CURRENT ISSUES IN SYMBIOTIC NITROGEN FIXATION

Developments in Plant and Soil Sciences

VOLUME 72

The titles published in this series are listed at the end of this volume.

Current Issues in Symbiotic Nitrogen Fixation

Proceedings of the 15th North American Symbiotic Nitrogen Fixation Conference, held at North Carolina, USA, August 13–17, 1995

Edited by

G. H. ELKAN and R. G. UPCHURCH

Partly reprinted from *Plant and Soil*, Volume 186, No. 1 (1996)

KLUWER ACADEMIC PUBLISHERS
DORDRECHT / BOSTON / LONDON

Library of Congress Cataloging-in-Publication Data

ISBN 0-7923-4367-0

Published by Kluwer Academic Publishers,
P.O. Box 17, 3300 AA Dordrecht, The Netherlands

Kluwer Academic Publishers incorporates
the publishing programmes of
D. Reidel, Martinus Nijhoff, Dr W. Junk and MTP Press.

Sold and distributed in the U.S.A. and Canada
by Kluwer Academic Publishers,
101 Philip Drive, Norwell, MA 02061, U.S.A.

In all other countries, sold and distributed
by Kluwer Academic Publishers,
P.O. Box 322, 3300 AH Dordrecht, The Netherlands

Printed on acid-free paper

Printed in the Netherlands

Contents

* Chapters indicated with an asterisk are reprinted from *Plant and Soil* Volume 186, No. 1 (1996)

Preface

Twenty-eight years ago, 35 scientists met at Iowa State University for the first American *Rhizobium* Conference, now known as the North American Symbiotic Nitrogen Fixation Conference. This series of conferences provides a forum for the exchange of current information regarding all of the phases of symbiotic N_2 fixation. The fifteenth biennial conference was convened from August 13-17, 1995, at North Carolina State University and brought together over 200 scientists from 25 countries to discuss their current research in 47 lectures and 96 posters. Subject matter ranged from very applied to the most basic and the format allowed for an informal forum for scientists and graduate students to exchange data and ideas concerning aspects of symbiotic nitrogen fixation. The sessions covered such diverse topics as ecology of rhizobia, environmental parameters affecting the infection and nodulation process, the nature of specificity, the biochemistry of host plants and microsymbionts, and chemical signalling between symbiotic partners.

This volume presents a summary of the current research efforts in symbiotic nitrogen fixation as presented during this conference. Because of the diversity of research in this field, no attempt was made to group the papers which represent a collection in the areas of research areas of physiology and metabolism, genetics, evolution, taxonomy, and ecology.

The editors hope that this volume serves to further disseminate the progress reported and will serve to stimulate additional research in this exciting and unique field.

Gerald H. Elkan
Robert G. Upchurch

Acknowledgements

A conference such as this one, with no internal organization or staff, is possible only because of the dedicated work of many, often unrecognized, heroes. First of all, we would again give our thanks to the sponsors listed previously who helped us financially. The plenary speakers and session chairs are vital to the success of the conference. We especially would like to thank the following organizations for their generous support:

Agrium Inc., Saskatchewan, Canada
CelPril Industries, Manteca, California
LiphaTech, Inc., Milwaukee, Wisconsin
Morgan Scientific Co., Andover, Massachusetts
North Carolina Biotechnology Center, Research Triangle Park, North Carolina
North Carolina State University, Raleigh, North Carolina
North Carolina State Agricultural Research Service, Raleigh, North Carolina
    Department of Crop Science
    Department of Microbiology
    Department of Soil Science
United States Department of Agriculture
    Agriculture Research Service
United States Department of Agriculture
    Cooperative States Research Service
    National Research Initiative
University of Minnesota, St. Paul, Minnesota
    College of Agriculture
Urbana Laboratories, St. Joseph, Missouri

We are also indebted to the reviewers of the papers which appear in the conference proceedings. They kept well to the narrow time frame that we could all allow, were careful in their evaluations, and generally very helpful in suggesting improvements.

This conference benefitted immensely from the hard work and computer desktop publishing skills of Betty Foley of the Microbiology Department. She worked many excess hours to design forms, organize the mass mailings, and doing numerous tasks voluntarily outside of her normal duties. Thomas J. Schneeweis worked diligently with all facets of this conference. His organizational skills and ideas and attention to detail surely prevented many mishaps.

After the first mailings, we received a response from Dr. Ira Baldwin, University of Wisconsin, one of the pioneers of nitrogen fixation (Fred, Baldwin, and McKoy - among many other contributions), who is now 100 years of age. Truly he is the last surviving pioneer of this field. He regretted not being 80 years old again so he could attend the conference, but requested that all outputs be sent to him. On behalf of the participants, I want to acknowledge Dr. Baldwin's longtime contributions.

Plant and Soil **186:** 1–7, 1996.

Advances in the positional cloning of nodulation genes in soybean

Gustavo Caetano-Anollés and Peter M. Gresshoff
*Plant Molecular Genetics, Institute of Agriculture and Center for Legume Research, The University of Tennessee, Knoxville, TN 37901, USA**

Key words: arbitrary primers, BAC, BSA, DAF, genetic markers, nodulation, positional cloning, symbiosis genes, soybean, YAC

Abstract

The positional cloning of genes involved in plant control of infection and nodule formation in legumes requires the development of improved tools for the analysis of large genomes and cloning of high molecular weight DNA. We have used bulked segregant analysis (BSA) and DNA amplification fingerprinting (DAF) with arbitrary oligonucleotide primers to detect polymorphisms linked to genes important to nodulation in soybean (*Glycine max* L. Merrill). Three loci controlling legume nodulation (*nts-1*, *rj1* and *rj6*) and one involved in early nodule development (*enod2*) were studied. Wild-type Bragg and EMS-induced mutants defective in autoregulation (nts382) or nodulation (nod49 and nod139) were crossed with *G. soja* (Sieb. and Zucc.), for the analysis of segregants in F_2 and F_3 populations. DNA pools from wild-type and mutant individuals, or DNA pools based on RFLP pattern, were screened with sets of structured (mini-hairpin) and unstructured primers, identifying polymorphic DNA. DAF with mini-hairpin primers, template endonuclease cleaved DAF (tecMAAP), and arbitrary signatures from amplification profiles (ASAP) were especially successful in detecting linked polymorphisms. Putative markers were confirmed by individual analysis of segregants, and some of them converted into sequence-characterized amplified regions (SCAR). These markers will be used for high density mapping of the relevant genomic regions and as landmarks for linking cloned soybean DNA from yeast and bacterial artificial chromosome (YAC and BAC) libraries. At present, the partial YAC library (avg. insert size=200 kb) represents about 20% of the soybean genome. YAC endclones were isolated using vectorette PCR, and consisted of unique or repeated DNA. Together with YAC-specific signatures generated with mini-hairpin primers, they will be used in the construction of contigs for positional cloning.

Introduction

The symbiosis between leguminous plants and rhizobia is a complex interplay between organisms that leads to the formation of a nitrogen-fixing organ, the root nodule. Nodulated legumes, such as soybean (*Glycine max* L. Merril), have the ability to host bacteria which fix atmospheric nitrogen for direct utilization by the plant. However, legumes seldomly fulfill their symbiotic potential because applied or residual nitrate in the soil inhibit the symbiosis. The analysis of developmental processes in legumes has been distinctly slow, and research on the role of the plant on nodulation and nitrogen fixation has been no exception. With only few natural variants in soybean, pea, alfalfa and clover, induced mutagenesis was the approach selected to increase the number of symbiotically defined loci, predominantly in soybean and pea, but also in other legumes such as french bean, chickpea, and sweet clover (reviewed in Caetano-Anollés and Gresshoff, 1991). As the result of a concerted genetic and physiological approach we have isolated unique soybean mutants that nodulate in the presence of fixed nitrogen (*nts*, nitrate-tolerant symbiosis mutants)(Carroll et al., 1985a, 1985b, 1986) and mutants blocked in the early stages of nodule development which fail to nodulate at all (homologous to *rj1* and *rj6*)(Carroll et al.,1986).

The lack of direct methods of mutant isolation in soybean, such as insertion mutagenesis by *Agrobacterium* T-DNA or transposable elements, prompted us to choose the strategy of positional cloning for the isolation of genes responsible for these symbiotic phenotypes (Figure 1). This has been done through a search

* FAX No: +14239742765

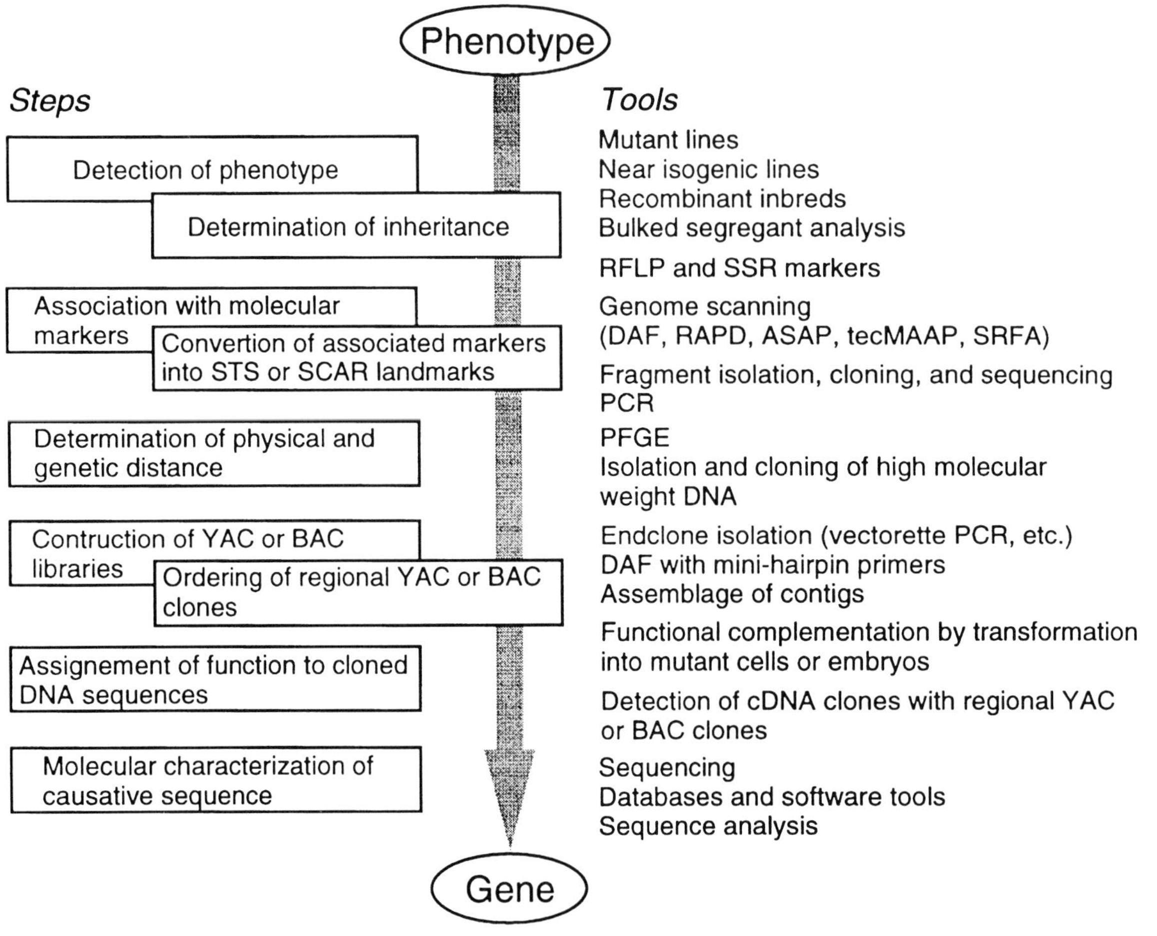

Figure 1. From mutant phenotype to causative gene sequence. A synopsis of steps and tools required in positional cloning.

of molecular markers tightly linked to symbiotic loci and construction of yeast artificial chromosome (YAC) and bacterial artificial chromosome (BAC) libraries containing large soybean DNA fragments. This map-based cloning approach has been spectacularly successful in the isolation of resistance genes in plants (Bent et al., 1994; Martin et al.,1993; Mindrinos et al., 1994). However, it has not been applied yet to any legume or broad-acreage crop plant. It should be noted that to overcome the complexities of positional cloning, *Lotus japonicus* and *Medicago truncatula* have been proposed as model plants. These legumes have been chosen mainly because of their small diploid genomes and the possibility of transformation, regeneration and insertional T-DNA mutagenesis. The isolation of symbiotic genes may help elucidate signal transduction events that determine symbiosis, and provide strategies for targeted improvement of agronomic and symbiotic traits.

Progress in map-based cloning of symbiotic loci

The challenge

The soybean genome is over a billion base pairs (1090 Mb), which is about seven times larger in size than that of *Arabidopsis thaliana* and about twice that of rice. The genome is organized in 20 chromosomes (2n=40), and contains a single nucleolar organizer region, duplicate genomic regions, and abundant (35%) repetitive DNA. Several soybean maps are available. For example, the USDA-ARS map (Iowa State University) contains 23 restriction fragment length polymorphism (RFLP) linkage groups, with over 500 linked RFLP,

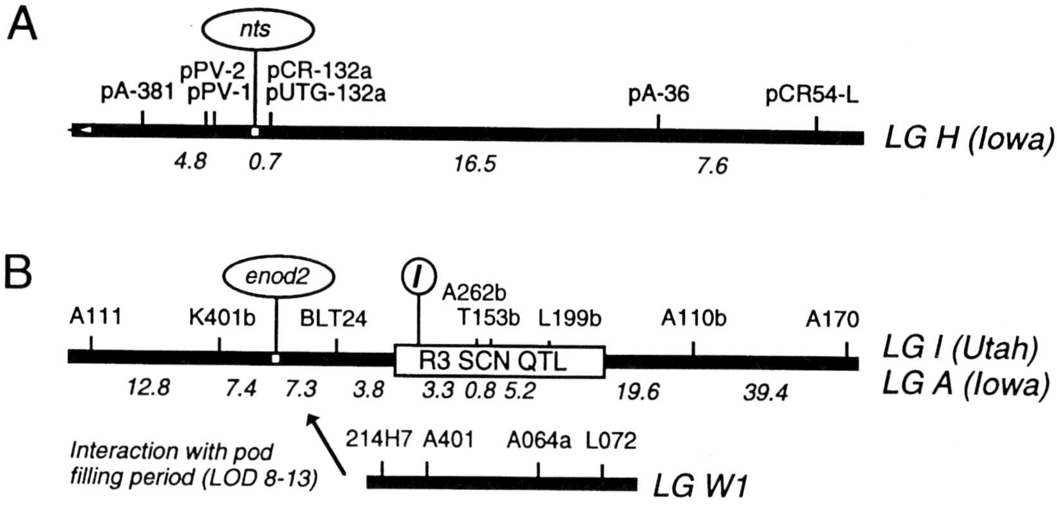

Figure 2. Molecular map of symbiotic loci. **A**. Map of the nitrate tolerant symbiosis (*nts-1*) region. pCR54-L is a SCAR marker (Kolchinsky et al., 1996). pPV-1 and pPV-2 are random genomic clones from *P. vulgaris* mapped by Shoemaker and colleagues between pA-381 and pA-132. About 10 additional *P. vulgaris* markers (E Vallejos, Gainesville, FL) are presently being mapped in the soybean region (A Filatov, unpublished). The arrow indicates proximity to the telomeric region. **B**. Map of the early nodulin *enod2* gene region. *enod2* maps closely to a quantitative trait loci (QTL) determining race 3 soybean cyst nematode resistance and to *gene I*, determinant of black seed coloration.

morphological and enzyme markers covering 3,000 cM (Shoemaker et al., 1992). Similarly, an immortal map based on recombinant inbreds from an F_2 population derived from a cross between soybean cultivars Minsoy and Noir 1 currently encompasses 1,550 cM in 31 linkage groups (Lark et al., 1993). Finally, a map based on simple sequence repeat (SSR) loci is now in the process of being generated (Rongwen et al., 1995). Despite abundance of genetic markers, a number of marker loci tightly linked to the targeted symbiotic genes had to be identified in order to initiate the positional cloning approach.

Mapping of the supernodulation nts-1 *locus*

The supernodulation *nts* trait segregates as a single recessive Mendelian locus, and appears to be determined by a shoot-mediated systemic control mechanism (Caetano-Anollés and Gresshoff, 1991). Wild-type Bragg soybean and ethyl methane sulfonate (EMS)-induced mutants defective in the autoregulatory control of nodule mass and number (nts382 (*nts-1*)) or nodulation (nod49 (*rj1*) and nod139 (*rj6*)) were crossed with *G. soja* PI468397 to produce F_2 and F_3 segregating populations. Screening of a set of molecular markers used to construct the USDA-ARS soybean RFLP map showed that marker pUTG-132a (derived from pA-132) was tightly linked to the *nts-1* locus (Landau-Ellis et al., 1991; Landau-Ellis and Gresshoff, 1992) (Figure 2A). The RFLP marker expresses a DNA polymorphism resulting from an 832 bp deletion in *G. soja* that resides within a 1.7 kb subcloned fragment (Kolchinski et al., 1996). Sequencing of the fragment permitted design of oligonucleotide primers for polymerase chain reaction (PCR) amplification, permitting the conversion of the RFLP into a sequence-tagged site (STS). Sequence analysis of the STS obtained from other soybean varieties (Peking, Enrei, Minsoy, Noir 1 and DPS3589) showed an unusual sequence conservation. PCR screening of a larger *G. soja* × nts382 F_2 segregating population (representing 330 plants), placed pUTG-132a as close as 0.7 cM away from *nts-1* (Figure 2A). Fluorescent in situ hybridization (FISH) of the pUTG-132a probe to soybean metaphase chromosomes revealed a single chromosomal location (P Keim, unpublished).

Physical mapping of the pA-36 marker region located 15 cM away from *nts*-1 showed an equivalence of 500 kb per cM (Funke et al., 1993). This suggests the *nts-1* region reflects average recombination frequencies, and supports the feasibility of genome walking based on contiguous segments (contigs) assembled from YAC or BAC libraries. The presence of repeated DNA (Funke et al., 1993) and the perils of chimerism in cloned DNA, warrant the construction of a high density map centered on *nts*-1. The availability of a large number of closely linked molecular markers will allow the use of a direct "genome landing" approach, whereby two flanking markers assign preferably one YAC or BAC clone to the region containing the symbiotic locus.

Bulked segregant analysis (BSA) (Michelmore et al., 1991) was used to identify markers associated to symbiotic loci with the sole knowledge of phenotype. Individuals in a segregating population that express or fail to express a particular trait are pooled, and DNA polymorphisms between the pooled samples are identified, generally using genome scanning techniques. DNA polymorphisms can be directly revealed without the need of cloned and characterized probes, radiolabeling, or prior knowledge of DNA sequence, by DNA amplification with arbitrary oligonucleotide primers. These primers initiate DNA polymerase-mediated amplification of discrete portions of a target nucleic acid molecule and produce characteristic profiles with which to identify and isolate molecular markers for genome mapping and general DNA fingerprinting (Caetano-Anollés, 1993). Of the different genome scanning techniques that take advantage of this strategy (Caetano-Anollés et al., 1991; Welsh and McClelland, 1990; Williams et al., 1990), DNA amplification fingerprinting (DAF) (Caetano-Anollés et al,, 1991) uses very short primers and generates complex amplification patterns. DNA pools from segregants homozygous for pUTG-132a and flanking markers pA-381 and pA-36 were analyzed by genome scanning with 181 unstructured primers (Kolchinsky et al., 1996). This BSA strategy maximizes the window for detection of potential polymorphisms. Two DNA polymorphisms were identified, cloned, sequenced, and converted into PCR-based STS markers, sequence characterized amplified regions (SCARs) defined by primers complementary to the ends of the isolated amplification markers (Paran and Michelmore, 1993). Only one of them was truly polymorphic and mapped 7.6 cM from marker pA-36, outside the cluster (Figure 2A). Overall results suggest that the *nts*-1 region is unusually conserved.

There is also a number of "fingerprint tailoring" strategies capable of increasing the information content of fingerprints or the ability to distinguish nucleic acids. For example, DAF can be used with extraordinarily stable mini-hairpin primers harboring a "core" arbitrary sequence at the 3' terminus (Caetano-Anollés and Gresshoff, 1994). Mini-hairpin primers increase the detection of polymorphic DNA and direct the controlled amplification of small template molecules, generating "sequence signatures" from plasmids, YACs, cloned DNA and PCR amplified fragments. Similarly, mini-hairpin primers or primers complementary to SSRs can also generate arbitrary sequence signatures from amplification profiles (ASAP) capable of distinguishing closely related nucleic acids (Caetano-Anollés and Gresshoff, 1996). We applied these and other fingerprint tailoring strategies to unravel new markers close to the *nts*-1 locus. While an extended screening of phenotypic pools of the *nts*-1 locus with 120 octamers did not produce additional polymorphisms, 64 mini-hairpin primers detected two clear polymorphisms, and ASAPs with several mini-hairpins and simple-sequence repeat primers revealed additional putative markers (Caetano-Anollés and Gresshoff, 1995). Furthermore, the use of DAF in conjunction with restriction endonuclease digestion of template DNA (tecMAAP) enhanced detection of polymorphic DNA and distinguished the EMS-induced mutant *nts*382 from its wild-type parent (Caetano-Anollés et al., 1993). Pre-digestion of template DNA with three endonucleases prior to amplification with octamer primers resulted in the detection of 42 polymorphisms, 14 of which segregated at 100% with the supernodulation phenotype in *G. soja* × *G. max* mutant derived F_2 populations. It is likely that these are valuable markers close to the *nts*-1 locus and their cloning and further characterization will facilitate the construction of a high density map around the region. Such map will allow the ordering of regional YACs. Several of the putatively linked markers are now being converted into SCARs. Of particular interest is the possible linkage of the histone H3 gene close to pA-132 (R Shoemaker, pers. commun.), and the possible placement of several clustered RFLP clones from *Phaseolus vulgaris* in the region (A Filatov, unpublished).

Mapping of the non-nodulation rj1 *and* rj6 *loci, and the early nodulin* enod2

The *rj1* and *rj6* non-nodulation traits segregate as single recessive Mendelian characters, and are necessary for nodule meristem development (*rj1*) and nodule initiation (*rj6*). A group of 64 mini-hairpin and 40 unstructured primers were used to screen bulked DNA from F_2 segregating populations (S Abbitt, unpublished). Only two mini-hairpins detected polymorphisms between the bulks, but showed about 36% recombination with the non-nodulation loci. As with *nts*, this region appears also highly conserved.

BSA was used to find RFLP molecular markers linked to *enod2* in soybean (F Ghassemi, unpublished). The *enod2* gene was mapped in linkage group I of the

Utah soybean map (Lark et al., 1993) using Minsoy × Noir 1 recombinant inbred lines, and was linked to the seed coat color gene *I*, and a quantitative trait locus (QTL) controlling race 3 cyst nematode resistance (Figure 2B). To generate additional markers linked to *enod2*, DNA pools from a *G. soja* × *G. max* cv. Bragg F_2 population from plants homozygous for parental RFLP patterns were amplified with more than 50 primers (8-12 nt in length). Three DNA polymorphisms were identified with two octamers and a decamer primer. These putative markers, when confirmed and mapped, will be used for positional cloning of cyst nematode genes.

Construction of soybean YAC and BAC libraries

In order to enable the positional cloning approach we had to initiate the construction of genomic libraries. High molecular weight DNA was isolated from agarose-embedded and lysed soybean mesophyll protoplasts, partially digested with *Eco*RI, and size-selected by pulsed field gel electrophoresis (PFGE) to enrich for large molecules (>250 kb). The isolated DNA was ligated into pYAC4 vector, transformed into yeast, and selected by colony color and uracyl auxotrophy (Funke et al., 1994). Candidate colonies were karyotyped by PFGE to detect additional chromosomes, and gels were blotted and probed with vector and total genomic DNA. Strong hybridization to total genomic DNA occurred in about half of clones, probably due to the abundance of repeated DNA sequences in the soybean genome. Weakly hybridizing signals were assumed to result from inserts containing predominantly unique DNA. At present, our partial YAC library represents about 20% of the soybean genome arrayed in microtiter plates. Insert size ranges 50-960 kb with an average of 200-250 kb. YAC inserts were stable but the extent of chimerism of our library was not examined. Chimerism is generally caused by the highly recombinogenic yeast, the presence of repeated DNA in soybean, or by physical factors during transformation and spheroplast recovery. Finally, endclones were isolated from YAC insert DNA, by restriction, vectorette-ligation and PCR amplification driven by primers complementary to vector and vectorette sequences. Endclones hybridized to restricted DNA from *G. max* and *G. soja*, showing either repeated or unique patterns. Repeated sequences appeared long and interspersed. Endclones can be use to link overlapping clones, once an initial YAC has been anchored on the molecular map through an RFLP, SCAR or microsatellite marker.

BACs have also been constructed with soybean and *Lotus japonicus* DNA (S Pillai, unpublished). The bacterial F factor is here used to clone size-selected PFGE into *Escherichia coli*. The system promises decreased recombination and chimerism, but produces clones with smaller insert sizes (50-80 kb).

Conclusions

Genome scanning techniques have been used extensively as amplification-based DNA diagnostic tools in breeding and general DNA fingerprinting of bacteria, fungi, plants and animals (Caetano-Anollés, 1993). The coupling of these techniques with BSA is now producing markers around defined genetic regions. Derived SCAR and STS markers closely linked to symbiotic loci will be used to screen soybean YAC and BAC libraries under construction and anchor initial contigs in a genome walking approach. Insert endclones will be isolated by vectorette PCR and used to further arrange regional overlaps. Assignments will be confirmed by DAF profiling with mini-hairpin primers and by Southern hybridization. Alternatively, the construction of a high density map centered on the symbiotic loci will allow a genome landing approach.

The analysis of the soybean genome is progressing rapidly. The construction of genetic maps (Lark et al., 1993; Shoemaker et al., 1992), the elucidation of a pachytene karyotype (Singh and Hymowitz, 1988), and *Agrobacterium tumefaciens* and biolistic transformation of soybean (Hinchee et al., 1988; McCabe et al., 1988), provided essential advances. Further advances in cell culture techniques, such as the use of protoplasts and microprojectile bombardment of immature embryos (Bond et al., 1992; Finer and McMullen, 1991), and *A. rhizogenes* transformation (Bond, 1993; Cheon et al., 1993; Miao and Verma, 1993) will make gene transfer more routine for soybean. Such developments will facilitate the culmination of our positional cloning efforts.

The cloning of nodulation genes will provide insight into the role of the plant in the symbiotic process. It will also reveal whether functional gene homologues exist in other legumes and non-legume plants. This will help identify the evolutionary relationship of the symbiotic genes to other genes used for normal cellular processes. The molecular understanding of the differences between plants capable or incapable of entering into symbiosis may open an avenue to the eventual extension of nodulation to other plants by genetic engineering.

Acknowledgements

Positional cloning projects are funded by the United Soybean Board (ASA), the Human Frontier Science Program, the Tennessee Soybean Board, the NSF-International Program, NATO and the Ivan Racheff endowment. We thank colleagues and students for support and unpublished data.

References

Bent A F, Kunkel B N, Dahlbeck D, Brown K L, Schmidt R, Giraudat J, Leung J and Staskawicz B J 1994 RPS2 of *Arabidopsis thaliana*: A leucine-rich repeat class of plant disease resistance genes. Science 265, 1856–1860.

Bond J E 1993 Transformation as a tool to study the genetics of nodulation in *Glycine max*. PhD Thesis. The University of Tennessee, Knoxville, USA.

Bond J E, McDonnell R, Finer J and Gresshoff P M 1992 Construction and use of low cost micro-projectile gene gun for gene transfer in plants. Tenn. Farm Home Sci. 162, 4–14.

Caetano-Anollés G 1993 Amplifying DNA with arbitrary oligonucleotide primers. PCR Methods Applic. 3, 85–94.

Caetano-Anollés G and Gresshoff P M 1991 Plant genetic control of nodulation. Annu. Rev. Microbiol. 45, 345–382.

Caetano-Anollés G and Gresshoff P M 1994 DNA amplification fingerprinting using arbitrary mini-hairpin oligonucleotide primers. Bio/Technology 12, 1011–1026.

Caetano-Anollés G and Gresshoff P M 1996 Generation of sequence signatures from DNA amplification fingerprints with mini-hairpin and microsatellite primers. Biotechniques 20, 1044–1056.

Caetano-Anollés G, Bassam B J and Gresshoff P M 1991 DNA amplification fingerprinting using very short arbitrary oligonucleotide primers. Bio/Technology 9, 553–557.

Caetano-Anollés G, Bassam B J and Gresshoff P M 1993 Enhanced detection of polymorphic DNA by multiple arbitrary amplicon profiling of endonuclease digested DNA: identification of markers linked to the supernodulation locus of soybean. Mol. Gen. Genet. 241, 57–64.

Carroll B J, McNeil D L and Gresshoff P M 1985a A supernodulation and nitrate tolerant symbiotic (*nts*) soybean mutant. Plant Physiol. 78, 34–40.

Carroll B J, McNeil D L and Gresshoff P M 1985b Isolation and properties of soybean mutants which nodulate in the presence of high nitrate concentrations. Proc. Natl. Acad. Sci. USA 82, 4162–4166.

Carroll B J, McNeil D L and Gresshoff P M 1986 Mutagenesis of soybean (*Glycine max* (L.) Merr.) and the isolation of non-nodulating mutants. Plant Sci. 47, 109–114.

Cheon C I, Lee N G, Siddique A B M, Bal A K and Verma D P S 1993 Role of plant homologs of Rap1p and Rap7p in the biogenesis of the peribacteroid membrane, a subcellular compartment formed (de novo) during root nodule symbiosis. EMBO J. 12, 4125–4135.

Finer J J and McMullen M D 1991 Transformation of soybean via particle bombardment of embryogenic suspension culture tissue. In Vitro Cell Dev. Biol. 27, 175–182.

Funke R, Kolchinsky A and Gresshoff P M 1993 Physical mapping of a region in the soybean (*Glycine max*) genome containing duplicated sequences. Plant Mol. Biol. 22, 437–446.

Funke R, Kolchinsky A and Gresshoff P M 1994 High EDTA concentrations cause entrapment of small DNA molecules in the compression zone of pulsed field gels, resulting in smaller than expected insert sizes in YACs prepared from size selected DNA. Nucleic Acids Res. 22, 2708–2709.

Hinchee M A W, Conner-Ward D V, Newell C A, McDonell R E, Sato S J, Gasser C S, Fischoff D A, Re D B, Fraley R T and Horsch R B 1988 Production of transgenic soybean plants using *Agrobacterium*-mediated DNA transfer. Bio/Technology 6, 915–922.

Kolchinsky A, Landau-Ellis D and Gresshoff P M 1996 Genome stability of the diploidized DNA region around the supernodulation (*nts 1*) locus of soybean. Mol. Gen. Genet.

Landau-Ellis D, Angermüller S, Shoemaker R and Gresshoff P M 1991 The genetic locus controlling supernodulation co-segregates tightly with a cloned molecular marker. Mol. Gen. Genet. 228, 221–226.

Landau-Ellis D and Gresshoff P M 1992 Supernodulating soybean mutant alleles nts382 and nts1007 show no recombination with the same RFLP marker supporting complementation data. Mol. Plant-Microbe Interact. 5, 428–429.

Lark K G, Wisemann J M, Mathews B F, Palmer R, Chase K and Macalma T 1993 A genetic map of soybean (*Glycine max* L.) using an interspecific cross of two cultivars Minsoy and Noir 1. Theor. Appl. Genet. 86, 901–906.

Martin G B, Brammonschenkel S, Chungwogse J, Frary A, Ganal M W, Spivey R, Wu T, Earle E D and Tanksley S D 1993 Map-based cloning of a protein kinase gene conferring disease resistance in tomato. Science 262, 1432–1436.

McCabe D E, Swain W F, Martinelli B J and Christou P 1988 Stable transformation of soybean (*Glycine max*) by particle acceleration. Bio/Technology 6, 923–926.

Miao G H and Verma D P S 1993 Soybean nodulin 26 gene - encoding a channel protein - is expressed only in the infected cells of nodules and is regulated differently in roots of homologous and heterologous plants. Plant Cell 5, 781–794.

Michelmore R W, Paran I and Kesseli R V 1991 Identification of markers linked to disease resistance genes by bulked segregant analysis: a rapid method to detect markers in specific genomic regions using segregating populations. Proc. Natl. Acad. Sci. USA 88, 9828–9832.

Mindrinos M, Katagiri F, Yu G L and Ausubel F M 1994 The *Arabidopsis thaliana* disease resistance gene RPS2 encodes a protein containing a nucleotide-binding site and leucine-rich repeats. Cell 78, 1089–1099.

Paran I and Michelmore R W 1993 Development of reliable PCR markers linked to downy mildew resistance genes in lettuce. Theor. Appl. Genet. 5, 985–993.

Rongwen J, Akkaya M S, Bhagwat A A, Lavi U and Cregan P B 1995 The use of microsatellite DNA markers in soybean genotype identification. Theor. Appl. Genet. 90, 43–48.

Shoemaker R C, Guffy R D, Lorenzen L and Specht J E 1992 Molecular genetic mapping of soybean: map utilization. Crop Sci. 32, 1091–1098.

Singh R J and Hymowitz T 1988 The genomic relationship between *Glycine max* (L.) Merr. and *Glycine soja* (Sieb. and Zucc.)

as revealed by pachytene chromosome analysis. Theor. Appl. Genet. 76, 705–711.

Welsh J and McClelland M 1990 Fingerprinting genomes using PCR with arbitrary primers. Nucleic Acids Res. 18, 7213–7218.

Williams J G C, Kubelik A R, Livak K J, Rafalski J A and Tingey S V 1990 DNA polymorphisms amplified by arbitrary primers are useful genetic markers. Nucleic Acids Res. 18, 6531–6535.

Section editor: H Lambers
Guest editors: G.H. Elkan and R.G. Upchurch

Plant and Soil **186:** 9–20, 1996.

Molecular analysis of actinorhizal symbiotic systems: Progress to date

Beth C. Mullin[1] and Svetlana V. Dobritsa[2]
[1]*Department of Botany and The Center for Legume Research, The University of Tennessee, Knoxville, TN 37996, USA and* [2]*Institute of Biochemistry and Physiology of Microorganisms, Russian Academy of Sciences, Pushchino, Moscow Region 142292, Russia**

Key words: actinorhizal symbiosis, ecology, *Frankia*, gene expression, molecular analysis

Abstract

The application of molecular tools to questions related to the genetics, ecology and evolution of actinorhizal symbiotic systems has been especially fruitful during the past two years. Host plant phylogenies based on molecular data have revealed markedly different relationships among host plants than have previously been suspected and have contributed to the development of new hypotheses on the origin and evolution of actinorhizal symbiotic systems. Molecular analyses of host plant gene expression in developing nodules have confirmed the occurrence of nodulin proteins and in situ hybridization techniques have been successfully adapted to permit the study of the spatial and temporal patterns of gene expression within actinorhizal nodules. The use of heterologous probes in combination with nucleotide sequence analysis have allowed a number of *nif* genes to be mapped on the *Frankia* chromosome which will ultimately contribute to the development of hypotheses related to *nif* gene regulation in *Frankia*. The use of both 16S and 23S rDNA nucleotide sequences has allowed the construction of phylogenetic trees that can be tested for congruence with symbiotic characters. In addition the development of *Frankia*-specific gene probes and amplification primers have contributed to studies on the genetic diversity and distribution of *Frankia* in the soil.

Introduction

The Tenth International Conference on the Biology of *Frankia* and Actinorhizal Plants was held in Davis, CA., August 6–11, 1995. The number of papers and posters presented at this meeting dealing with molecular analysis of *Frankia* and its host plants had increased dramatically since the last International Conference held in 1993. In this short review we have attempted to draw together results of molecular analyses of actinorhizal symbiotic systems reported in Davis as well as elsewhere over the past two or three years. Readers are referred to the excellent review of Benson and Silvester (1993) for more general information on actinorhizal symbiotic systems.

Host plant systematics and phylogenetics

Approximately 194 plant species distributed among 24 genera and eight families of dicotyledonous angiosperms are known to participate in nitrogen-fixing symbiosis with *Frankia* (Benson and Silvester, 1993). Traditional taxonomic schemes based on morphological characters assign the actinorhizal host plant families to eight higher plant orders distributed among four of the five subclasses of angiosperms (Cronquist, 1988). Such taxonomic schemes led to the belief that many of these plants are only very distantly related to one another. The only two features previously known to unite actinorhizal host plants were the ability to be nodulated and possession of a woody or suffrutescent habit. If one looks beyond traditional published taxonomic schemes at the rationale presented by plant systematists for the placement of several of the actinorhizal orders, it becomes evident that the placement of some has been based more on interpretation, expediency and intuition rather than on concrete evidence. For a taxonomic scheme to be complete, all orders need to be placed somewhere, no matter how tenuous is the placement. For some time it has been evident from taxonomic descriptions of many actinorhizal genera that molecular analyses would be required to resolve phy-

* FAX No: + 1423 974 0978. E-mail: bmullin@utk.edu

logenetic affinities. One of the first such studies was done to help resolve the phylogenetic affinities of the Datiscaceae (Swensen et al., 1994).

By 1993 the database of *rbcL* (ribulose bisphosphate carboxylase oxygenase large subunit gene) nucleotide sequences had grown to the point where it was possible to construct a global molecular phylogeny of the angiosperms (Chase et al., 1993). A major finding of this analysis was that taxa placed within subclass Dilleniidae on the basis of morphological characters did not form a natural group, and instead belonged to other subclasses. A second major finding of this analysis was that the so-called "higher hamamelids" of the Hamamelidae grouped with plants in the Rosidae. Although only a few actinorhizal genera were represented in the global molecular phylogeny of Chase et al. (1993), it was evident from this analysis that at least some actinorhizal plants were more closely related to one another than had previously been thought.

The addition of more actinorhizal taxa and their putative relatives to the *rbcL*, database has led to more refined phylogenetic trees that group all known actinorhizal taxa and their nonsymbiotic relatives, as well as legumes and *Parasponia*, within a single clade. This symbiotic nitrogen-fixing clade falls within one of four major clades within the Rosidae (Soltis et al., 1995; Swensen, 1996; Swensen and Mullin, 1996). This remarkable finding indicates that only one small group of angiosperms possesses the genetic predisposition to host nitrogen-fixing symbionts. In the case of legumes the symbiont is either *Rhizobium*, *Bradyrhizobium* or *Azorhizobium*; in the case of *Parasponia* it is *Bradyrhizobium*; and in the case of the actinorhizal plants the symbiont is the actinomycete *Frankia*.

In both legumes and actinorhizal plants, a variety of evidence points to the occurrence of multiple origins of nodulation within each group (Doyle, 1994; Sprent, 1994; Swensen, 1996). Hypotheses for the evolution of multiple origins of nodulation are further supported by molecular phylogenies that identify a single clade whose members appear to possess the genetic predisposition to be nodulated. Within this clade subclades whose members have the ability to be nodulated are interspersed with subclades whose members are not known to be nodulated. The identification of a single angiosperm clade containing plants with the genetic predisposition to be nodulated will undoubtedly lead to speculations, hypotheses and new experiments designed to identify the unique molecular, biochemical, physiological and developmental traits of members of this clade that would impart to them the ability to be nodulated.

Host plant gene expression

Actinorhizal nodules arise from lateral root primordia and maintain the structure of a modifed lateral root having a central vascular bundle and peripheral cortical tissue. *Frankia* infects host plant roots by either root hair infection (Berry et al., 1986) or intercellular penetration through the root epidermis (Liu and Berry, 1991; Miller and Baker, 1986). In plants infected via root hairs such as *Alnus*, root hair deformation occurs within the first few hours after inoculation and is the first visible sign of impending nodule formation. Progress is being made on the identification of root hair deforming factors (dnFs) which, in the case of *Casuarina glauca*, are released by *Frankia* cultures in response to treatment with extracts of *Casuarina* seeds (Schwencke and Selim, 1995). Using a different system Solheim et al. (personal communication) found that production by *Frankia* of a root hair deforming factor did not require induction by host plant compounds.

Root hair deformation is followed by penetration of the root hair by *Frankia* filaments and the initiation of cortical cell divisions leading to the formation of a structure called the prenodule. The colonization of prenodule tissue is followed by the induction of a lateral root primordium in the pericycle. As the lateral root primordium grows through the infected prenodule tissue, the postmeristematic cells of the primordium are themselves penetrated by the microsymbiont which rapidly proliferates within host cells and begins to fix nitrogen. In host plants infected via intercellular penetration, a prenodule is not formed and frankiae first penetrate host cells as the lateral root pushes through the infected cortical tissue. In both cases infection by *Frankia* alters the normal developmental pathway leading to lateral roots in such a way that they become nitrogen-fixing root nodules (Newcomb and Wood, 1987).

There is ample cytological evidence for differential gene expression throughout actinorhizal nodule development. In all cases studied to date micrographic evidence points to elevated levels of metabolic activity in developing root nodules, as well as to the existence of several distinct tissue and cell types. Throughout the infection process and symbiosis the microsymbiont remains separated from the host cytoplasm by modified plant cell wall material referred to as encapsulation material. The synthesis and export of encapsulation material represent major metabolic activities of infected cells which show extensive Golgi activity compared to non-infected cells. Proteins unique to the symbiotic state have been isolated from actinorhizal nodules. Hemoglobins, first reported to be in actinorhizal nodules by Tjepkema (1983), have been isolated and purified from nodules of *Casuarina* (Fleming et al., 1987), as well as from nodules of *Alnus* (Suharjo and Tjepkema, 1995) and *Myrica* (Tjepkema, personal communication). Séguin and Lalonde (1993) demonstrated changes in total polypeptide patterns in developing *Alnus* nodules. The nodule-specific proteins of either plant or bacterial origin are referred to by the authors as actinorhizins. At the 10th International Conference on *Frankia* and Actinorhizal Plants (August 6–11, 1995, Davis, CA), participants were in agreement that host plant proteins expressed in nodules, but not roots, of actinorhiral plants should be called nodulins rather than actinorhizins, to emphasize similarities with legume symbioses.

Only within the past three or four years has differential gene expression in the actinorhizal symbiosis been documented by molecular studies (Mullin et al., 1993). After several years of unsuccessful attempts at RNA isolation and cDNA synthesis, there are now several research groups that have successfully isolated mRNA from actinorhizal host plant species and constructed and screened cDNA libraries. Five approaches, all involving cDNA synthesis, have been used successfully to identify RNA transcripts that are expressed specifically or at altered levels in actinorhizal nodules. Differential screening of nodule cDNA libraries with root and nodule cDNA has resulted in the isolation of a number of putative nodule-specific cDNA clones. The construction of subtraction libraries has resulted in the isolation of several additional putative nodule-specific clones and the screening of nodule cDNA libraries using heterologous probes for genes expected to be involved in nodule metabolism has resulted in the isolation of additional nodule-specific or - enhanced cDNAs. In at least one case nodule-specific cDNA clones were isolated by antibody cross reactivity to specific clones in an expression library (Jacobsen-Lyon et al., 1995). More recently, Carlson et al. (1995) and Twigg (personal communication) have used reverse transcriptase-based PCR (RT-PCR) to construct cDNAs specific for targeted actinorhizal mRNAs.

Actinorhizal nodules have an indeterminate growth pattern with four major zones of metabolic activity within the nodules as defined by Ribeiro et al. (1995): Zone 1 is the nodule meristem and contains uninfected meristematic and postmeristematic cells; Zone 2 contains postmeristematic cells, some of which have become infected with endophyte filaments not yet engaged in nitrogen fixation; Zone 3 contains infected and uninfected cells and is the major region of nitrogen-fixing activity; Zone 4 contains cells in various stages of senescence. It has been possible by in situ hybridization to determine the time and location of expression of several actinorhizal mRNAs. In *Alnus glutinosa* infected cells in Zone 2 have been found to express high levels of mRNA for two subtilisin-like serine proteases, Ag12-1 and Ag12-2, that may be involved in protein processing extracellularly (Ribeiro et al., 1995). The derived amino acid sequence of *ag*12 indicates the presence of a signal sequence likely to target the protein to the extracellular space where it may play a role in processing proteins in the encapsulation material surrounding *Frankia* filaments. Expression of *ag*12 was also detected in Zone 3 but at much lower levels and only in infected cells with low levels of *nif*H gene expression. A cDNA (pAgNg118-203) encoding a putative nodule-specific cysteine proteinase (AgNOD-CP1) has been isolated from an *A. glutinosa* nodule cDNA library (Goetting-Minesky and Mullin, 1994) by differential screening of the library with root and nodule cDNA. Northern blot analysis has shown that this sequence is expressed in greatest amounts late in nodule development, indicating that this proteinase may be involved in protein turnover in Zone 4 (Pawlowski, personal communication). Northern blot analysis also demonstrated cross hybridization between a probe made from pAgNg118-203 and RNA from developing alder fruits (Pawlowski, personal communication). Southern hybridization of pAgNg118-203 to restriction digested alder genomic DNA revealed several hybridizing bands indicating that the AgNOD-CP1 may be a member of a multigene family (Goetting-Minesky and Mullin, 1994). Therefore it is not known at this time whether AgNOD-CP1 itself or a similar member of the cysteine proteinase gene family is involved in fruit development in *Alnus*.

Zone 2 is also the site of expression in *Alnus* nodules of a small unique glycine/histidine-rich protein (AgNOD-GHRP) that is expressed only in infected cells and has a putative signal sequence that might direct the protein to the extracellular space (Twigg, 1993; Twigg et al., 1995b; Pawlowski and Guan, personal communication). Although expressed in the same zone in the nodule the putative signal sequence of AgNOD-GHRP differs from that found in Ag12, the serine protease described above. The cDNA clones for AgNOD-GHRP, pAgNt84 and pAg164, isolated independently in two different laboratories, share high sequence similarity in the 5' untranslated region, the coding region and the 3' untranslated region. They are clearly homologs and differ largely on the basis of insertions or deletions rather than single base changes. The glycine/histidine-rich region is confined to a central portion of the protein and is similar to the metal-binding domains of both *Escherichia coli* and *Saccharomyces cerevisiae* metal-binding proteins (Conklin et al., 1992; Wülfing et al., 1994). If AgNOD-GHRP is a metal-binding protein, it is the first such glycine/histidine-rich metal-binding protein described in plants and may play a role in providing metal ions to *Frankia* during its rapid growth in the zone of infection. The cDNA for another putative metal-binding protein, metallothionein, has been isolated from *C. glauca* nodules and studies show that it is expressed in infected cells primarily in Zone 3 of *Casuarina* nodules, as well as in the pericycle (Bogusz et al., 1995; Duhoux, personal communication).

Also expressed in infected cells of Zone 3 of *C. glauca* nodules are two hemoglobin genes and a chalcone synthase gene that are expressed in uninfected roots as well (Bogusz et al., 1995; Duhoux, personal communication). In a separate study three symbiotic hemoglobin cDNAs were isolated from a *Casuarina* nodule cDNA expression library which was screened with antibody raised against *Casuarina* nodule hemoglobin (Jacobsen-Lyon et al., 1995). These cDNAs which share greater than 90%, derived amino acid sequence identity belong to a multigene family (cashb-sym) and are expressed only in nodule tissue. They share 53% derived amino acid sequence identity with a single copy non-symbiotic hemoglobin gene (cashb-nonsym) that is expressed in very low levels in nodule tissue as well as in root tissue unrelated to symbiosis. The promoter regions of both symbiotic and non-symbiotic *Casuarina* hemoglobins have been identified and have been tested for activity in a transgenic GUS fusion *Lotus* system. In *Lotus* each promoter directs the expression of GUS in a tissue specific pattern that precisely mirrors its activity in *Casuarina* nodules. The cashb-sym promoter regions contain nearly perfect copies (CAAGATn$_7$CTCTT) of the leghemoglobin nodulin box consensus sequence, AAAGATn$_6$CTCTT (Jacobsen-Lyon et al., 1995).

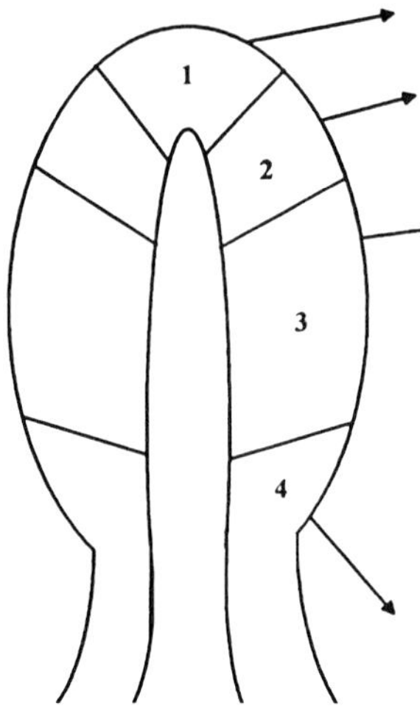

Figure 1. Host proteins whose location is predicted from expression of their mRNAs in developing actinorhizal nodules. The four zones of activity depicted are as described in Ribeiro et al. (1995). The asterisk (*) indicates nodule-specific proteins.

A number of other cDNAs that represent genes demonstrated to be expressed in Zone 3 or likely to be expressed in this zone have also been isolated. cDNAs coding for a putative fatty acid reductase and for a glutamine synthetase (GS) have been isolated by differential screening of an *A. glutinosa* nodule cDNA library (Guan et al., 1995; Pawlowski, personal communication), and a second GS cDNA has been isolated from this same host species by subtractive hybridization (Twigg et al., 1995a). Screening of an *Alnus incana* nodule cDNA library by heterologous probing resulted in the isolation of cDNA coding for PEP carboxylase, and amplification from the library using degenerate primers resulted in the isolation of fructokinase and glutamate synthase cDNAs, all three representing genes expressed at elevated levels in nodules (Lundquist and Vance, 1995, personal communication). cDNAs for two additional metabolic enzymes, sucrose synthase and enolase, have been isolated as well (Solheim, personal communication). Figure 1 is a summary of host proteins whose location is predicted from expression of their mRNAs in developing actinorhizal nodules.

To facilitate an understanding of the role in symbiosis and mechanisms of regulation of genes involved in the establishment and maintenance of actinorhizal nodules, plant transformation and regeneration systems need to be developed for a number of actinorhizal plants. Transformation of *C. glauca* root tissue with *Agrobacterium rhizogenes* to produce hairy roots that appear to support normal nodule formation and function as reported by Diouf et al. (1995) is an exciting step forward. This group has also developed a stable transformation and regeneration protocol for transformation of *Allocasuarina verticillata* using *Agrobacterium tumefaciens* (Franche et al., 1995; Duhoux, personal communication).

Gene expression in *Frankia*

Many standard tools used by microbial geneticists for analysis of gene function and regulation have not yet been adapted for use with *Frankia*. There are no reports of conjugation, transduction or reliable and stable transformation systems, no phage or replicating plasmid vectors and no standard mutation selection protocols. Frankiae grow slowly in culture forming mycelial mats and are difficult to recover from protoplast preparations (Normand et al., 1987; Tisa Ensign, 1987). It has been suggested that endogenous restriction/modification systems might lead to the degradation of DNA electroporated into *Frankia* protoplasts and that this might be a major factor underlying the difficulty in achieving stable transformation of *Frankia* cells. However it was shown by Cournoyer and Normand (1990) that plasmid DNA introduced into protoplasts is not rapidly degraded and that instead it apparently does not replicate at a sufficient rate to be maintained in the culture. In an effort to characterize cis-acting elements that might control gene expression in *Frankia*, Wigington et al. (1995) have screened sequences 5' to known *Frankia* genes looking for promoter activity using both in vitro and in vivo assays systems.

It has been possible using immunological methods to study levels of expression of *nif* gene protein products in actinorhizal nodules in response to changing environmental conditions (Lundquist, 1993), and more recently to study RNA transcript levels as well. In situ hybridization to *Alnus* nodule sections, for example, has demonstrated that *nif*H is expressed at high levels in Zone 3 of nodules (Pawlowski et al., 1995). Using a *nif*H probe in combination with other probes hybridized to adjacent tissue sections has made it possible to map gene expression precisely to fixing or non-fixing cells or tissues.

Using a combination of heterologous probing, restriction mapping and nucleotide sequence analysis, genetic maps of *nif* genes in at least two *Frankia* strains have been constructed. The order of *nif* genes in *Frankia* strain FaC1 as determined by analysis of the cosmid clone pFAH1 is orfA-orfB-*nif*V-H-D-K-E-N-X-W-B (An et al., 1995a,b). This order is reflected in the following gene order presented by Harriott et al. (1995) for *Frankia* strain CpI1, *nif*H-D-K-3kb-X-orf3-orf1-W-Z-B-orf2-U. Nucleotide sequence analysis of this *nif* region, as well as of 16S and 23S rDNA from *Frankia*, has provided data that can be used for building phylogenies, as well as for designing probes and primers for ecological studies as described below.

Recent advances in *Frankia* taxonomic and ecological studies

Molecular taxonomy and phylogeny

The genus *Frankia* (Actinomycetales) is presently defined by morphological features, such as sporangia and vesicle formation; cell chemistry, including cell wall type III, phospholipid type I and the presence of the diagnostic sugar 2-*O*-methyl-mannose; and the ability to fix nitrogen and to enter into symbiotic relationships with certain plant hosts (for recent reviews, see Benson and Silvester, 1993; Lechevalier, 1994), though molecular analyses have placed some non-infective and non-nitrogen-fixing isolates from actinorhizal root nodules within the phylogenetic radiation of authentic *Frankia* strains (Hönerlage et al., 1994; Mirza et al., 1994b; Nick et al., 1992). Conserved regions of the 16S ribosomal RNA (rRNA) molecule have been used to investigate quantitative evolutionary relationships of *Frankia* with other bacteria and to position the family Frankiaceae, including the genera *Frankia* and *Geodermatophilus*, in the phylogenetic tree of the actinomycetes (Hahn et al., 1989b).

Frankia species have been difficult to delineate by classical phenotypic methods. Therefore, some researchers (Lechevalier, 1994) believe that progress in classification of these difficult-to-grow organisms is only possible through molecular studies. Others (Benson and Silvester, 1993) argue that physiological studies, though slowly proceeding, are more relevant to the long-term objectives of understanding the biology of the symbiosis and thus should be given at least equal consideration in the development of taxonomic schemes.

Studies on total genomic DNA-DNA relatedness (Akimov and Dobritsa, 1992; Fernandez et al., 1989) have contributed to substantial progress in developing species concepts. At least 17 genomic species have been delineated among *Frankia* strains compatible with the plant genus *Alnus* and the genera within families Myricaceae, Elaeagnaceae and Casuarinaceae. Since infective *Frankia* isolates are available from members of less than half of the known actinorhizal plant genera and only non-infective isolates, so-called "atypical" isolates (unable to reinfect the source plant species), or no isolates have been obtained from the others (Benson and Silvester, 1993), much greater diversity of *Frankia* strains may exist.

Studies on the structure of the genus *Frankia* are presently focused on two genomic regions of conserved nature: the ribosomal RNA (*rrn*) operon and the nitrogenase region. Both contain highly conserved and more variable regions which can be used to design primers and probes for strain comparison. Detailed analysis of the *rrn* operon in *Frankia* strains allowed Normand et al. (1992a) to find regions with a high proportion of divergence in the intergenic spacer (IGS) between 16S and 23S *rrn* genes, as well as in the 23S and 16S genes, all of which can be used to study phylogenetic relationships within the genus.

Sequence analysis of different regions of the 16S rDNA gene is being extensively used to study phylogenetic relationships among *Frankia* strains. Although first studies included reverse transcriptase sequencing of the rRNA molecule (Hahn et al., 1989a,b), amplification of the 16S rRNA gene by the polymerase chain reaction (PCR) from pure cultures or directly from environmental samples has greatly facilitated evolutionary studies. Sequence information from a growing number of *Frankia* strains is now available and has been used to characterize and identify strains and to determine phylogenetic relationships (Hahn et al., 1989a,b; Mirza et al., 1992, 1994a,b; Nazaret et al., 1991; Nick et al., 1992; Normand et al., 1992a). It has become a standard procedure to determine partial nucleotide sequences of the 16S rRNA gene for placing new *Frankia* isolates (Ganesh et al., 1994) in the phylogenetic tree. This approach is especially valuable when "atypical" strains are studied. Sequence analysis of "atypical" nodule isolates has resulted in the placement of some of the strains within the *Frankia* phylogenetic clade and, moreover, has provided information about some uncultured *Frankia* strains (Mirza et al., 1992, 1994a,b,c; Nazaret et al., 1991; Nick et al., 1992; Simonet et al., 1994) i.e. strains which have never been isolated from the nodules.

A 268-bp segment spanning the variable domains V3 and V4 of the 16S rRNA gene has been sequenced in a number of strains (Nazaret et al., 1991) and specific sequences have been determined to be characteristic of some of the described *Frankia* genomic species. This study has shown, however, that among nine genomic species described by Fernandez et al. (1989) only seven are characterized by unique sequences within the 268-bp fragment, with two genospecies in the Elaeagnaceae host compatibility group being identical. This same region has also been found (Simonet et al., 1994) not to be divergent enough to allow discrimination between two clearly different groups of *Alnus*-compatible strains, Sp(+) and Sp(–). Also, Mirza et al. (1994c) showed that partial sequences of 16S rDNA of *Coriaria* and *Datisca* endophytes were identical in one region (domain V6), but different in another, more variable, region (domain V2), thus suggesting that the conclusion about remarkable homogeneity of *Coriaria* endophytes of different geographical origins made by Nick et al. (1992) on the basis of sequences obtained only in the V6 region may be questionable. The examples show that short stretches of rDNA do not always reflect overall genomic diversity and may not be sufficient to draw firm taxonomic or evolutionary conclusions.

Many differences within a short (248-bp) sequence of the 16S rRNA gene have been found to result in the acquisition or loss of restriction enzyme recognition sites, which have been used to demonstrate the possibility of PCR-based RFLP classification of *Frankia* strains into the described genomic groups (McEwan et al., 1994). Sequencing of PCR products amplified from a different variable region covering domains V1 and V2 of the 16S rRNA gene from a wide range of *Frankia* strains resulted in the design of a *Frankia* genus-specific oligonucleotide primer (Simonet et al., 1994). Primers for specific amplification of DNA in the 16S rRNA gene region have also been developed for Frankiaceae (Simonet et al., 1991), *Alnus*/*Casuarina*- and *Elaeagnus*-compatible strains (Bosco et al., 1992), *Coriaria* and *Coriaria*/*Datisca* nodule endophytes (Mirza et al., 1994c; Nick et al., 1992).

Recently the phylogenetic relationships among *Frankia* strains have been studied by comparing sequences in a highly variable region of 23S rDNA, with several phylogenetic subgroups delineated (Hönerlage et al., 1994). Although the strains used and the resulting trees obtained in this study and in the study performed by Nazaret et al. (1991) on 16S rDNA sequences are different enough to preclude unequivocal comparisons, both studies indicate that *Alnus*-compatible strains are more diverse as compared with

Elaeagnus- and *Casuarina*- compatible strains and that *Casuarina* strains group together with some typical *Alnus* strains, while *Elaeagnus* strains branch separately.

Nitrogen fixation (*nif*) genes of *Frankia* have been demonstrated to have a conserved character and can therefore be used to investigate phylogenetic relationships with other nitrogen-fixing microorganisms (Normand and Bousquet, 1989; Normand et al., 1992b). Different restriction sites within these genes can provide specific hybridization patterns of digested total DNA with *nif* probes (Simonet et al., 1994), which may be used in taxonomic studies but, due to conserved nature of the genes, only distantly related strains can usually be discriminated. Among diazotrophs, the *nifD-K* intergenic spacer has been found to be markedly less conserved than the neighboring nitrogenase genes. Within this region a sequence conserved among the *Frankia* strains studied, but absent in other diazotrophs, has been detected (Nalin et al., 1995). This has allowed the design of a primer which specifically amplifies the *nifD-K* intergenic spacer sequences from all Nif$^+$ and some of the Nif$^-$ *Frankia* strains tested. A PCR-RFLP typing method was developed for *Frankia* strains compatible with *Elaeagnus* and *Casuarina* host plants using primers flanking the *nifD-K* intergenic spacer (Jamann et al., 1993), with resulting PCR fragments being long enough (about 1.4 kb) to allow RFLP analysis, which resulted in strain groups that correlated well with the described genomic species. These methods may be useful for the rapid assigning of new strains or nodule endophytes to known genomic groups.

For discrimination between *Frankia* strains within a genomic group, highly sensitive PCR techniques which give multiple PCR fragments were shown to be most suitable. None of the following techniques has allowed discrimination between the closely related *Casuarina*- infective *Frankia* strains comprising genomic species 9 of Fernandez et al. (1989): DNA/DNA hybridization (Fernandez et al., 1989), RFLP analysis of the *nif* genes (Nazaret et al., 1989), partial sequencing of the 16S and 23S rDNA (Hönerlage et al., 1994; Nazaret et al., 1991), PCR amplification of the *nifH/nifD* region, followed by hybridization with specific probes, as well as RFLP analysis of the IGS between the 16S and 23S rRNA genes (Maggia et al., 1992), which is known to be the most variable region of the *rrn* operon (Normand et al., 1992a). Discrimination has been achieved, however, by application of the RAPD-PCR (Sellstedt et al., 1992) and rep-PCR using REP and ERIC primers (Dobritsa et al., 1995). Since these techniques do not require any knowledge of genomic sequences of the organisms and have been shown recently (McEwan and Wheeler, 1995) to work on intact *Frankia* cells without isolation of DNA, they may be of great use for discriminating between strains in root nodules as well as pure cultures.

Ecology

Frankia occupy two distinct ecological niches, the root nodule and the soil, and are difficult to study in either environment. They are also difficult to isolate from root nodules and cultivate in vitro, and only one successful attempt to isolate *Frankia* from soil has been reported (Baker and O'Keefe, 1984). Ecological research on soil populations of *Frankia* has long been based on plant bioassays using various actinorhizal host plants "traps" to show the presence of *Frankia* strains in particular soils, to estimate their relative abundance, as well as to isolate the microorganism from actinorhizal root nodules to study in pure culture. Consequently, the majority of in vivo studies have dealt with those *Frankia* that are both compatible with host plants and physiologically capable of infecting and subsequently forming root nodules and thus may comprise only a fraction of the total *Frankia* population in soils. For in vitro studies the fraction has been even smaller, because of the bias towards the selection of strains that are more easily isolated from nodules and are amenable to growth in pure culture.

Recent advances in molecular detection and identification methods at the DNA/RNA level (for reviews see Akkermans et al., 1991, 1994) have allowed nodulation and/or isolation and cultivation steps to be circumvented and have provided more reliable means to estimate *Frankia* populations in the environment both quantitatively and qualitatively. Many of the techniques described above for identification and taxonomic analysis of *Frankia* strains have been developed with the primary purpose of studying *Frankia* directly in root nodules and in soil. A common detection strategy for *Frankia* in environmental samples described in a number of reports includes the following steps: DNA/RNA extraction, in vitro amplification of a particular DNA/RNA fragment using PCR, direct sequencing of the PCR product or, alternatively, its cloning followed by sequencing. These analyses are followed by comparison of the sequences with those in a databank, development of a specific oligonucleotide probe, and its hybridization to nucleic acids extracted from environmental samples.

16S rRNA or *rrn* genes and *nif* genes have served as targets for designing specific probes and PCR primers to be used in ecological studies. Variable regions found in 16S rRNA of closely related organisms have indicated sufficient variation to design different probes of interest for *Frankia*. rRNA gene sequences were found to be suitable targets for developing oligonucleotide probes and/or PCR primers specific for the genus (Hahn et al., 1990a; Simonet et al., 1994) or for different groups of strains, including different host compatibility groups or their combinations (Bosco et al., 1992; Simonet et al., 1991), spore-positive and spore-negative strains (Simonet et al., 1994), effective and non-effective strains (Hahn et al., 1989a), and uncultured endophytes of *Datisca* and/or *Coriaria* (Mirza et al., 1994c; Nick et al., 1992). The probes and primers were demonstrated to be useful tools in ecological studies for detection and identification of the microorganism in soil and actinorhizae. In particular, the progress in population studies of uncultured strains has clearly demonstrated the advantages of molecular methods developed for rapid and direct detection and identification of *Frankia* on the basis of specific nucleotide sequences in the DNA or RNA over traditional approaches in assessing genetic diversity of these microorganisms in the environment.

The availability of unique strain-specific probes is still questionable, since sequence information is available for only a limited number of *Frankia* strains and only a limited number of experiments have been carried out with these probes. Oligonucleotide probes which reportedly are strain-specific were developed against the *nifH* mRNA sequences from two *Almus*-infective strains (Simonet et al., 1990). However, the probes were tested only in model experiments with axenic plants and were shown to be useful to detect and localize the strains within root nodule sections by in situ hybridization (Prin et al., 1993). Probes against mRNA are believed to be useful for discrimination between nucleic acids from living and dead cells, which is not possible on the basis of rRNA or DNA target sequences. In situ hybridization of *Alnus* and *Coriaria* root nodules (Mirza et al., 1994d; Prin et al., 1993) with specific probes which only hybridize to the mRNA of the *nifH* gene was demonstrated and thus may be used to detect *Frankia* gene expression under environmental conditions.

For broad population studies, probes and primers able to interact specifically with any *Frankia* genome are required but often turn out not to be specific enough in complex environments like soil and root nodules. Hahn et al. (1990a) reported the development of a *Frankia* genus-specific oligonucleotide probe targeted to 16S, rRNA that theoretically enables quantitative detection of total *Frankia* populations and is useful for detecting different *Frankia* strains in pure cultures, in mixtures with other microorganisms and in soil. This probe has been found, however, to lack absolute specificity for *Frankia* as was shown by its hybridization with RNA from two actinomycetes other than *Frankia* on the one hand, and the failure to hybridize with RNA from some *Frankia* strains under high stringency conditions, on the other hand. Similarly, highly efficient rep-PCR using REP and ERIC primers which are specific to short repetitive sequences in the genomes of prokaryotes and were previously reported (De Bruijn, 1992) not to amplify plant DNA from legume nodules have been shown to amplify actinorhizal plant DNA under conditions optimal for *Frankia* DNA amplification (Dobritsa et al., 1995).

Until now, no marker is available to trace the fate of a particular *Frankia* strain introduced to the environment, for example, that used for inoculation of a host plant to improve its symbiotic nitrogen fixation. For strains bearing plasmids, plasmid DNA was tested as a probe-target system in hybridization experiments on crushed nodules (Simonet et al., 1988). However, the hybridization signals were found to be weak, even though nodules are enriched for *Frankia* as compared with soils which generally contain much lower populations of *Frankia*. The application of oligonucleotide probes to identify introduced strains in soil remains restricted, due to their low specificity for strains. At least with root nodules, PCR amplification of the *rrn* genes followed by sequencing of the DNA still remains a method of choice to confirm strain identity (Mirza et al., 1994a). It should be noted that in case of well characterized strains sequences of only 150 nucleotides are sufficient to identify strains or discriminate between closely related strains as was shown with root nodule endophytes of *Datisca* and *Coriaria* (Mirza et al., 1994a,b).

The application of oligonucleotide probes for the detection of specific *Frankia* strains not only depends on specificity of the probes but also on the development of reliable isolation methods for target sequences. In the last few years, a number of procedures have been described to extract pure DNA or rRNA from environmental samples used for *Frankia* ecological studies, including soil (Hahn et al., 1991a; Picard et al., 1992) and actinorhizal root nodules (Baker and Mullin, 1994; Hahn et al., 1990b; McEwan et al., 1994; Simonet et al., 1994). In particular, 50 ng of DNA per mg of nodule tissue could be extracted within a few hours from individual nodule lobes as small as 3 to 25 mg (fresh weight) by a cetyl trimethyl ammonium bromide extraction procedure (Baker and Mullin, 1994), which may allow rapid screening of multiple samples for population studies.

It has been shown that use of DNA probes which hybridize to the *Frankia* genome or of PCR primers suitable for amplification of *Frankia* DNA allow comparison of *Frankia* sequences in DNA preparations obtained from root nodules without interference from plant DNA. Thus, diversity in RFLP and PCR patterns within and among endophyte populations of *Ceanothus*, a plant genus from which no infective pure-cultured *Frankia* strains have been isolated, was demonstrated by hybridization with DNA sequences from a *Frankia* strain and with a *nifHD* probe from *Klebsiella* (Baker and Mullin, 1994), as well as by rep-PCR with prokaryotic BOX primers (Murry et al., 1995).

Numerous studies on *Frankia* populations in soils using plant bioassay methods have found populations of infective *Frankia* to range from <1 to several thousand nodulation units (NU) per gram of soil. The PCR used to selectively amplify segments of the 16S rRNA gene has provided a quantitative estimate of the number of *Frankia* genomes (i.e. amplifiable target sequences), or genomic units (GU), present in a given soil (Myrold and Huss-Danell, 1994; Picard et al., 1992). As was shown by using DNA extraction directly from soil and subsequent nested PCR with a set of primers based on *Frankia*-specific regions in the 16S rRNA (Hahn et al., 1989a), numbers of GU greatly exceeded numbers of NU determined in parallel plant bioassays using *Alnus* spp. as the plant (Myrold and Huss-Danell, 1994). In addition, *Frankia* GU which are supposed to represent the total *Frankia* population in soil were unaffected, unlike NU, by soil amendment, host plant presence, or seasonal factors. Since little basic knowledge on *Frankia* population dynamics in soil and little information about specificity of the primers are yet available, it is difficult to interpret these data that contradict generally accepted points of view about seasonal dynamics of *Frankia* in soil and the effect of the presence of host plants on *Frankia* populations, based on plant baiting experiments. Such studies, if confirmed by others and extended to additional test sites, have the potential to expand greatly the knowledge of the population dynamics of *Frankia* in the soil.

Acknowledgements

Support for this work has been provided by United States Department of Agriculture National Research Initiative Competitive Grants Program (Awards 93-37305-9082 and 95-37305-2086 to BCM) and the Cooperation in Applied Science and Technology Program of the National Research Council of the US National Academy of Sciences and the Lexemual and Ester Hesler Visiting Professor Endowment Fund awards to SVD. We thank our many colleagues for sharing with us unpublished research results. Thanks are given also to Paula Goetting-Minesky and to Eunice Turner for help in preparing this manuscript.

References

Akimov V N and Dobritsa S V 1992 Grouping of *Frankia* strains on the basis of DNA relatedness. System. Appl. Microbiol. 15, 372–379.

Akkermans A D L, Hahn D and Mirza M S 1991 Molecular ecology of *Frankia*: Advantages and disadvantages of the use of DNA probes. Plant and Soil 137, 49–54.

Akkermans A D L, Mirza M S, Harmsen H J M, Blok H J, Herron P R, Sessitsch A and Akkermans W M 1994 Molecular ecology of microbes: A review of promises, pitfalls and true progress. FEMS Microbiol. Rev. 15, 185–194.

An C S, Hong J S and Mullin B C 1995a *nif*-gene organization and nucleotide sequence of *nif*K and *nif*E from *Frankia* FaC1 strain. 10th Int. Conf. on *Frankia* and Actinorhizal Plants, Davis, CA, August 6–11 (Abstract). Dept. of Environmental Horticulture, UC Davis, Davis, CA, USA.

An C S, Oh B S, Twigg P G and Mullin B C 1995b Nucleotide sequence of *nif*V from *Frankia* FaC1 strain. 10th Int. Conf. on *Frankia* and Actinorhizal Plants, Davis, CA, August 6–11. (Abstract). Dept. of Environmental Horticulture, UC Davis, Davis, CA, USA.

Baker D and O'Keefe D 1984 A modified sucrose fractionation procedure for the isolation of frankiae from actinorhizal root nodules and soil samples. Plant and Soil 78, 23–28.

Baker D D and Mullin B C 1994 Diversity of *Frankia* nodule endophytes of the actinorhizal shrub *Ceanothus* as assessed by RFLP patterns from single nodule lobes. Soil Biol. Biochem. 26, 547–552.

Benson D R and Silvester W B 1993 Biology of *Frankia* strains, actinomycete symbionts of actinorhizal plants. Microbiol. Rev. 57, 293–319.

Berry A M, McIntyre L and McCully M E 1986 Fine structure of root hair infection leading to nodulation in the *Frankia-Alnus* symbiosis. Can. J. Bot. 64, 292–305.

Bogusz D, Gherbi H, Nassar A, Franche C and Duhoux E 1995 Characterization of transcripts expressed in actinorhizal nodules of *Casuarina glauca*. 10th Int. Conf. on *Frankia* and Actinorhizal Plants, Davis CA, August 6–11. (Abstract). Dept. of Environmental Horticulture, UC Davis, Davis, CA, USA.

Bosco M, Fernandez M P, Simonet P, Materassi R and Normand P 1992 Evidence that some *Frankia* sp. strains are able to cross boundaries between *Alnus* and *Elaeagnus* host specificity groups. Appl. Environ. Microbiol. 58, 1569–1576.

Carlson D, Kuebler H and Twigg P 1995 Using the reverse transcriptase-based polymerase chain reaction (RT-PCR) to amplify a glutamine synthetase cDNA from nodules of *Elaeagnus augustifolia* 10th Int. Conf. on *Frankia* and Actinorhizal Plants, Davis CA, August 6–11. (Abstract). Dept. of Environmental Horticulture, UC Davis, Davis, CA, USA.

Chase M, Soltis D E, Olmstead R G, Morgan D, Les D H , Mishler B D, Duvall M R, Price R, Hills H G, Qiu Y, Kron K A, Rettig J H, Conti E, Palmer J D, Manhart J R, Sytsma K J, Michaels H J, Kress W J, Donoghue J D, Clark W D, Hedrén M, Gaut B S, Jansen R K, Kim K-J, Wimpee C F, Smith J F, Furnier G R, Straus S H, Xiang Q, Plunkett G M, Soltis P S, Swensen S M, Eguiarte L E, Learn G H Jr, Barrett S C, Graham S and Albert V A 1993 Phylogenetics of seed plants: an analysis of nucleotide sequences from the plastid gene *rbc*L. Ann. Missouri Bot. Garden 80, 528–580.

Conklin D S, McMaster J A, Culbertson and Kung C 1992 COT1, a gene involved in cobalt accumulation in *Saccharomyces cerevisiae*. Molec. Cell. Biol. 12, 3678–3688.

Cournoyer B and Normand P 1990 Uptake of macromolecules by *Frankia*. *In* Nitrogen Fixation: Achievements and objectives. Eds. P M Gresshoff, L E Roth, G Stacey and W E Newton. Chapman and Hall, New York, USA. 256 p.

Cronquist A 1988 The evolution and classification of flowering plants. New York Botanical Garden, Bronx, New York, USA.

De Bruijn F J 1992 Use of repetitive (repetitive extragenic palindromic and enterobacterial repetitive intergenic consensus) sequences and the polymerase chain reaction to fingerprint the genomes of *Rhizobium meliloti* and other soil bacteria. Appl. Environ. Microbiol. 58, 2180–2187.

Diouf D, Gherbi H, Prin Y, Franche C, Duhoux E and Bogusz D 1995 Hairy root nodulation of *Casuarina glauca* - a system for the study of symbiotic gene-expression in an actinorhizal tree. Molec. Plant Microbe Interact. 8, 532–537.

Dobritsa S V, Baker D D and Mullin B C 1995 DNA amplification fingerprinting of *Frankia* genomes using ERIC and REP primers. 10th Int. Conf. on *Frankia* and Actinorhizal Plants, Davis, CA, August 6–11 (Abstract). Dept. of Environmental Horticulture, UC Davis, Davis, CA, USA.

Doyle J J 1994 Phylogeny of the legume family: An approach to understanding of the origins of nodulation. Annu. Rev. Ecol. Syst. 25, 325–349.

Fernandez W P, Meugnier M, Grimont P A D and Bardin R 1989 Deoxyribonucleic acid relatedness among members of the genus *Frankia*. Int. J. Syst. Bacteriol. 39, 424–429.

Fleming A I, Wittenberg J B, Wittenberg B A, Dudman W F and Appleby C A 1987 The purification, characterization and ligand-binding kinetics of hemoglobins from root nodules of the non-leguminous *Casuarina glauca* - *Frankia* symbiosis. Biochem. Biophys. Acta 911, 209–220.

Franche C, Ndiaye A, Bogusz D, Gherbi H, Diouf D, Auguy F and Duhoux E 1995 Genetic transformations of *Allocasuarina verticillata* by *Agrobacterium tumefaciens*. 10th Int. Conf. on *Frankia* and Actinorhizal Plants, Davis, CA, August 6–11. (Abstract). Dept. of Environmental Horticulture, UC Davis, Davis, CA, USA.

Ganesh G, Misra A K, Chepelon C and Normand P 1994 Morphological and molecular characterization of *Frankia* sp. isolates from nodules of *Alnus nepalensis* Don. Arch. Microbiol. 161, 152–155.

Goetting-Minesky M P and Mullin B C 1994 Differential gene expression in an actinorhizal symbiosis: Evidence for a nodule-specific cysteine proteinase. Proc. Natl. Acad. Sci. USA 91, 9891–9895.

Guan C, Akkermans A D L, van Kammen A, Bisseling T and Pawlowski K 1995 Expression of a homologue of fatty acid reductase in actinorhizal nodules of *Alnus glutinosa* and in *Arabidopsis thaliana*. 10th Int. Conf. on *Frankia* and Actinorhizal Plants, Davis, CA, August 6–11. (Abstract). Dept. of Environmental Horticulture, UC Davis, Davis, CA, USA.

Hahn D, Dorsch M, Stackebrandt E and Akkermans A D L 1989a Synthetic oligonucleotide probes for identification of *Frankia* strains. Plant and Soil 118, 211–219.

Hahn D, Lechevalier M P, Fischer A and Stackebrandt E 1989b Evidence for a close phylogenetic relationship between members of the genera *Frankia*, *Geodermatophilus*, and "*Blastococcus*" and emendation of the family Frankiaceae. Syst. Appl. Microbiol. 11, 236–242.

Hahn D, Kester R, Starrenburg M J C and Akkermans A D L 1990a Extraction of ribosomal RNA from soil for detection of *Frankia* with oligonucleotide probes. Arch. Microbiol. 154, 329–335.

Hahn D, Starrenburg M J C and Akkermans A D L 1990b Oligonucleotide probes that hybridize with rRNA as a tool to study *Frankia* strains in root nodules. Appl. Environ. Microbiol. 56, 1342–1346.

Harriott O T, Hosted T J and Bensen D R 1995 Sequences of *nif*X, *nif*W, *nif*Z, *nif*B and two ORF in the *Frankia* nitrogen fixation gene cluster. Gene 161, 63–67.

Hönerlage W, Hahn D, Zepp K, Zeyer J and Normand P 1994 A hypervariable 23S rRNA region provides a discriminating target for specific characterization of uncultured and cultured *Frankia*. Syst. Appl. Microbiol. 17, 433–443.

Jacobsen-Lyon K, Jensen E O, Jorgensen J-E, Marcker K A, Peacock W J and Dennis E S 1995 Symbiotic and non-symbiotic haemoglobin genes in *Casuarina glauca*. Plant Cell 7, 213–223.

Jamann S, Fernandez M P and Moiroud A 1992 Genetic diversity of Elaeagnaceae-infective *Frankia* strains isolated from various soils. Acta Ecol. 13, 395–405.

Jamann S, Fernandez M P and Normand P 1993 Typing method for N_2-fixing bacteria based on PCR-RFLP-application to the characterization of *Frankia* strains. Molec. Ecol. 2, 17–26.

Lechevalier M P 1994 Taxonomy of the genus *Frankia* (*Actinomycetales*). Int. J. Syst. Bacteriol. 44, 1–8.

Liu Q and Berry A M 1991 The infection process and nodule initiation in the *Frankia-Ceanothus* root nodule symbiosis: A structural and histochemical study. Protoplasma 163, 82–92.

Lundquist P-O 1993 Dynamics of nitrogenase in the *Frankia-Alnus incana* symbiosis. PhD Dissertation, University of Umeå, Sweden.

Lundquist P-O and Vance C P 1995 Cloning of genes involved in nitrogen and carbon metabolism of *Alnus incana* root nodules. 10th Int. Conf. on *Frankia* and Actinorhizal Plants, Davis, CA, August 6–11. (Abstract). Dept. of Environmental Horticulture, UC Davis, Davis, CA, USA.

Maggia L, Nazaret S and Simonet P 1992 Molecular characterization of *Frankia* isolates from *Castuarina equisetifolia* root nodules harvested in West Africa (Senegal and Gambia). Acta Ecol. 13, 453–461.

McEwan N R, Wheeler C T and Milner J J 1994 Strain discrimination of cultured and symbiotic *Frankia* by RFLP-PCR. Soil Biol. Biochem. 26, 541–545.

McEwan N R and Wheeler C T 1995 PCR on the gram-positive organism *Frankia* without prior DNA extraction. TIG 11, 168.

Miller I M and Baker D D 1986 Nodulation of actinorhizal plants by *Frankia* strains capable of both root hair infection and intercellular penetration. Protoplasma 131, 82–91.

Mirza M S, Akkermans W M and Akkermans A D L 1994a PCR-amplified 16S rRNA sequence analysis to confirm nodulation of *Datisca cannabina* L. by the endophyte of *Coriaria nepalensis* Wall. Plant and Soil 160, 147–152.

Mirza M S, Hameed S and Akkermans A D L 1994c Genetic diversity of *Datisca cannabina*-compatible *Frankia* strains as determined by sequence analysis of the PCR-amplified 16S rRNA gene. Appl. Environ. Microbiol. 60, 2371–2376.

Murza M S, Hahn D and Akkermans A D L 1992 Isolation and characterization of *Frankia* strains from *Coriaria nepalensis*. Syst. Appl. Microbiol. 15, 289–295.

Mirza M S, Hahn D, Dobritsa S V and Akkermans A D L 1994b Phylogenetic studies on uncultured *Frankia* populations in nodules of *Datisca cannabina*. Can. J. Microbiol. 40, 313–318.

Mirza M S, Pawlowski K, Hafeez F Y, Chaudhary A M and Akkermans A D L 1994d Ultrastructure of the endophyte and localization of *nifH* transcripts in root nodules of *Coriaria nepalensis* Wall by in situ hybridization. New Phytol. 126, 131–136.

Mullin B C, Goetting-Minesky M P and Twigg P 1993 Differential gene expression in the development of actinorhizal root nodules. *In* New Horizons in Nitrogen Fixation. Eds. R. Palacios, J Mora and W E Newton. pp 309–314. Kluwer Academic Publishers, Boston, USA.

Murry M A, Konopka A S, Pratt S D and Davis S D 1995 The use of PCR-based typing methods to assess the diversity and phylogenetic relationships of nodule endophytes of *Ceanothus*. 10th Int. Conf. on *Frankia* and Actinorhizal Plants, Davis, CA, August 6–11. (Abstract). Dept. of Environmental Horticulture, UC Davis, Davis, CA, USA.

Myrold D D and Huss-Danell K 1994 Population dynamics of *Alnus*-infective *Frankia* in a forest soil with and without host trees. Soil Biol. Biochem. 26, 533–540.

Nalin R, Domenach A-M and Normand P 1995 Molecular structure of the *Frankia* spp. *nifD-K* intergenic spacer and design of *Frankia* genus compatible primer. Molec. Ecol. 4, 483–491.

Nazaret S, Cournoyer B, Normand P and Simonet P 1991 Phylogenetic relationships among *Frankia* genomic species determined by use of amplified 16S rDNA sequences. J. Bacteriol. 173, 4072–4078.

Nazaret S, Simonet P, Normand P and Bardin R 1989 Genetic diversity among *Frankia* isolated from *Casuarina* nodules. Plant and Soil 118, 241–247.

Newcomb W and Wood S M 1987 Morphogenesis and fine structure of *Frankia* (*Actinomycetales*): The microsymbiont of nitrogen-fixing actinorhizal root nodules. *In* International Review of Cytology, Vol. 109. Eds. G H Bourne, K W Joen and M Friedlander. pp 1–88. Academic Press, New York, USA.

Nick G, Paget E, Simonet P, Moiroud A and Normand P 1992 The nodular endophytes of *Coriaria* spp. form a distinct lineage within the genus *Frankia*. Mol. Ecol. 1, 175–181.

Normand P and Bousquet J 1989 Phylogeny of nitrogenase sequences in *Frankia* and other nitrogen-fixing microorganisms. J. Mol. Evol. 29, 436–447.

Normand P, Cournoyer B, Somonet P and Nazaret S 1992a Analysis of a ribosomal RNA operon in the actinomycete *Frankia*. Gene 111, 119–124.

Normand P, Gouy M, Cournoyer B and Simonet P 1992b Nucleotide sequence of *nifD* from *Frankia alni* strain ArI3: phylogenetic inferences. Mol. Biol. Evol. 9, 495–506.

Normand P, Simonet P, Prin Y and Moiroud A 1987 Formation and regeneration of *Frankia* protoplasts. Physiol. Plant. 70, 259–266.

Pawlowski K, Akkermans A D L, van Kammen A and Bisseling T 1995 Expression of *Frankia nif* genes in nodules of *Alnus glutinosa*. Plant and Soil 170, 371–376.

Picard C, Ponsonnet C, Paget E, Nesme X and Simonet P 1992 Detection and enumeration of bacteria in soil by direct DNA extraction and polymerase chain reaction. Appl. Environ. Microbiol. 58, 2717–2722.

Prin Y, Mallein-Gerin F and Simonet P 1993 Identification and localization of *Frankia* strains in *Alnus* nodules by in situ hybridization of *nif* H mRNA with strain-specific oligonucleotide probes. J. Exp. Bot. 44, 815–820.

Ribeiro A, Akkermans A, van Kammen A, Bisseling T and Pawlowski K 1995 A nodule-specific gene encoding a subtilisin-like protease is expressed in early stages of actinorhizal nodule development. Plant Cell 7, 785–794.

Schwencke J and Selim S 1995 Deforming factors of *Casuarina* root hairs (dnF) are induced in *Frankia* by flavonoids from *Casuarina* seeds. 10th Int. Conf. on *Frankia* and Actinorhizal Plants, Davis, CA, August 6–11. (Abstract). Dept. of Environmental Horticulture, UC Davis, Davis, CA, USA.

Séquin A and Lalonde M 1993 Modification of polypeptide patterns during nodule development in the *Frankia-Alnus* symbiosis. Symbiosis 15, 135–149.

Sellstedt A, Wullings B, Nyström U and Gustafsson P 1992 Identification of *Casuarina-Frankia* strains by use of polymerase chain reaction (PCR) with arbitrary primers. FEMS Microbiol. Lett. 93, 1–6.

Simonet P, Bosco M, Chapelon C, Moiroud A and Normand P 1994 Molecular characterization of *Frankia* microsymbionts from spore-positive and spore-negative nodules in a natural alder stand. Appl. Environ. Microbiol. 60, 1335–1341.

Simonet P, Grosjean M-C, Misra A K, Nazaret S, Cournoyer B and Normand P 1991 *Frankia* genus-specific characterization by polymerase chain reaction. Appl. Environ. Microbiol. 57, 3278–3286.

Simonet P, Le N T, du Cros E T and Bardin R 1988 Identification of *Frankia* strains by direct DNA hybridization of crushed nodules. Appl. Environ. Microbiol. 54, 2500–2503.

Simonet P, Normand P, Moiroud A and Bardin R 1990 Identification of *Frankia* strains in nodules by hybridization of polymerase chain reaction products with strain-specific oligonucleotide probes. Arch. Microbiol. 153, 235–240.

Soltis D E, Soltis P S, Morgan D R, Swensen S M, Mullin B C, Dowd J M and Martin P G 1995 Chloroplast gene sequence data suggest a single origin of the predisposition for symbiotic nitrogen fixation in angiosperms. Proc. Natl. Acad. Sci. USA 92, 2647–2651.

Sprent J I 1994 Evolution and diversity in the legume-*Rhizobium* symbiosis: Chaos theory? Plant and Soil 161, 1–10.

Swensen S 1996 The evolution of actinorhizal symbiosis: evidence for multiple origins of the symbiotic association. Am. J. Bot. (*In press*).

Suharjo U K S and Tjepkema T D 1995 Occurrence of hemoglobin in the nitrogen-fixing root-nodules of *Alnus glutinosa*. Physiol. Plant. 95, 247–252.

Swensen S and Mullin B 1996 Phylogenetic relationships among actinorhizal plants: The impact of molecular systematics and implications for the evolution of actinorhizal symbiosis. Physiol. Plant. (*In press*).

Swensen S M, Mullin B C and Chase M 1994 Phylogenetic affinities of the Datiscaceae based on analysis of nucleotide sequences from the plastid *rbcL* gene. Syst. Bot. 19, 157–168.

Tisa L S and Ensign J C 1987 Formation and regeneration of protoplasts of the actinorhizal nitrogen-fixing actinomycete *Frankia*. Appl. Environ. Microbiol. 53, 53–56.

Tjepkema J D 1983 Hemoglobins in the nitrogen-fixing root nodules of actinorhizal plants. Can. J. Bot. 61, 2924–2929.

Twigg P G 1993 Isolation of a nodule-specific cDNA encoding a putative glycine-rich protein from *Alnus glutinosa*. PhD Dissertation University of Tennessee, Knoxville, TN, USA.

Twigg P G, Kuebler H and Mullin B C 1995a NS85: A nodule enhanced cDNA from *Alnus glutinosa* encodes a glutamine synthetase. 10th Int. Conf. on *Frankia* and Actinorhizal Plants, Davis, CA, August 6–11. (Abstract). Dept. of Environmental Horticulture, UC Davis, Davis, CA, USA.

Twigg P G, Pawlowski K and Mullin B 1995b NS84: A nodule-specific cDNA from *Alnus glutinosa* encoding a putative glycine-rich protein. 10th Int. Conf. on *Frankia* and Actinorhizal Plants, Davis, CA, August 6–11. (Abstract). Dept. of Environmental Horticulture, UC Davis, Davis, CA, USA.

Wigington J, Bock J V, Battershell T and Johnson J D 1995 Cloning and characterization of a *Frankia* ArI5 DNA fragment with promoter activity. 10th Int. Conf. on *Frankia* and Actinorhizal Plants, Davis CA, August 6–11. (Abstract). Dept. of Environmental Horticulture, UC Davis, Davis, CA, USA.

Wülfing C, Lombardero J and Pluckthun A 1994 An *Escherichia coli* protein consisting of a domain homologous to FK506-binding proteins (FKBP) and a new metal binding motif. J. Biol. Chem. 269, 2895–2901.

Section editor: H. Lambers
Guest editors: G.H. Elkan and R.G. Upchurch

Plant and Soil **186:** 21–28, 1996.

Diversity of Rhizobia isolated from various *Hedysarum* species

B. Dov Kishinevsky[1], Dipankar Sen[2] and Guang Yang[3]
[1]*Department of Agronomy and Natural Resources, ARO, The Volcani Center, Bet Dagan, 50250, Israel**,
[2]*Department of Soil and Crop Sciences, Texas A and M University, College Station, Texas 77843, USA and*
[3]*S. Tolkowsky Laboratory, ARO, The Volcani Center, Bet Dagan, 50250, Israel*

Key words: Hedysarum spp., plasmid profile, RAPD, *Rhizobium*, serogroup

Abstract

Cultural and physiological properties, serology, plasmid profiles and infective traits were determined for 23 strains of rhizobia isolated from various *Hedysarum* species: *H. coronarium* (common name: sulla) (16), *H. carnosum* (1), *H. alpinum* (3), *H. mackenzii* (2) and *H. pallens* (1) from Portugal, Spain, Tunisia, Alaska and Israel. Strains isolated from *H. alpinum*, *H. mackenzii* and *H. pallens* have slow growth rates on yeast-extract mannitol medium and were unable to nodulate *H. coronarium* plants, whereas the latter were effectively nodulated by all sixteen fast growing strains from sulla. Regardless of the country of origin all *H.coronarium* strains fell into one serogroup and were not serologically related with strains of other *Hedysarum* species. The RAPD (random amplified polymorphic DNA) fingerprinting method which was carried out on five *H. coronarium* and three *H. alpinum* strains allowed distinction to be made among serologically related rhizobia. No particular plasmid profile pattern was observed in relation to the host or geographical origin of the strains.

Introduction

The genus *Hedysarum* L. includes 70 - 100 species of widely distributed herb legumes, especially in the Mediterranean, temperate Europe, North and South Africa, Asia Minor, Siberia and in North America from Arizona into Alaska and the arctic regions of Canada (Allen and Allen, 1981). Some *Hedysarum* spp. are known to have good forage value in semiarid conditions and are successful crops for soil improvement and honey production in North Africa and the near and middle east (Kernick, 1978). *H. coronarium* (common name: sulla) root nodules and their ability to fix N_2 from the atmosphere were reported by Rodriguez-Navarro et al. (1991) and Squartini et al. (1993). The extent of N_2 fixation for *H. coronarium* has been estimated by Gurfel et al. (1982) who reported that in greenhouse experiments the amount of N fixed in plants inoculated by effective strains of *Rhizobium* ranged from 63.2 to 81.2 mg/plant as compared to 1.8 mg/plant for the uninoculated treatment. The input of fixed N under field conditions was 141 kg ha^{-1} $year^{-1}$. Little information is, however, available regarding N_2-fixing capabilities of other *Hedysarum* species. Even less is known of the rhizobia responsible for this fixation. According to Allen et al. (1964) and Roizin (1959) profusely nodulated plants of *H. alpinum* and *H. mackenzii* are found in the soils of Alaska and the arctic regions of the former USSR. Hely and Ofer (1972) reported the nodulation of *H. spinosissimum (pallens)* in Israel. Root nodules are also known in *H. boreale, H. capitatum, H. caucasicum, H. iomiticum, H. mongolicum* and *H. obscurum* (Allen and Allen, 1981). Since scant information is available on *Hedysarum* rhizobia we undertook this study to examine the characteristics of rhizobia isolated from the same or different species of *Hedysarum.* The bacteria were evaluated in terms of their symbiotic effectiveness, growth rates, plasmid profiles, serological, biochemical and infective properties. The RAPD (random amplified polymorphic DNA) fingerprinting method was also used for discriminating among serologically related strains.

* FAX No: +97239669642

Materials and methods

Origin and maintenance of rhizobia

A list of strains of rhizobia included in this work, their parent hosts, countries of origin and the donor institutes are presented in Table 1. Stock cultures were maintained on yeast extract mannitol (YEM) agar.

Growth rate

For growth experiments all strains of rhizobia were cultured in 250 mL Erlenmeyer flasks containing 100 mL of liquid YEM media (pH 7). The inoculated flasks were incubated at 28 °C on a gyratory shaker (New Brunswick Scientific Instrument Co., USA) at 140 rev min^{-1}. Counts of viable rhizobia were carried out according to Vincent (1970). For the growth media tests *Rhizobium leguminosarum* bv. *viceae* strain 250V and *Bradyrhizobium* sp. (*Arachis*) strain 280A, from the culture collection of the Volcani Center, were used as standard fast and slow-growing strains respectively.

Colony characteristics

Observations on growth rate, size and morphology of colonies were made for cultures grown on YEM agar at 27 °C for up to 10 days. The same conditions were used to assess the ability of strains to grow on peptone glucose agar.

Temperature range of growth

Plates of YEM agar were inoculated and incubated for 20 days at 5 °C and 15 °C, for 10 days at 27 °C and for 7 days at 37 °C, 40 °C and 42 °C.

Change of pH on YEM

Bromthymol blue incorporated into the agar medium (Vincent, 1970) was used to demonstrate the change of pH.

Tolerance to pH and NaCl

Growth was assessed in YEM broth (without $CaCO_3$) adjusted to pH 4.0, 5.0, 7.0, 8.0, 9.0 and 10.0 by addition of sterile acid or alkali. Inoculated tubes were incubated for ten days at 27 °C and then scored for growth. Growth in the presence of 1% and 2% NaCl on YEM agar was determined after 10 days of incubation at 27 °C.

Serological characteristics

Four antisera against strains of rhizobia isolated from *H. coronarium* (2), *H. alpinum* (1) and *H. mackenzii* (1) were produced during the course of this study. The antisera were prepared according to the immunization schedule adopted in previous work (Kishinevsky and Gurfel, 1980). The indirect enzyme-linked immunosorbent assay (ELISA) of Engvall and Perlmann (1971) with the modifications of Kishinevsky and Maoz (1983) and Fuhrman and Wollum II (1985) was followed as described by Kishinevsky et al. (1984).

Plasmid profiles

All strains, except those isolated from Alaska, were grown on tryptone-yeast extract medium (TY) (Beringer, 1974) and then subcultured for overnight growth in peptone-yeast extract-tryptone medium (HP) (Hynes et al., 1985) until a turbidity of 40 to 70 Klett units (red filter) was obtained. The Alaskan isolates grew poorly in TY and showed almost no growth in HP. They were grown on YM broth to a turbidity of 30 to 50 Klett units for developing their plasmid profiles. Plasmid profiles were determined by horizontal agarose gel electrophoresis following the Eckhardt (1978) method as modified by Hynes et al. (1985) except that sodium dodecylsulfate (SDS) was incorporated in 0.7% agarose gels instead of adding to a second set of wells in the gels. Plasmid sizes were determined by comparing their mobility (Casse et al., 1979) with three plasmids of known sizes from *R. leguminosarum* bv. *phaseoli* strain DB1 (Berryhill et al., 1985).

RAPD fingerprinting

Rhizobium cells were grown on a gyratory shaker to stationary phase in 50 mL of YEM broth, and chromosomal DNA was extracted following the procedure described by Webb and Wilson (1991). RAPD markers were amplified on templates containing DNA from rhizobia, using single oligonucleotide primers. The 10-mer oligonucleotides were from kit D of Operon Technologies Inc., USA. The polymerase chain reaction (PCR) was carried out in 25 μL containing 10 mM Tris-HCl pH 9.0, 0.1% Triton X-100, 2.0 mM $MgCl_2$, 0.2 mg mL^{-1}BSA, 0.3 mM each of dCTP, dGTP, dATP and dTTP (Boehringer Mannheim, pH 7.0), 0.25 units Taq DNA polymerase (Appligene, France), 1.6 μg mL^{-1} template DNA and 0.5 μM primer. Each amplification reaction contained a single unique random primer and was overlaid with 40 μL mineral oil. For all data used in the present analysis PCR were carried out using a thermal reactor (Hybaid, USA) with the following profile: 1) 96 °C for 5 min × 1 cycle;

Table 1. Rhizobia used in this study

Strain	Number of strains	Parent host	Country of origin	Donor institute
3-3, 1b, 13b	3	*H. coronarium*	Portugal	Institute of Agriculture, Portugal
851, 853, 857	3	*H. coronarium*	Spain	Ciudad University, Spain
73, 588, 611, 616, 635, 636, 637, 793, 802, 803	10	*H. coronarium*	Tunisia	NifTAL Project Hawaii, USA
614	1	*H. carnosum*	Tunisia	NifTAL Project Hawaii, USA
1332, 1334, 1335	3	*H. alpinum*	Alaska, USA	NifTAL Project Hawaii, USA
1337, 1338	2	*H. mackenzii*	Alaska, USA	NifTAL Project Hawaii, USA
W4	1	*H. pallens*	Israel	Volcani Center, Israel

2) 96 °C for 50 sec, 37 °C for 1 min, 72 °C for 2 min × 40 cycles; 3) 72 °C for 5 min × 1 cycle.

Greenhouse experiment

Nodulation tests were carried out in the greenhouse using sterilized soil in tin-coated metal containers of 3.0-L capacity. Each container held 3.5 kg of sandy nitrogen-poor soil. Ten surface-sterilized seeds were planted in each pot and inoculated with 5 mL of the appropriate strain of rhizobial broth culture containing 5×10^8 cells mL^{-1}. After addition of inoculum, the soil was covered with paraffin-coated sand as described by Vincent (1970), and 7 days after sowing and inoculation the seedlings were thinned to five per pot. Effectiveness study was done only with *H. coronarium* in which an uninoculated control was included. These plants were harvested 12 weeks after planting. Plant roots were observed for nodulation and dry weight of the tops was determined.

Results

Comparison of cultural and physiological properties

All strains formed almost colorless circular and convex colonies on YEM agar but differed greatly in the amounts of extracellular gum. *H. coronarium* strains, irrespective of their origin, had colonies appearing after 2 to 4 days with a diameter of 3 mm or more at 7 days (Table 2). Similar growth parameters were observed for strain 614 isolated from *H. carnosum.* In contrast most of the Alaskan strains (1334 and 1335 of *H. alpinum* and 1337, 1338 of *H. mackenzii*) and strain W4 isolated from *H. pallens* showed slower growth with colonies appearing 5 to 6 days after plating with a diameter not exceeding 3 mm within 7 days at 27 °C. Only the strain 614 grew on glucose peptone agar. Similar growth patterns of the strains of rhizobia were observed in YEM broth (Figure 1). Maximum growth was observed after 4 - 8 days with *H. coronarium* and *H. carnosum* strains but with *H. alpinum*, *H. mackenzii* and *H. pallens* strains growth reached its peak only after 12 - 14 days. Growth characteristics of the standard strains 250V and 280A were similar to those of strain 853 (*H. coronarium*) and W4 (*H. pallens*) respectively and data are not presented. As shown in Table 2 only strain 614 was able to grow at 37 °C. This strain was also the only one which grew on YEM agar at 40 °C and 42 °C. Eight strains from *H.coronarium* and the *H. carnosum* strain 614 were able to grow at pH 10. All strains from Alaska and strain W4 isolated from *H. pallens* were found to be susceptible to high concentrations (1%-2%) of NaCl in YEM broth.

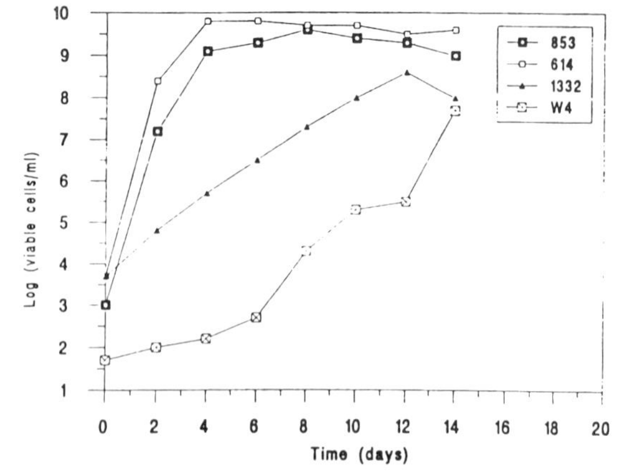

Figure 1. Comparison of growth rates of *Rhizobium* strains isolated from *H. coronarium* (853), *H. carnosum* (614), *H. alpinum* (1332) and *H. pallens* (W4)

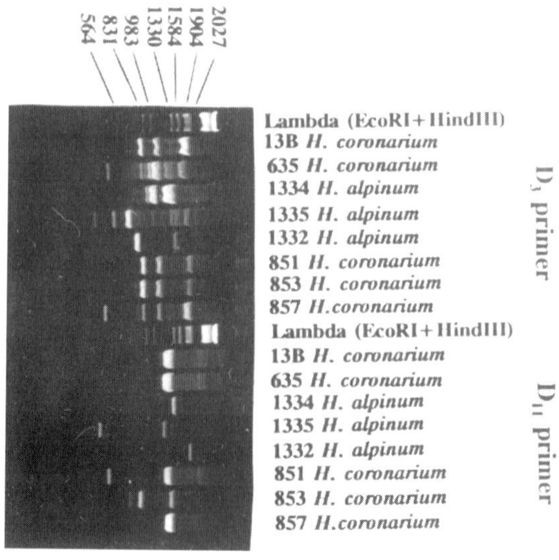

Figure 2. RAPD profiles generated with two primers applied to 8 *Rhizobium* strains from *H. coronarium* (5) and *H. alpinum* (3). The leftmost lane of each primer contain lambda DNA digested by EcoRI + Hind III and used as a molecular marker. Sizes are shown in bp.

With one exception all strains from *H. coronarium, H. carnosum* and *H. mackenzii* acidified the YEM medium. For two strains of *H. alpinum* a phenomenon of alkalization followed by acidification was observed. Strain W4 alkalized the YEM medium (Table 2).

Serological characteristics

The serological properties of the strains were evaluated using four antisera: two against strains 3-3 and 13b of *H. coronarium*, one against strain 1332 of *H. alpinum* and one against strain 1337 of *H. mackenzii*. Somatic cross-agglutination and ELISA analysis applied to 23 strains of rhizobia showed that the sulla-specific strains, irrespective of their origin, reacted with antisera against *H. coronarium* strains and none of them reacted with antisera against strains specific to *H. alpinum* or *H. mackenzii* (Table 3). Strain 614 from *H. carnosum* was serologically related to the *H. coronarium* strains. However, the ELISA readings for *H. coronarium* strains with antisera 3-3 and 13B were significantly higher than that of *H. carnosum* strain. None of *H. alpinum, H. mackenzii* or *H. pallens* strains reacted with antisera specific to *H. coronarium* strains. There were no ELISA reactions of *H. alpinum* strains with antibodies against *H. mackenzii* strain 1337 and vice versa. Moreover, *H. alpinum* strains form at least two serological groups. Strain W4 from *H. pallens* did not react with any of the antisera used.

Diversity within serogroups

RAPD test was carried out on five *Rhizobium* strains isolated from *H. coronarium* and three strains isolated from *H. alpinum*. Two primers, OPD-3 (D_3) and OPD-11 (D_{11}) that had yielded reproducible fragments in preliminary tests with the strains tested were used. The RAPD fingerprinting test detected diversity among strains (Figure 2) that were not distinguished by agglutination and ELISA methods (Table 3). For example primer D_3 distinguished 7 of the 8 strains tested. Two *H. coronarium* strains (851 and 853) produced similar 0.8 kb, 1.0 kb and 1.9 kb fragments. However, these two strains were not identical to one another when typed with primer D_{11} (Figure 2). *Rhizobium* strains isolated from *H. alpinum* (1332, 1334 and 1335) were distinguishable from one another by RAPD fingerprinting with each of the primers used. When different preparations of DNA from each of the eight strains were tested in parallel using the D_3 and D_{11} primers, the differences in RAPD profiles were highly reproducible.

Plasmid profile

Plasmids of sizes ranging from 22 to 885 Mdal were detected among isolates from *H. coronarium, H. carnosum, H. alpinum, H. mackenzii* (Figure 3). Most of the strains have one or two plasmids of sizes between

Table 2. Cultural and physiological characteristics of rhizobia isolated from different *Hedysarum* species

	Hedysarum sp.				
	coronarium	*carnosum*	*alpinum*	*mackenzii*	*pallens*
No. of strains from sp.	16	1	3	2	1
First colony appearance at 27 °C after:					
2-4 days	16[a]	1	-[b]	-	-
5-6 days	-	-	3	2	1
Colony diameter after 7days:					
< 3mm	-	-	2	2	1
> 3mm	16	1	1	-	-
Growth at 15 °C	16	1	3	2	-
Growth at 37 °C	-	1	-	-	-
Growth at pH:					
4.0	7	-	-	-	-
5.0	16	-	1	-	-
7.0	16	1	3	2	1
8.0	15	1	2	-	-
9.0	12	1	-	-	-
10.0	8	1	-	-	-
Tolerance of NaCl:					
1%	5	1	-	-	-
2%	-	1	-	-	-
Acid and alkaline reactions:					
Acidification	15	1	1	2	-
Alkalization followed by acidification	1	-	2	-	-
Alkalization	-	-	-	-	1

[a] Numbers indicate number of positive strains.
[b] All strains negative.

100 and 500 Mdal except strains 588 and 637 which have no plasmids and strains 853 and 857 which have four plasmids each. Strain 614, which is the only isolate from *H. carnosum* and showed a fast rate of growth, showed a very large plasmid of 885 Mdal size. The isolate from *H. pallens* was not tested.

Nodulation studies

Sixteen strains were tested on *H. coronarium* for their ability to nodulate this host and their effectiveness on it. All eleven sulla isolates were able to nodulate *H. coronarium* whereas the strains isolated from other *Hedysarum* species failed to form nodules on it. Eight sulla isolates (611, 802, 637, 588, 635, 636, 616, and 73) did not differ significantly with regard to their effect on *H. coronarium* top dry weight (Figure 4). We have not yet been able to test all the other species of *Hedysarum*. We could not obtain seeds for some of the species. *H. alpinum* did not grow under our greenhouse conditions. *H. pallens* was nodulated by W4, a strain isolated from the homologous host, but this strain has not yet been tested on other species of *Hedysarum*.

Table 3. Cross-agglutination and ELISA tests with rhizobia isolated from different *Hedysarum* species

Strain	Parent host	Somatic agglutination titre (S) and ELISA values (A_{405}) with antisera							
		H. coronarium				*H. alpinum*		*H. mackenzii*	
		13		3-3		1332		1337	
		S	A_{405}[a]	S	A_{405}	S	A_{405}	S	A_{405}
3-3	*H. coronarium*	1280[b]	2.0	2560	1.8	-	-	-	-
13B	"	2560	1.6	2560	1.6	-	-	-	-
1B	"	1280	1.4	1280	1.4	-	-	-	-
851	"	640	1.2	1280	1.5	-	-	-	-
853	"	1280	1.2	2560	1.5	-	-	-	-
857	"	1280	1.0	1280	1.5	-	-	-	-
73	"	2560	1.1	1280	1.5	-	-	-	-
588	"	640	0.9	640	1.2	-	-	-	-
611	"	2560	1.2	2560	1.5	-	-	-	-
616	"	2560	1.0	1280	1.0	-	-	-	-
635	"	1280	1.2	2560	1.4	-	-	-	-
636	"	640	1.1	640	1.2	-	-	-	-
637	"	320	0.8	640	1.2	-	-	-	-
793	"	2560	1.3	2560	1.3	-	-	-	-
802	"	640	0.7	1280	1.0	-	-	-	-
803	"	1280	0.9	1280	1.0	-	-	-	-
614	*H. carnosum*	320	0.6	160	0.4	-	-	-	-
1332	*H. alpinum*	-	-	-	-	2560	1.5	-	-
1334	"	-	-	-	-	-	-	-	-
1335	"	-	-	-	-	-	-	-	-
1337	*H. mackenzii*	-	-	-	-	-	-	2560	1.2
1338	"	-	-	-	-	-	-	2560	1.3
W4	*H. pallens*	-	-	-	-	-	-	-	NT

[a]Detecting antibody, 8 μg mL^{-1}, Antigen, 10^9 cells mL^{-1}. Dilution of enzyme-labeled protein a=1:4000. ELISA readings of $A_{405} < 0.09$ were considered as negative.
[b]All values represent the reciprocal of the highest serum dilution in which a strong agglutination occurred.
-, no antigen-antiserum reaction; NT, not tested.

Discussion

Twenty three strains of rhizobia isolated from five *Hedysarum* species (*H. coronarium, H. carnosum, H. alpinum, H. mackenzii* and *H. pallens*) are phenotypically heterogeneous with respect to their growth rates, tolerance of NaCl, serological properties and plasmid content. The strains isolated from *H. alpinum, H. mackenzii* and *H. pallens* grew slowly on YEM media like members of the genus *Bradyrhizobium*, whereas the isolates from *H. coronarium* and *H. carnosum* behaved as fast-growers. Although for some strains a phenomenon of alkalization followed by acidification was observed, the majority of rhizobia in both groups were acid producers like members of the genus *Rhizobium*. All strains isolated from Alaska, the isolate from nodules of *H. pallens* and most of the sulla-specific isolates were found to be susceptible to high concentrations (1-2%) of NaCl in YEM agar, a feature usually characterizing root-nodule bacteria with the exception of *R. meliloti* (Graham and Parker, 1964). Our study indicates that isolates from *Hedysarum* have some features in common with both the fast-growing and the slow-growing types of rhizobia: the genera *Rhizobium* and *Bradyrhizobium*. *R. fredi*, the fast-growing symbiont of soybeans, was also reported to have traits common to both the genera *Rhizobium* and *Bradyrhizobium* (Stowers and Eaglesham, 1984). We have not yet established the positive nodulation specificities for all the rhizobia used in this study, but assuming the authenticity of the rhizobial isolates from different host species, we suggest that rhizobia isolated from *H. coronarium* form a distinct nodulation group because none of the strains isolated from other *Hedysarum* species

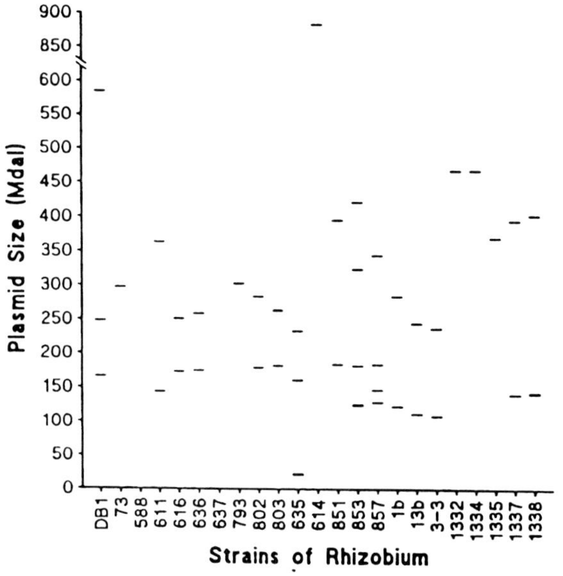

Figure 3. Plasmid profile types of *Rhizobium* strains isolated from different *Hedysarum* species.

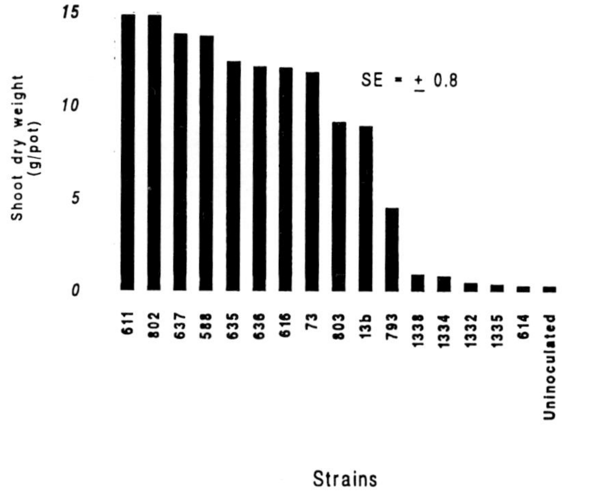

Figure 4. Response of sulla to inoculation with *Rhizobium* strains isolated from different *Hedysarum* species.

was able to nodulate sulla. No particular plasmid profile pattern was observed in relation to the host or geographical origin of the strains. However, all the plasmid-carrying sulla-specific strains, irrespective of their geographical origin showed a slow moving band of estimated mol. wt. of over 1000 Mdal, which may be a very large megaplasmid as reported by Mozo et al. (1988). This was not seen in other isolates, indicating again that these strains form a distinct group. Plasmids were detected in most of the strains tested in this study, but we have not yet determined whether they carry the symbiotic genes. It was reported, that in several species of *Rhizobium* the symbiotic genes are located on plasmids, whereas there is no evidence that plasmids in *Bradyrhizobium* carry symbiotic genes (Long, 1992). It should be noted that all strains isolated from *H. coronarium*, irrespective of their origin, fell into one serogroup and were not serologically related to strains isolated from *H. alpinum, H. mackenzii* and *H. pallens*. The latter, according to our results, may belong to a new species of the slow-growing type of root-nodule bacteria, but this awaits further investigations.

Acknowledgements

The authors are indebted to Drs P Somasegaran and R Bruce Martin (NifTAL Project, University of Hawaii, USA.), Prof A Garcia (Institute of Agriculture, Portugal) and Prof Santa Maria (Ciudad University, Spain) for providing strains of rhizobia, to Prof R W Weaver (Texas A and M University, USA) and Prof M Bar-Joseph (Volcani Center, Israel) for providing their laboratory facilities for developing the plasmid profiles and RAPD analysis, and to Dr Debora Gurfel (Volcani Center, Israel) for her able technical assistance.

References

Allen O N and Allen E K 1981 The Leguminosae. Univ. of Wisconsin Press, Madison,WI, USA. 812 p.

Allen E K, Allen O N and Klebesabel L J 1964 An insight into symbiotic nitrogen-fixing plant association in Alaska. *In* Alaska Science Conference, Science in Alaska. Anchorage, Alaska. Proceedings 14, 54–63.

Beringer J E 1974 R factor transfer in *Rhizobium leguminosarum*. J. Gen. Microbiol. 84, 188–198.

Berryhill D L, Shroeder M B and Obermiller T L 1985 Plasmid contents of commercial *Rhizobium leguminosarum* biovar. *phaseoli* strains. Research report #105, p 4, Agricultural Experiment Station, North Dakota State University, Fargo, USA.

Casse F, Boucher C, Julliot J S, Michel M and Denarie J 1979 Identification and characterization of large plasmids in *Rhizobium meliloti* using agarose gel electrophoresis. J. Gen. Microbiol. 113, 229–242.

Eckhardt T 1978 A rapid method for the identification of plasmid deoxyribonucleic acid in bacteria. Plasmid 1, 584–588.

Engvall E and Perlmann P 1971 Enzyme-linked immunosorbent assay (ELISA). Quantitative assay of immunoglobulin G. Immunochemistry 8, 871–874.

Fuhrmann J and Wollum II A G 1985 Simplified enzyme-linked immunosorbent assay for routine identification of *Rhizobium japonicum* antigens. Appl. Environ. Microbiol. 49, 1010–1013.

Graham P H and Parker C A 1964 Diagnostic features in characterization of the root-nodule bacteria of legumes. Plant and Soil 3, 383–396.

Gurfel D, Lobel R and Schiffmann J 1982 Symbiotic nitrogen-fixing activity and yield potential of inoculated *Hedysarum coronarium* in Israel. Isr. J. Bot. 31, 296–304.

Hely F W and Ofer I 1972 Nodulation and frequencies of wild leguminous species in the Northern Negev region of Israel. Aust. J. Agric. Res. 23, 267–284.

Hynes M F, Simon R and Puhler A 1985 The development of plasmid-free strains of *Agrobacterium tumefaciens* by using incompatibility with a *Rhizobium meliloti* plasmid to eliminate pAtC58. Plasmid 13, 99–105.

Kernick M D 1978 Indigenous arid and semi-arid forage plants of North Africa, the Near and Middle East. *In* Ecological Management of Arid and Semi-arid Rangelands in Africa and the Near and Middle East, vol IV, pp 599–619. FAO, Rome, Italy.

Kishinevsky B and Gurfel D 1980 Evaluation of enzyme-linked immunosorbent assay (ELISA) for serological identification of different *Rhizobium* strains. J. Appl. Bacteriol. 49, 514–526.

Kishinevsky B and Maoz A 1983 ELISA identification of *Rhizobium* strains by use of enzyme labeled protein A. Curr. Microbiol. 2, 45–49.

Kishinevsky B, Maoz A, Gurfel D and Nemas C 1984 A comparative evaluation of direct and indirect enzyme-linked immunosorbent assay (ELISA) in serological studies of *Rhizobium* strains. *In* Symbiotic Nitrogen Fixation, Vol 1. Ed. B S Ghai. USG Publishers and Distributors, Ludhiana, India.

Long S R 1992 Genetic analysis of *Rhizobium* nodulation. *In* Biological Nitrogen Fixation. Eds. G Stacey, R H Burris and H J Evans. pp 560–597. Chapman and Hall, New York.

Mozo T, Cabrera E and Ruiz-Argueso T 1988 Diversity of plasmid profiles and conservation of symbiotic nitrogen fixation genes in newly isolated *Rhizobium* strains nodulating sulla (*Hedysarum coronarium* L.). Appl. Environ. Microbiol. 54, 1262–1267.

Rodriguez-Navarro D N, Temprano F and Orive R 1991 Survival of *Rhizobium* sp. (*Hedysarum coronarium* L.) on peat-based inoculants and inoculated seeds. Soil Biol. Biochem. 23, 375–379.

Roizin M B 1959 The root nodules of leguminous plants of the Kola Peninsula, Bot. Zh. 44, 467–474 (*In Russian*).

Squartini A, Dazzo F B, Casella S and Nuti M P 1993 The its drought-tolerant host *Hedysarum coronarium*. Symbiosis 15, 227–238.

Stowers M D and Eaglesham A R J 1984 Physiological and symbiotic characteristics of fast-growing *Rhizobium japonicum*. Plant and Soil 77, 3–14.

Vincent J M 1970 A manual for the practical study of the root nodule bacteria. Blackwell Scientific Publications, Oxford, UK.

Webb K E and Wilson R 1991 Extraction, purification and assay of DNA. *In* Genetic Manipulation: Techniques and Application. Ed. J Grange, M A Fox and N Morgan. pp 1–14. Technical series, Blackwell Scientific Publications, Oxford, UK.

Section editor: H Lambers
Guest editors: G.H. Elkan and R.G. Upchurch

Plant and Soil **186:** 29–38, 1996.

Diazotrophic endophytes: progress and prospects for nitrogen fixation in monocots

Eric W. Triplett
Department of Agronomy and the Center for the Study of Nitrogen Fixation, University of Wisconsin-Madison, Madison, WI 53706–1597, USA *

Key words: Acetobacter, Azoarcus, Hebaspirillum, Klebsiella, maize, sugar cane

Abstract

The development of nitrogen fixation in maize can be considered the "holy grail" of nitrogen fixation research. As nitrogen fertilization is one of the highest costs of corn production, the development of a symbiosis between diazotrophic bacteria and corn would be of enormous economic value. Such a discovery would also improve human health as it would decrease the amount of nitrate in ground water as well as in corn cultured for human consumption. Several proposals have been made toward this end. These include: a) the transfer of root nodulation genes from a legume to maize; b) the expression of the bacterial *nif* regulon in maize organelles; and c) the development of corn lines with the ability to accept fixed nitrogen from diazotrophs in the rhizosphere. All of these proposals have enormous technical problems to overcome such that the development of nitrogen-fixing corn in the near term has been considered unlikely. An alternative and less-technically challenging approach may be a thorough study of non-pathogenic bacterial endophytes that already inhabit the corn plant.

The discovery of a nitrogen-fixing bacterial-sugar cane association by Döbereiner and coworkers in Brazil illustrates the enormous potential of endophytic bacteria to enhance grass biomass in the absence of nitrogen fertilizer. Döbereiner and coworkers have discovered diazotrophic strains of *Acetobacter diazotrophicus* and *Herbaspirillum seropedicae* in lines of sugar cane that were bred in the absence of nitrogen fertilizer. The Brazilian group has also demonstrated that sugar cane plants infected with these diazotrophs are capable of deriving all of their nitrogen needs from N_2.

Recently, the presence of non-pathogenic endophytic bacteria in corn has been shown. Based on this evidence and using the sugar cane paradigm as an example, investigators are working toward the discovery and analysis of diazotrophic endophytes in corn which includes the search for corn germplasm that would readily benefit from an association with these bacteria. Several diazotrophic endophytes have been identified in grass species that are members of the α-, β-, and γ-subclasses of the proteobacteria. Our understanding of the ability of these bacteria to enhance the growth of grasses through nitrogen fixation is only beginning to be explored but this approach is thought to be far less technically challenging than are other proposals to develop ‘nitrogen fixation in maize.

Abbreviations: VAM – vesicular-arbuscular mycorrhizae.

Justification for the development of nitrogen-fixing corn

The development of nitrogen-fixing corn has often been touted as one of the ultimate goals of nitrogen fixation research. Many research proposals to granting agencies interested in funding nitrogen fixation research, often as part of their justification for funding, state that the proposed work can eventually lead toward the "holy grail" of nitrogen fixing corn. The economic and human health benefits of nitrogen-fixing corn are enormous. Nitrogen fertilization of non-leguminous crops such as corn is one of the most expensive inputs in agriculture. Eliminating nitrogen fertilization of corn will also reduce our dependence on fossil fuels as they are necessary for the production of nitrogen fertilizer. From health and environmental perspectives, the qual-

* FAX No: 608–262–5217. E mail: ewtriple@facstaff.wisc.edu

ity of ground water is adversely affected by nitrogen fertilization of corn. Nitrate in ground water eventually enters our rivers, lakes, and streams. In these waterways, excess nitrate and other nutrients causes eutrophication followed by oxygen 'depletion. This causes massive death of fish and other wildlife who depend on these sources of fresh water.

This review will briefly describe past proposals to develop nitrogen-fixing corn whose success would require overcoming major technical obstacles. However, the primary focus of this review is to describe a new approach now being examined by several laboratories that avoids these technical problems and instead relies on the ability of endophytic diazotrophs to provide the fixed nitrogen necessary for plant growth. My confidence that success can be achieved in this work is based on the recent discovery 'of nitrogen 'fixation by bacteria in sugar cane as described by the work of Döbereiner, Boddey, and coworkers (Boddey et al., 1991: Cavalcante et al., 1988; Lima et al., 1987; Urquiaga et al., 1992) as well as on our recent discovery of non-pathogenic, endophytic diazotrophs in asymptomatic corn (Palus et al., 1996). The readers of this review are encouraged to examine reviews by Döbereiner (1992), Döbereiner et al., (1993), and Boddey (1995). This review will not discuss the vast literature on cyanobacterial-grass associations or root-associations in which the bacteria are not endophytes. These topics have been reviewed very well elsewhere (Döbereiner 1992; Elmerich et al., 1992; Glick, 1995; Hoflich et al., 1994; Okon and Labandera-Gonzalez., 1994).

Recent proposals for the development of nitrogen-fixing corn

Transforming corn plants with nif genes

Ever since the discovery of the nitrogen fixation (*nif*) regulon in *Klebsiella* in the mid-1970s, which was soon followed by *nif* gene analysis in other bacteria, many investigators have speculated how this knowledge could be used to develop nitrogen-fixing monocots. Small attempts have been made at introducing *nif* genes into plants by transformation (Dixon et al., 1993; Zilberstein et al., 1988). Zilberstein et al. (1988) used an *Agrobacterium* transformation system to insert *nifH* and *nifD* from *K. pneumoniae* into tobacco. Stable expression of *nifH* was observed in the transgenic lines. However, there are major hurdles to be overcome in order to achieve nitrogen fixation in plants via the introduction of bacterial *nif* genes. The first one of these is the oxygen sensitivity of the two proteins which comprise nitrogenase, dinitrogenase reductase and dinitrogenase. The second problem is to use properly regulated promoters for the expression of these bacterial genes in a eukaryotic cell. And third, despite the enormous advances in our understanding of the synthesis and regulation of the nitrogenase complex (Allen et al., 1994), there remain unknown steps in the maturation of nitrogenase whose 'identity may be required to permit successful expression of the nitrogenase complex in plants. The solution of each of these problems individually may not be very difficult but success requires solutions to all three problems.

Inducing 'nitrogen-fixing nodules on corn and other grasses

Our understanding of how nitrogen-fixing root nodules are formed on certain legumes has blossomed enormously in recent years. The isolation and characterization of the bacterial genes involved in this process (*nod* and *no1*), the plant excretion of small molecules that induce these bacterial genes, the description of the wide array of Nod factors produced by the nodulation gene products in the root nodule bacteria, and the isolation, characterization, and genetic mapping of host genes involved in nodule formation have all been reviewed recently by many authors (Carlson et al., 1994; Dénarié et al., 1992; Franssen et al., 1992; Garvin and Weeden., 1994; Kiss et al., 1993; Nodari et al., 1993; Stacey et al., 1995; Torres et al., 1993; Verma and Miao, 1992). This knowledge is being applied by a number of laboratories in attempt to encourage root nodulation of non-legumes by strains of *Rhizobium and Bradyrhizobium.* Indeed, the effective root nodulation of one non-legume in the elm family, *Parsponia*, by broad host-range strains of *Bradyrhizobium* is a shining example of how these bacteria can nodulate at least one non-leguminous plant species.

Attempts to induce nodules on non-legumes are described well in reviews by Cocking et al. (1993) and Ridge et al. (1993b). Two approaches have been used to induce nodule-like structures on the roots of *Brassica*, wheat, and rice. Cocking and coworkers (Al-Mallah et al., 1990; Cocking, et al., 1990) have used cellulases and pectolyases to degrade the cell walls of root tip cells. Following such treatments with rice and wheat plants cultured in a nitrogen-free medium and inoculated with strains of *Rhizobium* or *Bradyrhizobium,* small nodule-like structures were formed on the roots. About 1% of the plant cells in these structures are infected with bacteria. The amount of acetylene reduction observed was near the lowest detectable level. Rolfe and colleagues have induced similar nodule-like structures by treatment of rice and wheat roots with 2, 4-dichlorophenoxyacetic acid (Ridge et al., 1992, 1993a).

To date, no one has been able to demonstrate that nitrogen-fixing root nodules can be induced on grass species by inoculation with *Rhizobium* or *Bradyrhizobium*. De Bruijn et al. (1995) have shown the problems associated with studying these systems. As the receptors for Nod factors in the host legume are unknown, it is difficult to determine the reasons for these problems. We also do not know whether these grasses are capable of synthesizing sufficient quantities of haemoglobin at the right time and place to permit the proper oxygen balance between bacterial respiration and nitrogenase protection in a grass root nodule. Monocot homologs or analogs of the nodule-specific proteins from legumes necessary for the establishment of the symbiosis have not yet been described. All of these factors contribute to making this approach extremely difficult.

Development of corn lines with the inability to accept fixed-nitrogen from diazotrophs in the rhizophere

Ela et al. (1982) described an approach of breeding corn lines with the ability to accept ammonia excreted from diazotrophs that naturally inhabit the rhizosphere. The work by Ela et al. (1982) has not been repeated and it showed nitrogenase activity only by acetylene reduction without any ^{15}N analysis. The hurdles to this approach are also daunting. First, the diazotroph to be used for this work must be genetically altered to excrete large quantities of ammonia rather than use the ammonia for its own metabolism. Second, the diazotroph chosen must be capable of surviving and competing in the rhizosphere niche while at the same time sacrificing much of its energy 'supply for the reduction of N_2 and the excretion of ammonia. This is a penalty not paid by other soil microorganisms. And third, the ammonia excreted by the bacteria must be absorbed quickly by the plant prior to its utilization by soil microorganisms. Ammonia is very rapidly metabolized in soil by bacteria and fungi. Once again, each of these problems may not be difficult to solve individually but taken together they are a formidable barrier to the development of a corn cropping system that does not need nitrogen inputs.

Diazotrophic endophytes in sugar cane

An alternative to the above approaches for the development of nitrogen-fixing cereal crops has been discovered in sugar cane by Döbereiner and coworkers at EMBRAPA in Brazil. Four million hectares in Brazil are used for the culture of sugar cane (*Saccharum* spp.) The sucrose extracted from the cane is used in the production of ethanol. Ethanol fuels about one-half of the automobiles in Brazil. Unlike sugar cane development in Hawaii, most sugar cane lines in Brazil have been bred in the absence of nitrogen fertilizer. In many areas of Brazil, no nitrogen fertilizer is used in the culturing of sugar cane and yet high yields of this crop are obtained.

Döbereiner and coworkers were curious about the source of nitrogen for these plants and began looking for diazotrophs in the rhizophere that could supply fixed N to the sugar cane. Diazotrophs were found in the rhizosphere but the nitrogenase activities of these isolates were too low to account for the nitrogen needs of the plant. Then the stem and root tissue of the sugar cane were examined for endophytic bacteria. A new diazotroph among the α-subclass of the proteobacteria, *Acetobacter diazotrophicus*, was isolated from surface-sterilized stem and root tissue of sugar cane by using a semi-solid nitrogen-free medium with sugar cane juice as carbon source (Cavalcante and Döbereiner, 1988; Gillis et al., 1989). This bacterium has a number of interesting properties some of which make it a likely candidate for 'providing fixed N to sugar cane. These properties include: a) an ability to grow on 10% sucrose as sole carbon source; b) a very low pH for optimum growth (pH 5.5) and an ability to grow at pH values as low as 3.0; c) NO_3^- and NH_4^+ show no or little inhibition of *Acetobacter* nitrogenase activity, respectively; and d) an aerobe capable of microaerobic nitrogen fixation (Boddey et al., 1991; Burris, 1994; Cavalcante and Döbereiner, 1988; Gillis et al., 1989; Stephan et al., 1991). These characteristics are very different from those of *Azospirillum* strains which are also known to be associated with grasses (Döbereiner, 1992).

This bacterium is found in large numbers ($10^3 - 10^7$ cells per g fresh weight) in root and stem tissue as well as in rhizosphere soil but not in non-rhizosphere soil (Döbereiner et al., 1988). The endophytic habitat of these bacteria has been confirmed using immunogold labelling in which the bacteria are first seen to colonize the epidermis of stems and roots followed by distribution in xylem vessels (James et al., 1994). *Acetobacter diazotrophicus* has also been isolated from sugar cane plants in Mexico, Cuba, and Australia (Caballero-Mellado and Martinez-Romero, 1994; Dong et al., 1994; Fuentes-Ramirez et al., 1993; Li and MacRae, 1991, 1992).

Elegant and thorough ^{15}N isotope dilution experiments and nitrogen balance experiments have conclusively shown that *Acetobacter*-infected sugar cane plants obtain up to 80% of their N from N_2 (Boddey et al., 1991; Döbereiner et al., 1993; Lima et al., 1987; Urquiaga et al., 1992). However, it is not certain whether *Acetobacter* or another diazotroph is responsible for the availability of fixed N for sugar cane. Other diazotrophs have been identified in sugar cane. In particular, *Herbaspirillum seropedicae* and *Acetobacter diazotrophicus* have been found in roots, stems, and leaves of sugar cane (Baldani et al., 1986, 1992; Döbereiner et al., 1993). As *H. seropedicae* cannot use sucrose as a carbon source and as sucrose is in abundance in sugar cane, the prediction could be made that a more successful diazotroph in sugar cane would be one, such as *A. diazotrophicus,* which thrives on sucrose as carbon source. However, the availability of sucrose to *Acetobacter* when those bacteria are found in the apoplast is unknown (Dong et al., 1994, 1995; James et al., 1994). In addition to the concerns about carbon utilization by *H. seropedicae*, this species is also known to be pathogenic on *Sorghum* and *Pennisetum* but not sugar cane and is closely related to the sugar cane pathogen *Pseudomonas rubrisubalbicans* (Döbereiner et al., 1993; Pimentel et al., 1991). The pathogenic nature of *H. seropedicae* would preclude its use as an inoculant for agriculture.

Clearly what is needed to resolve this question are *nif* mutants of the diazotrophic endophytes of sugar cane. The laboratory of Christina Kennedy has isolated and characterized the *nifA*, *nifHDK*, and *ntrBC* from *Acetobacter diazotrophicus* with the goal of producing *nif* mutants of this species by marker exchange (Meletzus et al., 1995; Sevilla and Kennedy, 1995). The effect of inoculation of mutant and wild-type strains of *A. diazotrophicus* on the nitrogen economy of sugar cane can then be ascertained.

Transmission in plants - mychorrizae and vegetative propagation

Three means by which *Acetobacter* cells enter the sugar cane plant have been proposed. As sugar cane is vegetatively propagated, much of the transmission of this organisms probably occurs when stem tissue is planted in soil (Döbereiner et al., 1990; Paula et al., 1990). The second transmission method is via an insect vector. Cells of *A. diazotrophicus* have been found in the pink sugar cane mealy bug which colonizes the plant tissue between the leaf sheath and stem (Ashbolt and Inkerman, 1990). Consistent with this notion are the observations that the honey dew excreted by this insect is rich in carbon sources and is very acidic (Ashbolt and Inkerman, 1990). *Acetobacter diazotrophicus* thrives under these conditions.

The third, and perhaps most intriguing method of transmission is via vesicular-arbuscular mycorrhizae (VAM) (Paula et al., 1990, 1991). The presence of bacteria in mycorrhizal spores has been known for many years (Mosse, 1962; Varma et al., 1981). The function of these bacteria, if any, within these spores is unknown. Recently, Paula et al. (1991) found the presence of several diazotrophic bacteria in spores of *Glomis clarum* including *A. diazotrophicus* and *Klebsiella* sp. Paula et al. (1991) obtained sterile spores of VAM and co-inoculated sweet potato, sweet sorghum and sugar cane with these spores and with *A. diazotrophicus* or with a mixture of diazotrophs including *A. diazotrophicus, Klebsiella* sp. and others. In those treatments which contained inoculation with both microorganisms, the number of diazotrophic cells within surface-sterilized roots and tops of sweet potato and sugar cane was significantly higher than in the uninoculated treatments or in the treatment inoculated with the diazotrophs only.

Isopi et al. (1995) have coinoculated sorghum plants with VAM and *A. diazotrophicus.* The bacterium decreased fungal infection of roots but the fungus increased bacterial infection of roots, stems, and leaves. The bacterium also increased N levels in the plant especially when co-inoculated with the fungus. The co-inoculation treatment also increased root length and branching.

The entry of bacteria into plants through mycorrhizal infection is very interesting and raises many questions. How do the bacteria enter the spores? Do the bacteria multiply in the fungal hyphae? Do the bacteria reduce N_2 in the hyphae? How common are diazotrophic bacteria in VAM? How many plant species benefit from infection with diazotrophic bacteria from VAM?

Unfortunately, the current agricultural practice of heavy phosphorous fertilization has probably resulted in a massive decline in VAM infection of crop plants. It is not known whether modern maize lines are commonly infected with mycorrhizae. Modern wheat lines are not dependent on mycorrhizal infection (Hetrick et al., 1992, 1993) probably as a result of being developed in the presence of phosphorous fertilization.

Cultivar dependence

Not all sugar cane lines are capable of forming an effective nitrogen-fixing symbiosis (Boddey et al., 1991; Urquiaga et al. 1992). Only those lines that were developed in the absence of high nitrogen fertilization appear able to develop this symbiosis. Thus, just as wheat cultivars developed in the presence of phosphorous fertilizer appear to have lost the ability to form mycorrhizal associations (Hetrick et al., 1992, 1993), sugar cane lines have lost the ability to develop nitrogen-fixing symbioses when nitrogen fertilization was used during their development. Indeed given that mycorrhizae have been implicated as one of the major transmission modes of diazotrophs into the sugar cane plants, it is possible that heavy fertilization during maize development may have caused the loss of the ability to be inhabited by mycorrhizal or actively nitrogen-fixing bacterial endophytes. This implies that looking for such nitrogen-fixing associations in maize or other monocots will only be successful if early lines of these plants are examined.

Diazotrophic endophytes in Kallar grass (*Leptochloa fusca* [L.] Kunth)

A new bacterial genus in the β-subclass of the proteobacteria, *Azoarcus*, was recently described by Reinhold-Hurek et al. (1993b) which includes members that are diazotrophic under microaerobic conditions (Reinhold-Hurek and Hurek, 1993). Many diazotrophs inhabit the rhizosphere of Kallar grass but it was the gram-negative rods of *Azoarcus* that predominates inside the roots of this grass species (Reinhold et al., 1986; Reinhold-Hurek et al., 1993b). Although *Azoarcus* strains are not capable of utilizing carbohydrates (Reinhold-Hurek et al., 1993b), they nonetheless possess several enzymes capable of degrading cellulose on their cell surface (Reinhold-Hurek et al., 1993a). Mutations in the endoglucanase gene causes reduced spreading of the bacterium within rice seedlings (Reinhold-Hurek et al., 1993c). The development of genus-specific probes for *Azoarcus* allows their rapid identification in Kallar grass Huek et al. (1993).

As is the case with *Acetobacter* in sugar cane, *Azoarcus* sp. strain BH72 resides in the xylem and not in living plant cells (Hurek et al., 1994a). Also, there is strong evidence that presence of *Azoarcus* cells in rice seedlings significantly promotes the growth of the plants (Hurek et al., 1994a). However, in contrast with the *Acetobacter*-sugar cane system, the growth benefit does not seem to be caused by bacterial nitrogen fixation (Hurek et al., 1994a). A *nifK* mutant of *Azoarcus* sp. strain BH72, which has a Nif^- phenotype in vitro, provides the same growth benefit to rice as does the wild-type strain (Hurek et al., 1994a). The ability of alternative nitrogenase systems in BH72 to provide fixed nitrogen to Kallar grass is unknown. No ^{15}N isotope dilution or nitrogen balance experiments have been done with this association. Addition of malate to the infected plants reverses the growth promotion caused by BH72 (Hurek et al., 1994a). Hurek et al. (1994b) have shown that under certain conditions BH72 has a very high respiration rate that can result in the development of an environment with a very low pO_2 and a high nitrogenase activity. All of these data taken together suggest that the role of *Azoarcus* in Kallar grass is neither diazotrophic or pathogenic but more data is needed to rule out the possibility of a diazotropric symbiosis.

Bacterial endophytes in corn

Fisher et al. (1992) have found that healthy corn plants in England possess endophytic bacteria and fungi. Many of the bacterial and fungal endophytes in maize have been identified. The most common bacterial endophytes in maize were identified using the Biolog system as strains of *Enterobacter agglomerans, Klebsiella terrigena, Pseudomonas corrugata, P. fluorescens, P. marginalis, P.* sp. 4, *P.* sp. 5, and *Vibrio* sp. 1. Strains of *E. agglomerans* and *K. terrigena* have been shown to be diazotrophs (for review see Young, 1992). Various strains of *Vibrio* and *Pseudomonas* have also been identified as diazotrophs (Young, 1992). Thus, it seems likely that diazotrophs can inhabit maize. However, the ability of the Biolog system to correctly identify strains of *Enterobacter* and *Klebsiella* is in doubt (Klingler et al., 1992). The primary objective of the work by Fisher et al. (1992, 1993) is to develop techniques to inoculate corn with genetically-engineered bacteria to provide pest control.

McInroy and Kloepper (1991, 1995) have also isolated and provided preliminary identification for bacterial endophytes of maize. These workers found 29 genera of bacteria in field-grown stems and roots of maize. Among these genera are several which include diazotrophic members such as *Bacillus, Enterobacter, Erwinia, Klebsiella,* and *Xanthobacter* (Young, 1992). The bacteria in this work were identified by gas chromatographic analysis of fatty acid methyl esters. This is a good and reliable method but it is difficult to assign strains at the species level using this method.

McInroy and Kloepper (1991, 1995) have also made four other interesting observations. First, the number of bacterial genera found in corn stems decreases with the age of the plant. At plant maturity, *Pseudomonas* and *Bacillus* strains are dominant. Second, corn tissue was found to contain about 10^4 bacteria per gram fresh weight of tissue. Third, surface-sterilized seeds possessed 10^5–10^6 bacteria per gram fresh weight. Fourth, plants grown under sterile conditions from surface-sterilized seeds also contained stem bacteria but at levels 100-fold lower than plants cultured in nonsterile soil. This suggests that the endophytic bacteria can be transmitted either through seed or soil. The observation of seed transmission is important in an annual such as corn and suggests a simple method of inoculation of plants via seed coating.

Discovery of diazotrophic endophytes in root and stems of maize

As many of the bacterial species identified in maize have been described as diazotrophs or are close relatives of diazotrophs, this begs the question of the possible diazotrophic role of these bacteria in maize particularly in light of the discovery of nitrogen-fixing associations in the closely-related grass, sugar cane. The extent of the close relationship between corn and sugar cane was recently described by Weerakoon et al. (1993).

Herbaspirillum seropedicae is a diazotroph which to date has been described as an maize endophyte in roots only (Baldani et al., 1986, 1992; Döbereiner et al., 1993). As described above, *H. seropedicae* is a pathogen on certain grass species (Döbereiner et al., 1993). Thus, its exploitation as a potential inoculant for corn is doubtful. Until recently, no diazotrophs have been isolated from maize stem tissue.

Diazotrophic endophytes isolated from maize stems by this laboratory

Based on the sugar cane paradigm, my laboratory has begun to analyze diazotrophs isolated from maize stem tissue in Wisconsin. This work will be summarized briefly here and is described in detail in this volume by Palus et al. (1996). We decided to attempt to isolate bacteria from the stems of sweet corn rather than field corn because sweet corn lines on average have a higher sugar content in the stem than do field corn lines although corn lines vary enormously in stem sugar content (Bertolini et al., 1993). Our next step was to decide which medium to use to isolate the diazotrophs. As *A. diazotrophicus* had been isolated from sugar cane plants over a wide geographical region in Latin America and Australia, we were expecting to find either *A. diazotrophicus* or a close relative in maize stem tissue. As a result, we used the LGI of Cavalcante and Döbereiner (1988) with the exception that the 0.02 g L^{-1} yeast extract was omitted. The apoplastic fluid from maize and teosinte stems was isolated as described by Dong et al. (1994). Two isolates from maize and one from teosinte were isolated from the apoplastic fluid that were capable of growth on semi-solid N-free LGI medium.

The diazotrophic nature of one of the maize isolates was characterized further. Strain zmvsy expressed nitrogenase activity microaerobically as measured by acetylene and $^{15}N_2$ reduction. A partial *nifH* sequence from zmvsy was PCR-amplified and sequenced using the universal *nifH* primers described by Ueda et al. (1995). A near full length 16S rDNA fragment using the universal eubacterial primers of Angert et al. (1993) was PCR-amplified and sequenced from all three isolates. These sequence data along with many physiological and nutritional characteristics suggest that all three of these isolates are members of *Klebsiella* sp. A species designation of these isolates would require more analysis. The number of these isolates in the corn plant appears to be about 10^3 cells per g fresh weight which is at the lower limit of the number of *A. diazatrophicus* found in sugar cane.

Preliminary evidence for the isolation diazotrophic endophytes from maize tissues by other laboratories

A flurry of activity in this area seems to be emerging from a number of laboratories around the world as illustrated by the presence of several abstracts on this topic at the 10th International Congress on Nitrogen Fixation at St. Petersburg, Russia in the spring of 1995. Strains of *Azotobacter chroococcum, Klebsiella terrigena* and others were described as inhabiting maize tissue at that meeting (Hallberg, 1995; Li et al., 1995; Raičević et al., 1995a, b).

Future directions

Clearly, much more work is necessary to determine whether any of the diazotrophs found in maize are capable of contributing to the nitrogen economy of corn. In our work, the *Klebsiella* isolates almost certainly do not contribute fixed N to the corn plant because the field from which these plants were collected was heavily fertilized with nitrogen. By analogy with the sugar cane system, the cultivar chosen to examine any benefit from these strains is crucial. It may be that none of the strains isolated to date are capable of contributing fully to the nitrogen needs of corn even if the appropriate cultivar was discovered. It may be necessary to isolate maize diazotrophic endophytes from many locations including the areas of maize origin in Mexico. Primitive maize lines of one of the relatives of maize, such as *Zea luxurians* or *Z. diploperennis,* may already be benefiting from a diazotrophic association that remains to be discovered. The presence of mycorrhizae may be required as well which would also suggest careful choosing of both cultivar and bacterial strain. During the course of this work careful nitrogen balance or ^{15}N isotope dilution studies must be conducted in order to verify the ability of these associations to fix nitrogen.

Although these may be high hurdles to overcome, none of them possess the technical challenges that are found in the more traditional approaches to obtaining nitrogen fixation in corn. To this end, the College of Agricultural and Life Sciences of the University of Wisconsin-Madison has funded a multi-disciplinary group of faculty to approach this problem in a systematic fashion. This group includes a sweet corn breeder (W F Tracy) to help us choose the appropriate corn lines and inoculate corn plants with any diazotrophs identified in this work; a plant pathologist and mycorrhizal expert (J L Parke) to help us determine the mode of transmission of these bacteria in maize; a *nif* biochemist (P W Ludden) to demonstrate the ability of these bacteria to fix nitrogen in vitro and in planta; a corn extension specialist (J G Lauer) who will effectively communicate our results to Wisconsin taxpayers; an agricultural economist (J -P Chavas) who will help justify our work by providing a detailed economic analysis of the benefits of nitrogen-fixing corn; and a bacteriologist (E W Triplett) who will continue to isolate and characterize more diazotrophs from maize tissue.

Conclusions

The presence of diazotrophic endophytes from the α-, β-, and γ-subclasses of the proteobacteria in three grass species have been described by several laboratories. The ability of these diazotrophs to provide fixed nitrogen to sugar cane has been demonstrated by Döbereiner, Boddey and coworkers. A number of diazotrophs have been identified in sugar cane but most of the evidence suggests that *Acetobacter diazotrophicus* is responsible for the fixed N provided to sugar cane. Strains of *Azoarcus* have been found in Kallar grass and rice by Reinhold-Hurek, Hurek and coworkers. These bacteria enhance the growth of rice seedlings but this enhancement is probably not the result of N_2 fixation. Using the sugar cane discovery as a guide, many investigators worldwide are now interested in studying diazotrophic endophytes in maize with the hope that a nitrogen-fixing association can someday be established in cultivated maize. Given the large technical hurdles to more traditional approaches to this problem, the development of a diazotrophic endophytic association in maize appears to be the most likely route to success in the development of a corn plant which does not require nitrogen fertilization for optimum growth and yield.

Acknowledgements

The author gratefully acknowledges the support of his laboratory from the following sources: USDA-NRI grant no 94–37305–0767, USDA risk assessment program grant no 94–33120–0433, USEPA cooperative agreement no CR–822882–01–0, and the College of Agricultural and Life Sciences Hatch project no 5201.

References

Allen R M, Chatterjee R, Madden M S, Ludden P W and Shah V K 1994 Biosynhesis of the iron-molybdenum cofactor of nitrogenase. Crit. Rev. Biotechnol. 14, 225–249.

Al-Mallah M K, Davey M R and Cocking E C 1990 Nodulation of oilseed rape (*Brassica napus*) by rhizobia. J. Exp. Bot. 41, 1567–1572.

Angert E R, Clements K D and Pace N R 1993 The largest bacterium. Nature 362, 239–241.

Ashbolt N J and Inkerman P A 1990 Acetic acid bacterial biota of the pink sugar cane mealybug, *Saccharococcus sacchari*, and its environs. Appl. Environ. Microbiol. 56, 707–712.

Baldani J I, Baldani V L D, Seldin L and DöberEiner J 1986 Characterization of *Herbaspirillum seropedicae* gen. nov., sp. nov., a root-associated nitrogen-fixing bacterium. Int. J. Syst. Bacteriol. 36, 86–93.

Baldani V L D, Baldani J I, Olivares F and Döbereiner J 1992 Identification and ecology of *Herbaspirillum seropedicae* and the closely related *Pseudomonas rubrisubalbicans*. Symbiosis 13, 65–73.

Bertolini M, Lorenzoni C, Marocco A and Maggiore T 1993 Soluble solids content in the stalk of maize (*Zea mays* L.) lines and hybrids. Maydica 38, 321–324.

Boddey R M 1995 Biological nitrogen fixation in sugar cane: a key to energetically viable biofuel production. Crit. Rev. Plant Sci. 14, 263–279.

Boddey R M, Urquiaga S, Reis V and Döbereiner J 1991 Biological nitrogen fixation associated with sugar cane. Plant and Soil 137, 111–117.

Burris R H 1994 Comparitive study of the response of *Azotobacter vinelandii* and *Acetobacter diazotrophicus* to changes in pH. Protoplasma 183, 62–66.

Caballero-Mellado J and Martinez-Romero E 1994 Limited genetic diversity in the endophytic sugarcane bacterium *Acetobacter diazotrophicus*. Appl. Environ. Microbiol. 60, 1532–1537.

Carlson R W, Price N P J and Stacey G 1994 The biosynthesis of rhizobial lipo-oligosaccharide nodulation signal molecules. Molec. Plant-Microbe Interact. 7, 684–695.

Cavalcante V A and Döbereiner J 1988 A new acid-tolerant nitrogen-fixing bacterium associated with sugarcane. Plant and Soil 108, 23–31.

Cocking E C, Al-Mallah M K, Benson E and Davey M R 1990 Nodulation of non-legumes by rhizobia. *In* Nitrogen Fixation: Achievements and Objectives. Eds. P M Gresshoff, L E Roth, G Stacey and W E Newton. pp 813–823. Chapman and Hall, New York, USA.

Cocking E C, Davey M R, Kothari S L, Srivastava J S, Jing Y X, Ridge R W and Rolfe B G 1993 Altering the specificity control of the interaction between rhizobia and plants. Symbiosis 14, 123–130.

De Bruijn F J, Ying Y and Dazzo F B 1995 Potential and pitfalls of trying to extend symbiotic interactions of nitrogen-fixing organisms to presently non-nodulated plants, such as rice. Plant and Soil 174, 225–240.

Dénarié J, Debellé F and Rosenberg C 1992 Signalling and host range variation in nodulation. Annu. Rev. Microbiol. 46, 497–531.

Dixon R, Vanderleyden J and Romero D 1993 Extension of nitrogen fixation to other crops. *In* New Horizons in Nitrogen Fixation. Eds. R Palacios, J Mora and W E Newton. pp 765–768. Kluwer Academic Publishers, Dordrecht, The Netherlands.

Döbereiner J 1992 History and new perspectives of diazotrophs in association with non-leguminous plants. Symbiosis 13, 1–13.

Döbereiner J, Pimentel J P, Olivares F and Urquiaga S 1990 Bactérias diazotróficas podem ser endófitas e/ou fitopathogěnicas? An. Acad. Bras. Ci. 62, 319.

Döbereiner J, Reis V and Lararini A C 1988 New N_2 fixing bacteria in association with cereals and sugarcane. *In* Nitrogen Fixation: Hundred Years After. Eds. H Bothe, F J de Bruijn and W E Newton. pp 717–722. Gustav Fischer, Stuttgart, Germany.

Döbereiner J, Reis V M, Paula M A and Olivares F 1993 Endophytic diazotrophs in sugar cane, cereals and tuber plants. *In* New Horizons in Nitrogen Fixation. Eds. R Palacios, J Mora and W E Newton. pp 671–676. Kluwer Academic Publishers, Dordrecht, The Netherlands.

Dong Z, Canny M J, McCully M E, Roboredo M R, Cabadilla C F, Ortega E and Rodés R 1994 A nitrogen-fixing edophyte of sugarcane stems. A new role for the apoplast. Plant Physiol. 105, 1139–1147.

Dong Z, Heydrich M, Bernard K and McCully M E 1995 Further evidence that the N_2-fixing endophytic bacterium from the intercellular spaces of sugarcane stems is *Acetobacter diazotrophicus*. Appl. Environ. Microbiol. 61, 1843–1846.

Ela S W, Anderson M A and Brill W J 1982 Screening and selection of maize to enhance associative bacterial nitrogen fixation. Plant Physiol. 70, 1564–1567.

Elmerich C, Zimmer W and Vieille C 1992 Associative nitrogen-fixing bacteria. *In* Biological Nitrogen Fixation. Eds. G Stacey, R H Burris and H J Evans. pp 212–258. Chapman and Hall, New York, USA.

Fisher P J, Broad S A, Clegg C D and Scott H M L 1993 Retention and spread of a genetically-engineered pseudomonad in seeds and plants of *Zea mays* L. - a preliminary study. New Phytol. 124, 101–106.

Fisher P J, Petrini O and Scott H M L 1992 The distribution of some fungal and bacterial endophytes in maize (*Zea mays* L.). New Phytol. 122, 299–305.

Franssen H J, Vijn I, Yang W C and Bisseling T 1992 Developmental aspects of the *Rhizobium*-legume symbiosis. Plant Molec. Biol. Internat. J. Molec. Biol. Biochem. Genet. Eng. 19, 89–107.

Fuentes-Ramirez L E, Jimenez-Salgado T, Abarca-Ocampo I R and Caballero-Mellado J 1993 *Acetobacter diazotrophicus,* an indoleacetic acid producing bacterium isolated from sugarcane cultivars of México. Plant and Soil 154, 145–150.

Garvin D F and Weeden N F 1994 Genetic linkage between isozyme, morphological, and DNA markers in tepary bean. J. Hered. 85, 273–278.

Gillis M, Kersters K, Hoste B, Janssens D, Kroppenstedt R M, Stephan M P, Teixeira K R S, Döbereiner J and de Ley J 1989 *Acetobacter diazotrophicus* sp. nov., a nitrogen- fixing acetic acid bacteria associated with sugarcane. Int. J. Syst. Bacteriol. 39, 361–364.

Glick B R 1995 The enhancement of plant growth by free-living bacteria. Can. J. Microbiol. 41, 109–117.

Hallberg T B 1995 Nitrogen-fixing bacteria associated with maizes native of Oaxaca. *In* Abstracts for the 10th International Congress on Nitrogen Fixation, St. Petersburg, Russia, no 638.

Hetrick B A D, Wilson G W T and Cox T S 1992 Mycorrhizal dependence of modern wheat varieties, landraces, and ancestors. Can. J. Bot. 70, 2032–2040.

Hetrick B A D, Wilson G W T and Cox T S 1993 Mycorrhizal dependence of modern wheat cultivars and ancestors: a synthesis. Can. J. Bot. 71, 512–518.

Hoflich G, Wiehe W and Kuhn G 1994 Plant growth stimulation by inoculation with symbiotic and associative rhizosphere microorganisms. Experientia 50, 897–905.

Hurek T, Burggraf S, Woese C R and Reinhold-Hurek B 1993 16S rRNA-targeted polymerase chain reaction and oligonucleotide hybridization to screen for *Azoarcus* spp., grass-associated diazotrophs. Appl. Environ. Microbiol. 59, 3816–3824.

Hurek T, Reinhold-Hurek B, Turner G L and Bergersen F J 1994a Augmented rates of respiration and efficient nitrogen fixation an nanomolar concentrations of dissolved O_2 in hyperinduced *Azoarcus* sp. strain BH72. J. Bacteriol. 176, 4726–4733.

Hurek T, Reinhold-Hurek B, van Montagu M and Kellenberger E 1994b Root colonization of *Azoarcus* sp. strain BH72 in grasses. J. Bacteriol. 176, 1913–1923.

Isopi R, Fabbri P, Delgallo M and Puppi G 1995 Dual inoculation of *Sorghum bicolor* (L.) Moench ssp bicolor with vesicular-arbuscular mycorrhizas and *Acetobacter diazotrophicus*. Symbiosis 18, 43–55.

James E K, Reis V M, Olivares F L, Baldani J I and Döbereiner J 1994 Infection of sugar cane by the nitrogen-fixing bacterium *Acetobacter diazotrophicus*. J. Exp. Bot. 45, 757–766.

Kiss G B, Csanadi G, Kalman K, Kalo P and Okresz L 1993 Construction of a basic genetic map of alfalfa using RFLP, RAPD, isozyme and morphological markers. Molec. Gen. Genet. 238, 129–137.

Klingler J M, Stowe R P, Obenhuber D C, Groves T O, Mishra S K and Pierson D L 1992 Evaluation of the Biolog automated microbial identification system. Appl. Environ. Microbiol. 58, 2089–2092.

Li F, Liu R, Wang W and Yan Z 1995 Research on distribution of 32P labelled nitrogen fixing bacteria in plant. *In* Abstracts for the 10th International Congress on Nitrogen Fixation, St. Petersburg, Russia, no 623.

Li R P and MacRae I C 1991 Specfic association of diazotrophic acetobacters with sugarcane. Soil Biol. Biochem. 23, 999–1002.

Li R and MacRae I 1992 Specific identification and enumeration of *Acetobacter diazotrophicus* in sugarcane. Soil Biol. Biochem. 24, 413–419.

Lima E, Boddey R M and Döbereiner J 1987 Quantification of biological nitrogen fixation associated with sugar cane using a ^{15}N aided nitrogen balance. Soil Biol. Biochem. 19, 165–170.

Mc Inroy J A and Kloepper J W 1991 Analysis of population densities and identification of endophytic bacteria of maize and cotton in the field. *In* Plant Growth-Promoting Rhizobacteria - Progress and Prospects. Eds. C Keel, B Koller and G Défago. pp 328–331. IOBC/WPRS Bull. XIV/8.

McInroy J A and Kloepper J W 1995 Survey of indigenous bacterial endophytes from cotton and sweet corn. Plant and Soil 173, 337–342.

Meletzus D, Pero R, Kennedy C, Teixeira K and Baldani I 1995 Identification of nifA and *ntrBC* genes in *Acetobacter diazotrophicus*. *In* Abstracts for the 10th International Congress on Nitrogen Fixation, St. Petersburg, Russia, no 125.

Mosse B 1962 The establishment of vesicular-arbuscular mycorrhizae under aseptic conditions. J. Gen. Microbiol. 27, 509–520.

Nodari R 0, Tsai S M, Guzman P, Gilbertson R L and Gepts P 1993 Toward an integrated linkage map of common bean. III. Mapping genetic factors controlling host-bacteria interactions. Genetics 134, 341–350.

Okon Y and Labandera-Gonzalez C A 1994 Agronomic applications of *Azospirillum* - an evaluation of 20 years worldwide field inoculation. Soil Biol. Biochem. 26, 1591–1601.

Palus J A, Borneman J, Ludden P W and Triplett E W 1996A diazotrophic bacterial endophytes isolated from stems of *Zea mays* L. and *Zea luxurians* Iltis and Doebley. Plant and Soil. 186.

Paula M A, Reis V M and Döbereiner J 1991 Interactions of *Glomus clarum* with *Acetobacter diazotrophicus* in infection of sweet potato (*Ipomoea batatas*), sugarcane (*Saccharum*), sweet sorghum (*Sorghum vulgare*). Biol. Fertil. Soils 11, 111–115.

Paula M A, Reis V M, Urquiaga S and Döbereiner J 1990 Esporos de fungo MNA *Glomus clarum* como veículo de infecçao de *Acetobacter diazotrophicus*. Anais Acad. Bras. Ciĕnc. 62, 318.

Pimentel J P, Olivares F, Pitard R M, Urquiaga S, Akiba F and Döbereiner J 1991 Dinitrogen fixation and infection of grass leaves by *Pseudomonas rubrsubalbicans* and *Herbaspirillum seropedicae*. Plant and Soil 137, 61–65.

Raičević V, Sarić M, Sarić Z and Bogdanović V 1995a *Azotobacter* movement in maize root. *In* Abstracts for the 10th International Congress on Nitrogen Fixation, St. Petersburg, Russia, no 607.

Raičevic V, Sarić M, Sarić Z and Bogdanović V 1995b Colonization and adsorption of some *Azotobacter* strains in maize root. *In* Abstracts for the 10th International Congress on Nitrogen Fixation, St. Petersburg, Russia, no 608.

Reinhold B, Hurek T, Niemann E-G and Fendrik I 1986 Close association of *Azospirillum* and diazotrophic rods with different root zones of Kallar grass. Appl. Environ. Microbiol. 52, 520–526.

Reinhold-Hurek B and Hurek T 1993 Capacities of *Azoarcus* sp., a new genus of grass-associated diazotrophs. *In* New Horizons in Nitrogen Fixation. Eds. R Palacios, J Mora and W E Newton. pp 691–694. Kluwer Acad. Publ., Dordrecht, The Netherlands.

Reinhold-Hurek B, Hurek T, Claeyssens M and van Montagu M 1993a Cloning, expression in *Escherichia coli*, and characterization of cellulolytic enzymes of *Azoarcus* sp., a root-invading diazotroph. J. Bacteriol. 175, 7056–7065.

Reinhold-Hurek B, Hurek T, Gillis M, Hoste B, Vancanneyt M, Kersters K and de Ley J 1993b *Azoarcus* gen. nov., Nitrogen-fixing proteobacteria associated with roots of kallar grass (*Leptochloa fusca* (L.) Kunth), and description of two species, *Azoarcus indignes* sp. nov. and *Azoarcus communis* sp. nov. Int. J. Syst. Bacteriol. 43, 574–584.

Reinhold-Hurek B, Hurek T, Maes T and van Montagu M 1993c Analysis of the role of endoglucanse in spreading of *Azoarcus* in grass seedlings. *In* New Horizons in Nitrogen Fixation. Eds. R Palacios, J Mora and W E Newton. p 256. Kluwer Academic Publ., Dordrecht, The Netherlands.

Ridge R W, Bender G L and Rolfe B G 1992 Nodule-like structures induced on the roots of wheat seedlings by the addition of the synthetic auxin 2, 4-dichlorophenoxyacetic acid and the effects of microorganisms. Aust. J. Plant Physiol. 19, 481–492.

Ridge R W, Ride K M and Rolfe B G 1993a Nodule-like structures induced on the roots of rice seedlings by addition of the synthetic auxin 2,4-dichlorophenoxyacetic acid. Aust. J. Plant Physiol. 20, 705–717.

Ridge R W, Rolfe B G, Jing Y and Cocking E C 1993b *Rhizobium* nodulation of non-legumes. Symbiosis 14, 345–357.

Sevilla M and Kennedy C 1995 Characterization of *nifHDK* in *Acetobacter diazotrophicus*. *In* Abstracts for the 10th International Congress on Nitrogen Fixation, St. Petersburg, Russia, no 124.

Stacey G, Sanjuan J, Luka S, Dockendorff T and Carlson R W 1995 Signal exchange in the *Bradyrhizobium*-soybean symbiosis. Soil Biol. Biochem. 27, 473–483.

Stephan M P, Oliveira M, Teixeira K R S, Martinez-Drets G and Döbereiner J 1991 Physiology and dinitrogen fixation of *Acetobacter diazotrophicus*. FEMS Microbiol. Lett. 77, 67–72.

Torres A M, Weeden N F and Martin A 1993 Linkage among isozyme, RFLP and RAPD markers in *Vicia faba*. Theor. Appl. Genet. 85, 937–945.

Ueda T, Suga Y, Yahiro N and Matsuguchi T 1995 Remarkable N_2-fixing bacterial diversity detected in rice roots by molecular evolutionary analysis of *nifH* gene sequences. J. Bacteriol. 177, 1414–1417.

Urquiaga S, Cruz K H S and Boddey R M 1992 Contribution of nitrogen fixation to sugar cane: Nitrogen-15 and nitrogen-balance estimates. Soil Sci. Soc. Am. J. 56, 105–114.

Varma A K, Singh K and Lall V K 1981 Lumen bacteria from endomycorrhizal spores. Curr. Microbiol. 6, 207–211.

Verma D P S and Miao G H 1992 Induction of nodulin genes and root nodule symbiosis. Semin. Ser. Soc. Exp. Biol. 49, 175–204.

Weerakoon L K, Weber D F and Schneerman M C 1993 Estimating genetic relatedness of maize to twenty-three plant species by RFLP analysis. Maydica 38, 231–237.

Young J P W 1992 Phylogenetic classification of nitrogen-fixing organisms. *In* Biological Nitrogen Fixation. Eds. G Stacey, R H Burris and H J Evans. pp 43–86. Chapman and Hall, New York, USA.

Zilberstein A, Koncz C, Kaufman R, Geva N, Zamir A and Schell J 1988 Introduction of *nifH* and *nifD* into the plant nuclear genome. *In* Nitrogen Fixation: Hundred Years After. Eds. H Bothe, F J de Bruijn and W E Newton. p 648. Gustav Fischer, Verlag, Stuttgart, Germany.

Section editor: H Lambers
Guest editors: G.H. Elkan and R.G. Upchurch

Plant and Soil **186:** 39–43, 1996.

Drought-avoidant soybean germplasm maintains nitrogen-fixation capacity under water stress

Robert P. Patterson and Colleen M. Hudak
Department of Crop Science, N.C. State University, P.O. Box 7620, Raleigh, NC 27695, USA and Food and Drug Protection Division, N.C. Department of Agriculture, P.O. Box 27647, Raleigh, NC 27611, USA **

Key words: acetylene reduction, drought, drought-avoidant germplasm, *Glycine max* (L.) Merrill, growth, nitrogen assimilation

Abstract

Inoculated soybeans (*Glycine max* L. (Merrill)) were grown in controlled environments to evaluate the relationship between genotype and plant water status on nodule function, nitrogen assimilation, growth rates, and seed yield. Plants were grown under well-watered (WW) and water-stressed (WS) conditions during the linear pod-filling growth stage in sand culture using N-free nutrient solution. Dry matter and N accumulation were greater for the drought-adapted Plant Introduction 416937 (PI) than for 'Forrest', a commercially adapted genotype of similar phenology. These differences are attributed to: (i) more favorable internal water balance throughout the pod-filling period (higher total leaf water potential), (ii) higher photosynthetic function (more total leaf area and higher net carbon exchange rates), and (iii) stronger nodule function (larger nodule mass, greater specific and total nodule activity, and thus more nitrogen assimilation) for the PI than for Forrest. While Forrest out yielded the PI under WW conditions, the percentage reduction in seed mass per plant was less for the PI than for Forrest when both genotypes were exposed to desiccating conditions. The inference is that soybean germplasm with the capacity to maintain tissue turgidity, and thus leaf and nodule function, during reproductively-imposed desiccation may reduce the extent to which yield is compromised during drought. These findings have implications for the role of symbiotic nitrogen fixation in conserving yield under dry weather conditions.

Abbreviations: DAE - Days After Emergence, NCE - Net CO_2 Exchange, PI - PI 416937, SNA - Specific Nodule Activity, TNA - Total Nodule Activity, WS - Water Stressed, WW - Well Watered.

Introduction

Water deficit stress is a major factor in limiting soybean (*Glycine max* (L.) Merrill) seed yield in many soybean-producing regions throughout the world. Plant desiccation during seed filling causes greater reductions in yield of determinate cultivars than at any other time (Doss et al., 1974; Sionit and Kramer, 1977) since moisture stress at this range of developmental stages (R3-R6) (Fehr and Caviness, 1977) results in both a reduction in duration of seed fill (Sionit and Kramer, 1977) and early senescence of leaves (Cure et al., 1982). The effects of water stress on soybean yield appear to be related to limited availability of photosynthate and nitrogen for translocation to developing seed (Boyer, 1976; Silvius et al., 1977). Poorly understood, however, are the effects of reproductively-mediated water deficit on availability of photosynthate for translocation to the root system for maintenance of nodule function and hence for nitrogen assimilation during seed fill.

We have evidence from research conducted under controlled climate conditions using adapted soybean cultivars that application of water deficit during pod filling reduces photosynthetic activity of leaves as well as nitrogen-fixation capacity of roots (Cure et al., 1985). More recently, we evaluated a drought-avoidant soybean Plant Introduction 416937 (PI) under field conditions (Hudak, 1993). Under drying soil conditions, PI was found to maintain a higher total leaf water potential than 'Forrest', an adapted germplasm of sim-

* FAX No: +19195157959. E-mail: bob-patterson@ncsu.edu

ilar phenology, and to experience less of a seed yield reduction. Results indicated that PI's maintenance of leaf turgidity was due to efficient extraction of soil water by a larger, fibrous root system. These findings suggest the possibility that soybean germplasm capable of maintaining leaf turgidity during drought may be able to extend nodule function further into the pod-filling period, thus ensuring a more satisfactory flow of nitrogen assimilate to the developing embryos than otherwise would be the case. Limited field data also indicate that, under desiccating soil conditions, PI has the capacity to maintain relatively high net carbon exchange rates in comparison to Forrest (Marlow, 1993). We therefore postulated that soybean germplasm capable of preserving leaf turgidity under drought may be able to maintain leaf function, and therefore nodule function, during the growth stage of greatest sensitivity to drought, namely the pod-filling period. Such a phenomenon would have clear implications for the genotype's capacity to extend the duration of seed fill, and thus conserve yielding ability under drought.

Table 1. Effect of watering protocol (WW = well watered and WS = water stressed) on total leaf water potential, net CO_2, exchange rate, and leaf area production throughout reproductive development of soybean genotypes (For = Forrest and PI = PI 416937)

Water Status	Genotype	Growth stage R1	R3	R4	R5	R7
Total leaf water potential (MPa)						
WW	For	-0.6	-0.7	-0.7	-0.9	-1.2
	PI	-0.5	-0.6	-0.5	-0.7	-1.1
WS	For	-0.6	-0.7	-1.4	-1.6	-1.8
	PI	-0.5	-0.5	-0.9	-1.0	-1.4
	$LSD_{(0.05)}$	0.2	0.1	0.2	0.2	0.3
Net CO_2 exchange rate (mg CO_2 m^{-2} s^{-1})						
WW	For	1.1	1.0	0.9	0.7	nil
	PI	1.0	1.1	1.0	0.8	nil
WS	For	0.8	0.7	0.4	0.2	nil
	PI	1.0	1.0	0.8	0.5	nil
	$LSD_{(0.05)}$	0.2	0.3	0.2	0.2	–
Leaf area (dm^2 $plant^{-1}$)						
WW	For	325	515	705	725	–[a]
	PI	505	620	825	840	–
WS	For	340	505	620	615	–
	PI	495	590	785	770	–
	$LSD_{(0.05)}$	32	34	37	36	–

[a] Not collected.

Materials and methods

The experiment was conducted in the phytotron at North Carolina State University. Soybean (PI 416937 and cv. Forrest) seed, inoculated with a commercial (Nitrogen)[1] preparation of *Bradyrhizobium japonicum*, were pre-germinated in moist vermiculite for 36 h at 25 °C. Primary roots of the germinated seedlings were dipped into a slurry of the commercial *Rhizobium* preparation. Seedlings were transplanted, three per pot, in 9 L plastic pots filled with steam-sterilized river-bottom sand. Seedlings were thinned to one per pot 14 days after emergence (DAE).

Plants were grown in two (replicated) controlled-environment chambers at day/night temperatures of 26/22° (+ 0.25°C). Day and night transitions of light and temperature were abrupt and coincident. During the 12-h day period, a combination of cool-white fluorescent and incandescent lamps provided a photosynthetic photon flux density of 670 to 735 μE m^{-2} sec^{-1} and photomorphogenic radiation of 10 to 13 W m^{-2}. Further details regarding the lighting regimen can be found in Russell (1979). Ambient CO_2 within growth chambers was monitored by infrared gas analysis and maintained at 350 to 400 μL L^{-1}. Relative humidity was 70% or greater at all times.[1]

Pots were watered within the first hour of light and again in the middle of the light period with 0.600 L of deionized water. During the last hour of the light period, the pots were flushed with 1.2 L of deionized water. After excess water had drained, 0.600 L of a nitrogen-free nutrient solution (Russell, 1979) was applied to each pot. Immediately prior to application of the nutrient solution, 0.010 L of fresh inoculum slurry was applied daily around the base of plants in each pot until 14 DAE.

The procedure of Sanchez-Diaz and Kramer (1971), utilizing double junction isopiestic thermocouple psychrometry, was used to monitor changes in total leaf electrochemical water potential (Ψw,) throughout the drying and recovery cycle. The stress cycle was ini-

[1] The use of trade names in this publication neither implies endorsement by the North Carolina Agricultural Service of the products named nor criticism of similar ones not mentioned.

Table 2. Effect of watering protocol (WW = well watered and WS = water stressed) on total plant dry matter[a] accumulation during reproductive growth and on final seed yield (R7 growth stage) of soybean genotypes (For = Forrest and PI = PI 416937)

Water status	Genotype	Growth stage			Seed yield	YSI[b]
		Rl	R4	R7[c]		
		Dry matter (g $plant^{-1}$)			(g $plant^{-1}$)	(% red.)
WW	For	35.1	49.8	55.8	38.4	–
	PI	37.5	53.9	69.4	35.2	–
WS	For	37.3	44.2	51.7	24.7	35.6
	PI	38.8	49.2	64.3	29.3	16.8
	$LSD_{(0.05)}$	3.2	3.3	3.7	3.3	

[a]Total dry matter = leaves + stems + petioles + nodules
[b]Yield stress Index, percentage reduction (red.) in yield in WS compared to WW.
[c]Fallen leaves were collected daily and added to the mass of the appropriate treatment.

tiated at R3 growth stage (Fehr and Caviness, 1977). Our intent was to maintain well-watered (WW) plants at Ψ_W of about -0.6 to -0.8 MPa, and water-stressed (WS) plants at a Ψw of about -1.1 to -1.6 MPa, during the drying phase. For WS plants, volume and concentration of nutrient solutions were adjusted to ensure consistent nutrient additions to each treatment combination. Our experience is that this approach affords us the best opportunity to simulate WW and WS conditions which normally prevail in the field. Stress was terminated at the end of the R5 stage: previously stressed plants were watered using the same protocol as for WW plants.

Measurement of net CO_2 exchange rate (NCE) was accomplished using a LiCor LI-6000 infrared gas analyzer equipped with a microcuvette suitable for measuring CO_2 flux of individual soybean leaflets. Both NCE and Ψw measurements were taken at the same node position between five and seven hours following the "lights on" phase of the diurnal cycle.

A modification of the acetylene reduction assay of Patterson et al. (1979) was used to estimate the potential for root nodule nitrogenase activity. Total nodule activity (TNA) was calculated based on production of ethylene, which was quantified by gas chromatography. Specific nodule activity (SNA) was obtained by dividing TNA by the mass (dry weight basis) of the nodules recovered from each root system.

Plants representing all four treatment combinations (Forrest WW and WS; PI WW and WS) were harvested at each of the growth stages Rl through R7. Four plants were sampled from each treatment combination at each harvest date. Immediately before harvest, tissue was excised for Ψ_W measurement from the terminal leaflet of uppermost fully developed leaves. At the time of harvest, intact root systems with attached nodules were excised and assayed as described above. Leaves and stems plus petioles were separated and frozen. After the incubation for acetylene reduction, roots and nodules also were separated and frozen. Following freeze-drying and weighing, the plant components were ground for analysis of total nitrogen by a modified Kjeldahl method (Nelson and Sommers, 1973).

The experiment was conducted as a completely randomized block with four single-plant replicates within each chamber. The positions of plants within a chamber were changed daily to minimize positioning effects. A factorial analysis of variance was used for each sampling date to evaluate the differences among responses to treatments.

Results

Application of water stress reduced Ψw for each genotype (Table 1). As expected (based on other phytotronic and field studies by our group), Ψ_W for Forrest fell more rapidly and fully under desiccating conditions than did the PI, underscoring the relatively greater capacity of the PI for preserving its tissue in a hydrated state. Application of water stress reduced photosynthetic capacity of leaves for both genotypes, but more so for Forrest than for PI (Table 1). This obser-

Table 3. Effect of watering protocol (WW = well watered and WS = water stressed) on total nodule dry weight, specific and total activity of nodules, and on total leaf N content throughout reproductive development of soybean genotypes (For = Forrest and PI = PI 416937)

Water status	Genotype	Growth stage R1	R3	R4	R5	R7
Nodule dry wt.(g plant^{-1})						
WW	For	4.32	4.65	4.42	3.69	2.05
	PI	6.45	6.80	6.94	5.71	3.84
WS	For	4.10	4.45	3.61	2.87	1.71
	PI	6.25	6.75	6.62	6.45	4.34
	$LSD_{(0.05)}$	0.69	0.71	0.70	0.72	0.73
Specific activity (μmol C_2H_4 g^{-1} h^{-1}						
WW	For	52.4	48.1	45.4	18.4	nil
	PI	53.9	50.2	52.5	27.3	nil
WS	For	49.2	46.2	41.1	11.1	nil
	PI	52.9	49.1	44.3	21.4	nil
	$LSD_{(0.05)}$	8.4	8.1	8.3	8.4	–
Total activity (μmol C_2H_4 plant^{-1} h^{-1})						
WW	For	226	224	201	68	nil
	PI	283	341	364	156	nil
WS	For	202	206	148	32	nil
	PI	331	331	293	138	nil
	$LSD_{(0.05)}$	13	12	14	17	–
Total leaf N (g plant^{-1})						
WW	For	1.25	1.08	0.85	0.69	0.24
	PI	1.30	1.15	0.91	0.75	0.45
WS	For	1.20	0.98	0.61	0.41	0.15
	PI	1.31	1.06	0.75	0.51	0.38
	$LSD_{(0.05)}$	0.21	0.18	0.20	0.17	0.19

vation is generally consistent with the findings of Marlow (1993), who examined the NCE capacity of PI under field conditions. The relatively greater leaf area production of the PI through growth stage R5 emphasizes the comparatively favorable photosynthetic surface presented by this genotype (Table 1).

Under WW conditions, the PI accumulated plant dry matter at a faster rate during reproductive growth than did Forrest, but Forrest partitioned a relatively higher proportion of its phytomass into seed mass (Table 2). While Forrest tended (numerically, not statistically) to out yield PI under WW conditions, PI out yielded Forrest under stress and also experienced a considerably less (16.8 vs. 35.6%) percentage yield reduction under stress. This finding is consistent with what we have observed under field conditions for these two genotypes (Sloane et al., 1990).

Total nodule mass for each genotype appeared to reach a maximum at the beginning of R3, but the decline in mass as pod filling continued was much greater for Forrest than for PI (Table 3). Regardless of water treatment, the exotic germplasm was able to allocate fixed carbon to nodules more effectively than was the case for Forrest, as indicated by consistently higher nodule mass for PI than for Forrest at each growth stage. Moreover, in comparison to Forrest, the PI was able to maintain the potential for SNA and TNA much later into reproductive development and also much more fully under drying conditions. While leaf N content for both genotypes dropped rapidly with the onset of maturation and water deficit, a faster decline was observed for Forrest than for PI (Table 3).

Discussion

A critically important period for fixation and assimilation of nitrogen in soybean production is during the interval between initial seed formation and the end of the linear seed-filling phase (Graham, 1979). The sensitivity of both soybean pod fill (Sionit and Kramer, 1977) and biological nitrogen fixation (Masyhudi and Patterson, 1991) to drought during the linear seed-filling phase suggests that water stress may impose a serious impediment to the genetic yield potential of a given genotype being realized. Both duration and rate of seed fill appear to be linked closely to the flow of N assimilate from source tissues to the developing embryos (Cure et al., 1983, 1985). Thus, the ability of a genotype to avoid drought could be pivotal in its capacity to fix, assimilate, and utilize N for seed growth during the critical linear filling phase.

The magnitude of acetylene reduction, NCE, leaf area production, and total plant dry matter production exhibited by the PI under stress (compared to Forrest) suggests that this exotic germplasm was capable of maintaining photosynthetic function sufficient to ensure an abundant supply of carbohydrates to the roots to establish and maintain potential for high rates of nitrogen fixation in the presence of desiccating growing conditions. Hudak (1993) has suggested that tissue of this PI is more hydrated under stress than are control genotypes such as Forrest due to a more efficient extraction of soil water.

Our observation of PI's seed yield being reduced relatively less under water stress (compared to Forrest) is consistent with the findings of Hudak (1993) in a field study and suggests that PI's capacity to avoid drought does not necessarily compromise its potential yielding ability. Part of the reason for this supposition resides in the preservation of potential for symbiotic N_2 fixation later into the pod-filling period for this drought adapted germplasm.

Photosynthetic and N_2 fixation rates for PI under stress were sufficient to maintain an adequate concentration of N in leaves throughout seed filling, but the leaf N concentration of stressed Forrest plants fell below the critical level for continued high rates of photosynthetic activity (Boote et al., 1978). As has been observed in the field (personal observations), Forrest leaves exhibit premature senescence under drought more readily than do PI leaves. We attribute at least some of Forrest's decline in leaf function under stress to more rapidly declining nodule function, resulting in less N assimilate available for transport from nodules to leaves. In conclusion, PI 416937 appears to possess the capacity to preserve nodule function under desiccating growing conditions and, therefore, provide a more favorable supply of N assimilate to leaves and developing pods for preservation of yield stability under stress.

Acknowledgements

Sincere appreciation is expressed to the N C Soybean Producers' Association and to the United Soybean Board of the American Soybean Association for providing partial funding for the conduct of this research.

References

Boote K J, Gallaher R N, Robertson W K, Hinson K and Hammond L C 1978 Effect of foliar fertilization on photosynthesis, leaf nutrition, and yield of soybeans. Agron. J. 70, 787–791.

Boyer J S 1976 Water deficits and photosynthesis. *In* Water Deficits and Plant Growth. Vol. IV. Soil Water Measurement, Plant Responses, and Breeding for Drought Resistance. Ed. T T Kozlowski. pp 151–190. Academic Press, New York, USA.

Cure J D, Patterson R P, Raper C D Jr and Jackson W A 1982 Assimilate distribution in soybeans as affected by photoperiod during seed development. Crop Sci. 22, 1245–1250.

Cure J D, Raper C D Jr, Patterson R P and Jackson W A 1983 Water stress recovery in soybeans as affected by photoperiod during seed development. Crop Sci. 23, 110–115.

Cure J D, Raper C D Jr, Patterson R P and Robarge W P 1985 Dinitrogen fixation in soybean in response to leaf water stress and seed growth rate. Crop Sci. 25, 52–58.

Doss B D, Pearson R W and Rogers H T 1974 Effect of soil water stress at various growth stages on soybean yield. Agron. J. 66, 297–299.

Fehr W R and Caviness C E 1977 Stages of soybean development. Iowa State Univ. Agric. and Home Econ. Stat. Special Report 80. Iowa Cooperative Extension Service, Ames, Iowa, USA.

Graham P H 1979 Influence of temperature on growth and nitrogen fixation in cultivars of *Phaseolus vulgaris* L. inoculated with *Rhizobium*. J. Agric. Sci. Camb. 93, 365–370.

Hardy R W F, Burns R C, Herbert R R, Holsten R D and Jackson E K 1965 Biological nitrogen fixation: A key to world protein. Plant and Soil 35, 561–590.

Hudak C M 1993 An evaluation of a soybean plant introduction for drought avoidance under field conditions. Doctorate thesis. N.C. State University, Raleigh, USA.

Marlow W S 1993 Canopy apparent photosynthesis and biological nitrogen fixation of a drought tolerant soybean genotype. Master's thesis. N.C. State University, Raleigh, USA.

Masyhudi M F and Patterson R P 1991 The effect of water stress on nitrogen absorption of soybean. Indones. J. Crop Sci. 6, 43–63.

Nelson D W and Sommers L E 1973 Determination of total nitrogen in plant material. Agron. J. 65, 109–112.

Patterson R P, Raper C D Jr and Gross H D 1979 Growth and specific nodule activity of soybean during application and recovery of a leaf moisture stress. Plant Physiol. 64, 551–556.

Russell S J 1979 Nitrogen-temperature interaction in soybeans [*Glycine max* L. (Merrill)]. Master's thesis. N.C. State University, Raleigh, USA.

Sanchez-Diaz M F and Kramer P J 1971 Behavior of corn and sorghum under water stress and during recovery. Plant Physiol. 48, 613–616.

Silvius J E, Johnson R R and Peters D B 1977 Effect of water stress on carbon assimilation and distribution in soybean plants at different stages of development. Crop Sci. 17, 713–716.

Sionit N and Kramer P J 1977 Effect of water stress during different stages of growth of soybean. Agron. J. 69, 274–278.

Sloane R J, Patterson R P and T E Carter Jr 1990 Field drought tolerance of a soybean plant introduction. Crop Sci. 30, 118–123.

Section editor: H Lambers
Guest editors: G.H. Elkan and R.G. Upchurch

Plant and Soil **186:** 45–52, 1996.

Phylogeny and taxonomy of rhizobia

J.P.W. Young
Department of Biology, University of York, P.O. Box 373, York YO1 5YW, UK *

Key words: Azorhizobium, Bradyrhizobium, Mesorhizobium, phylogeny, *Rhizobium Sinorhizobium*, species, taxonomy

Abstract

Rhizobia are bacteria that form nitrogen-fixing nodules on the roots, or occasionally the shoots, of legumes. There are currently more than a dozen validly named species, but the true number of species is probably orders of magnitude higher. The named species are listed and briefly discussed. Sequences of the small subunit ribosomal RNA (SSU or 16S rRNA) support the well-established subdivision of rhizobia into three genera: *Rhizobium, Bradyrhizobium*, and *Azorhizobium*. These all lie within the alpha subdivision of the *Proteobacteria*, but on quite distinct branches, each of which also includes many bacterial species that are not rhizobia. It has been clear for several years that *Rhizobium*, on this definition, is still too broad and is polyphyletic: there are many non-rhizobia within this radiation. Recently, therefore, it has been suggested that this genus should be split into four genera, namely *Rhizobium (R. leguminosarum, R. tropici, R. etli), Sinorhizobium (S. fredii, S. meliloti, S. teranga, S. saheli), Mesorhizobium (M. loti, M. huakuii, M ciceri, M. tianshanense, M. mediterraneum*), and a fourth, unnamed, genus for the current *R. galegae*. The evidence and pros and cons are reviewed.

Introduction

The taxonomy of the rhizobia is in a state of flux at the moment, with a flood of proposals for new names. This has been stimulated both by the isolation of symbionts from new sources and by advances in molecular methods, notably DNA sequencing and DNA-DNA hybridization, that have greatly increased the confidence with which we can assign strains to species and describe the relationships among species. This chapter sets out to describe the present situation and to explain how we got to this point and where things might go in the near future. The taxonomy has changed almost beyond recognition since the last account in Bergey's Manual (Jordan, 1984); there have been a number of more recent accounts (Elkan, 1992; Graham et al., 1991; Martinez-Romero, 1994; Young, 1991).

In this review I shall list, with brief comments, all the names of genera and species that are currently valid (end of 1995), and describe the evidence for their phylogenetic relationships. There are now over a dozen "good" species of *Rhizobium*, but in fact this is probably still orders of magnitude lower than the true number of rhizobial species. This might even be comparable with the number of legume species, currently estimated at up to 18,000, because only a small number of legume species have so far been systematically sampled for the diversity of their symbionts. Although it is true that most known rhizobial isolates are capable of nodulating more than one legume species, it is equally true that one legume species often yields more than one species of rhizobium symbiont.

* FAX No: + 441904432860. E-mail: JPY1@YORK.AC.UK

Phylogeny of rhizobia

For many years all bacteria that formed nitrogen-fixing root nodules on legumes were, by definition, assigned to the genus *Rhizobium*. However, two distinct groups of rhizobia, slow-growing and fast-growing on laboratory media, were recognised as early as 1932 by Fred et al., who also pointed out that some fast-growing rhizobia were closely related to agrobacteria. On the basis of the evidence available at the time, Norris (1965) suggested that slow-growers (which do not acidify the growth medium) tend to be associated with tropical legumes, and the acid-producing fast-growers with temperate legumes. This statement has been repeated by numerous authors since, even though counter-examples have steadily accumulated and it is now evident that the generalisation is of little value. Many temperate legumes e.g. *Lupinus* and *Coronilla* (Allen and Allen, 1981) have slow-growing rhizobia, many tropical legumes, e.g. *Leucaena, Acacia* and *Sesbania* (Lajudie et al., 1994; George et al., 1994) have fast-growing rhizobia, and indeed fast and slow rhizobia can frequently be isolated from the same species, e.g. *Glycine max* (Scholla and Elkan, 1994), or even the same plant, e.g. *Acacia* (Dreyfus and Dommergues, 1981) and *Prosopis* (Jenkins et al., 1987).

Modern molecular data allow us to see that the rhizobia are clearly polyphyletic, meaning that there is no branch of the evolutionary tree that carries all the rhizobia and no other bacteria. They have confirmed the distinctness of the classical "slow-growers", now the genus *Bradyrhizobium*, but have shown that the fast-growers are rather diverse.

Sequence comparisons of the small subunit of the ribosomal RNA (SSU rRNA) have become the standard method for assessing phylogenetic relationships among bacteria (Olsen et al., 1994), and such comparisons show the diversity of rhizobia very clearly (Sawada et al., 1993; Willems and Collins, 1993; Yanagi and Yamasato, 1993). All the rhizobia discovered so far are members of the alpha subdivision of the proteobacteria, but the branch that includes them all also carries other bacteria that are not root-nodule symbionts but have been assigned to at least 24 different genera (Maidak et al., 1994), some of which are shown in Figure 1.

One way to disentangle rhizobia from these genera is to recognise the distinct lineages as separate genera. The genera *Bradyrhizobium* and *Azorhizobium* are so clearly separate from *Rhizobium* that no one disputes their reality as independent evolutionary branches. The species remaining in *Rhizobium* are commonly known as the "fast-growing rhizobia", but they are still very heterogeneous so there are now moves to split up this branch, as described in the next section. Of course it is highly likely that, as more rhizobia are isolated and described, some will fall well outside the phylogenetic groups defined so far and further new genera will be needed.

There remain some non-rhizobia that are so closely related to rhizobia that it is not appropriate to "define away the problem" by creating new genera. In particular *Agrobacterium rhizogenes* (i.e. biovar 2) is so similar to *Rhizobium tropici* that they might be one species, and certainly cannot realistically be considered as separate genera. *Bradyrhizobium* has a number of odd bedfellows and it is not yet clear whether the different genera form distinct branches. These issues are addressed below.

While SSU rRNA sequences are a reliable guide to bacterial relationships at the level of genus and above, there are a number of reasons why they become less informative at the level of individual species. Ribosomal RNA is a rather conserved gene, so the number of informative sequence differences becomes very small when closely related species are compared. Differences between species may be no greater than variation within species, or even within genomes (e.g. Petersson et al., 1994), and relationships deduced from SSU sequences may be different from those implied by other genetic traits (Eardly et al., 1995), perhaps because of recombination. At this level, hybridization of total genomic DNA (reviewed by Grimont, 1988) provides a more credible estimate of overall genetic similarity. This approach is not without problems: there are many technical variants and results are not directly comparable between laboratories; multiple pairwise comparisons are required so large studies are laborious and full distance matrices are seldom obtained; hybridization declines to background levels at quite moderate levels of DNA sequence divergence; and the contributions of plasmid and chromosomal DNA cannot be distinguished. Nevertheless, this remains the best technique we have for deciding whether two strains belong to the same species, and the creation of a new species has always to be supported by low values of DNA-DNA hybridization with all previously-described species.

Taxonomy of the rhizobia

Branch 1: the genus Rhizobium

In any splitting of the genus, the name *Rhizobium* must, according to the rules of nomenclature, stay with *Rhizobium leguminosarum* as this is the type species and defines the genus. In addition to the named species described below there are quite a few strains that seem to belong to the *leguminosarum* species group but are distinct from the named species according to DNA-DNA hybridization and/or SSU rRNA sequence. One potential new species is formed by FL27 and some similar strains (Laguerre et al., 1993). Strain TAL1145, isolated from *Leucaena*, is also distinct in SSU sequence (George et al., 1994).

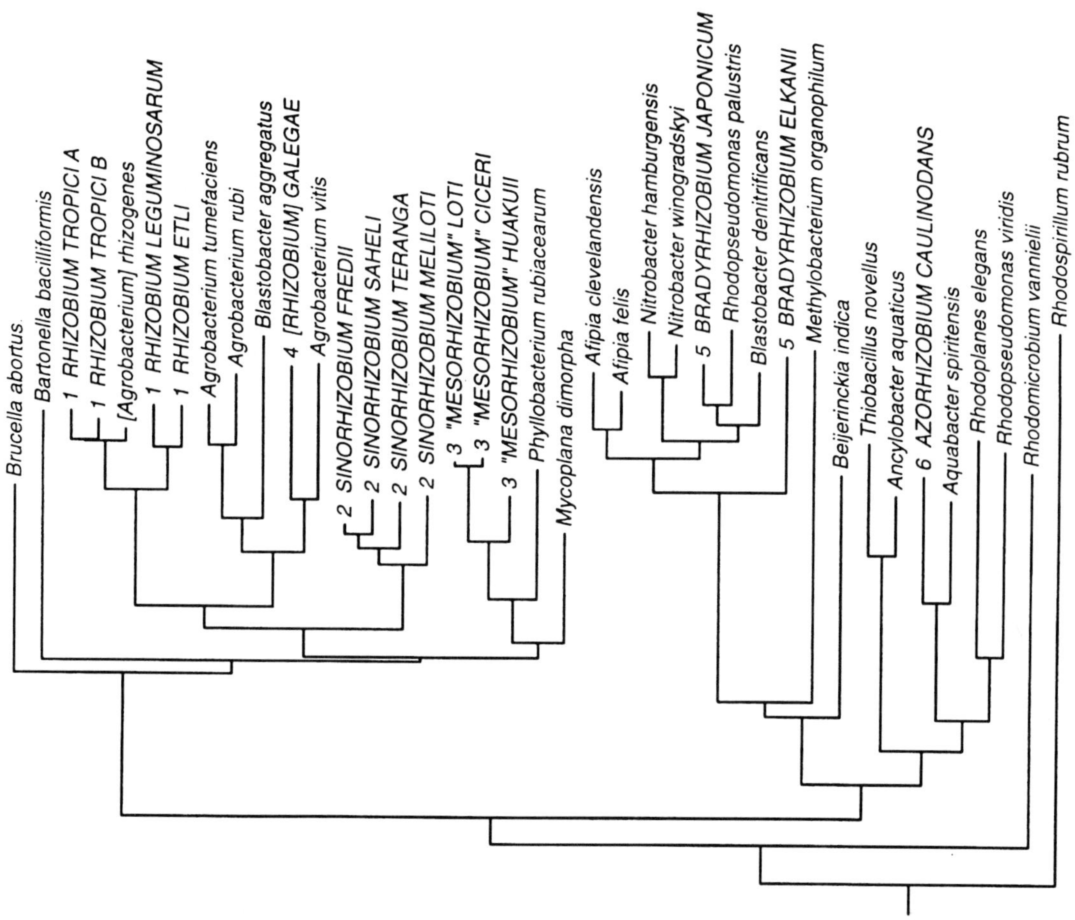

Figure 1. Phylogenetic tree of rhizobia and some related bacteria in the alpha subdivision of the *Proteobacteria*. For rhizobia, phylogenetic branches are numbered as in the text. The tree is constructed by the Neighbor-Joining method (Saitou and Nei, 1987) from SSU rRNA sequences. All the sequences used are almost full length and are derived from the type strain wherever possible. "*Mesorhizobium*" *tianshanense*, "*M.*" *mediterraneum* and *Bradyrhizobium liaoningense* are omitted as full sequences were not available. The Genbank/EMBL accession numbers, or other sources. are (top to bottom): X13695, Z11683, X67233, X67234, X67224, U29386, U28916, X67223 X67228, X73041, X67226, X67225, X67231, X68390, X68387, X67222, X67229, U07934, D12797, D12790, D12786, M69186, M65248, L11663, L11661, X66024, D25312, S46917, U35000, D32226, M59060, D32247, M62790, X67221, aqb.spirit (from Ribosomal Database Project, Maidak et al., 1994), D25311, D25314, M34127, D30778.

Rhizobium leguminosarum Frank 1889. This name originally described just the symbionts of the tribe Vicieae. However, it became clear that bacteria from clover nodules (*R. trifolii*) and *Phaseolus* bean nodules (*R. phaseoli*) could not be reliably distinguished from *R. leguminosarum* except by their host range, and the currently accepted position (proposed by Jordan, 1984) is that there are three biovars of the species *R. leguminosarum*. Biovar *viciae* nodulates the tribe Vicieae (*Pisum, Vicia, Lathyrus, Lens*), biovar *trifolii* nodulates *Trifolium*, and biovar *phaseoli* nodulates *Phaseolus* (but so also do *R. tropici* and *R. etli*, see below). The host ranges of the three biovars are quite distinct and seem to be mutually exclusive: strains that are good effective symbionts of both peas and clover (for example) have never been found. The nitrogen fixation and nodulation genes, including the host range determinants, are carried on plasmids, so the biovar designation is really a description of the plasmid rather than the chromosomal background of the strain.

Rhizobium tropici Martinez-Romero et al. 1991. The symbionts of *Phaseolus* originally called *R. phaseoli*, became *R. leguminosarum* biovar *phaseoli* according to Jordan's 1984 revision. However, it was clear even then that they were rather diverse and further evidence accumulated to indicate that, while some isolates from *Phaseolus* did indeed belong in *R. leguminosarum* (for example, Young, 1985), others were quite different. Extensive genetic studies at the Centro de Fijacion de Nitrogeno in Mexico identified two major groups, Type I and Type II. The Type II strains, which form effective nodules on *Phaseolus vulgaris, Leucaena esculenta*, and *L. leucocephala*, and have numerous distinctive phenotypic and genetic characteristics, have been renamed as *R. tropici* (Martinez-Romero et al., 1991). There are two distinct subgroups, A and B which are sufficiently different that they might be described as separate species (Geniaux et al., 1995; Martinez-Romero et al., 1991). The type strain (CIAT 899) is in subgroup B.

Rhizobium etli Segovia et al. 1993. The type I symbionts of *Phaseolus vulgaris* are also different from "typical" *R. leguminosarum* in genetic and phenotypic properties and have been reclassified as *R. etli.* They form an effective symbiosis only with *Phaseolus vulgaris*.

"Rhizobium"/[Agrobacterium] rhizogenes. Nomenclature in *Agrobacterium* is also changing. At one time the name *A. tumefaciens* was used for tumor-inducing strains (carrying a Ti plasmid), and *A. rhizogenes* for hairy-root strains (Ri plasmid). It was recognised that there were several distinct chromosomal species in *Agrobacterium*; these were identified as "biotypes" or "biovars". More recent proposals give these biovars proper recognition as species: biovar 1 strains becomes *A. tumefaciens* regardless of the type of plasmid (Ti or Ri), while biovar 2 is *A. rhizogenes*, again this may have either type of plasmid (Bouzar, 1994; Oyaizu and Sawada, 1994; Sawada et al., 1993). Basing the nomenclature on chromosomal relatedness in this way, rather than plasmid-encoded characters, is obviously desirable, but leaves a problem because *A rhizogenes* is only distantly related to the other *Agrobacterium* species but is very closely related to *R. tropici*. It should therefore be called *Rhizobium rhizogenes*. The only remaining question that needs investigation is whether *R. rhizogenes* is so close to *R. tropici* that these should be regarded as biovars of the same species (using biovars to describe plasmid-encoded characteristics, as in *R. leguminosarum*, rather than to describe chromosomal species, as was formerly the practice in *Agrobacterium*).

Branch 2: the genus Sinorhizobium

Chen et al. (1988), on the basis of a numerical taxonomic study, proposed that *R. fredii* and related soybean rhizobia should be moved to a new genus, *Sinorhizobum*, with two species *S. fredii* and *S. xinjiangensis*) because their phenotypic characteristics were so different from other rhizobia, including *R. meliloti*. However, this is not consistent with the molecular data, which clearly indicate a close relationship between *R. fredii* and *R. meliloti* (Jarvis et al., 1992), and *Sinorhizobium* as defined by Chen et al. (1988) was never widely accepted. Recently, however, the name has been revived by de Lajudie et al. (1994) with a wider definition that now includes the former *R. meliloti, R. fredii*, and two new species. *Sinorhizobium* in this new sense is well supported by the SSU rRNA data (Figure 1).

Sinorhizobium meliloti (Dangeard 1926) the former *Rhizobium meliloti*, nodulates *Melilotus, Medicago* and *Trigonella*, i.e. the tribe Trifolieae apart from *Trifolium* (which is nodulated by *R. leguminosarum* biovar *trifolii*) and *Parochetus* (nodulated by a distinct but unnamed bacterium). There are two clearly defined subgroups that could justifiably be split into separate species (Eardly et al., 1990).

Sinorhizobium fredii (Scholla and Elkan 1984) can form effective nodules on soybean (*Glycine max*), cowpea (*Vigna unguiculata*) and pigeon pea (*Cajanus cajan*), although some strains do not form effective nodules on US soybean cultivars, which have a narrow genetic base. Scholla and Elkan (1984) recognised two groups within the species on the basis of DNA-DNA hybridization and serology (chemovars *fredii* and *siensis*), and Chen et al. (1988), using phenotypic tests, also identified two groups and proposed two species, *S. fredii* and *S. xinjiangensis*. Unfortunately these two different bases for subdividing the species give contradictory classifications, so it seems best not to change the nomenclature until the situation is clearer, and indeed the name *S. xinjiangensis* has not been commonly used. Strain NGR234 is a widely-studied organism, originally isolated from *Lablab purpureus* but having a very wide host range, including the tropical tree *Parasponia* (which is in the Ulmaceae and is the only nonlegume known to form nitrogen-fixing nodules with rhizobia). Although NGR234 has never been formally assigned to any species, its SSU rRNA and *nod* gene sequences indicate a very close relationship with *S. fredii*, which is also supported by immunological studies (Jarvis et al., 1992).

Sinorhizobium saheli de Lajudie et al. 1994. The genus *Sinorhizobium* in its new sense was defined by de Lajudie et al. when they described two new species, *S. saheli* and *S. teranga*, isolated from *Acacia* and

Sesbania. Both species also nodulate a number of other tree legume genera.

Sinorhizobium teranga de Lajudie et al. 1994. See *S. saheli*.

Branch 3: the genus "Mesorhizobium"

This group of species is the most distant branch within the fast-growing rhizobia, and definitely merits a new genus. No name has yet been formally published, but at the meeting of the International Subcommittee on *Rhizobium* and *Agrobacterium* held in St Petersburg in June 1995 the name "*Mesorhizobium*" was put forward (Lindström et al., 1995), and it is likely that a formal proposal will be made soon (B D W Jarvis, pers. comm.). The name (which was originally suggested by W X Chen and colleagues) reflects the fact that many (but by no means all) strains in these species grow on laboratory medium at rates intermediate between typical "fast-growers" (*Rhizobium*) and "slow-growers" (*Bradyrhizobium*).

"Mesorhizobium" loti (Jarvis, Pankhurst and Patel 1982) forms effective nodules on *Lotus, Lupinus*, and *Anthyllis*, though not necessarily on all species. The same hosts can also be nodulated by *Bradyrhizobium* (see below). The nodulation and nitrogen fixation genes are usually chromosomal in *M. loti*, rather than plasmid-encoded as in most other fast-growing rhizobia.

"Mesorhizobium" huakuii (Chen et al., 1991) was isolated from *Astragalus sinicus*. It nodulates this and a few other hosts and is distinct from other named species.

"Mesorhizobium" ciceri (Nour et al., 1994) is very similar to "*M.*" *loti* in SSU rRNA sequence (Nour et al., 1995), but is distinguished by DNA-DNA hybridization, by phenotype, and by host range as most strains nodulate only chickpea.

"Mesorhizobium" tianshanense (Chen et al., 1995) describes a set of related strains from various legumes (including soybean) in Xinjiang province, China. Only a partial SSU rRNA sequence is available, and this is identical to that of "*M.*" *ciceri*. Furthermore the type strain shows more than 60% DNA-DNA relatedness to "*M.*" *huakuii*, so some further work on this genus is needed to clarify just how many distinct species there are.

"Mesorhizobium" mediterraneum (Nour et al., 1995) nodulates chickpea but is well differentiated from "*M.*" *ciceri* and other mesorhizobia by SSU rRNA sequence and other characteristics.

Branch 4: an unnamed genus

The species *R. galegae* has no very near known relatives. The closest SSU rRNA sequence belongs to *Agrobacterium vitis*, which appears to be closer to *R. galegae* than it is to *A. tumefaciens* and *A. rubi*. Perhaps a new genus could be proposed for *R. galegae* and *A. vitis*, but they are not sufficiently close to make this uncontroversial. An alternative might be to propose *Agrobacterium galegae*, but this would broaden the genus unacceptably. It would be good to wait until more relatives are discovered, but this leaves [*R.*] *galegae* without a satisfactory genus name, since it is clearly distinct from the new, narrowly defined *Rhizobium*.

[Rhizobium] galegae Lindström 1989 forms effective nodules on either *Galega orientalis* or *G. officinalis*. Strains from each host species form only ineffective nodules on the other *Galega* species, and no nodules on other legumes tested.

Branch 5: the genus Bradyrhizobium

Although the existence of two distinct groups of root nodule bacteria had been recognised since the early years of the century (Fred et al., 1932), the creation of the new genus *Bradyhizobium* Jordan 1982 was the first accepted change in rhizobial nomenclature for fifty years. Besides slow growth on standard media, there are a host of other physiological differences that distinguish *Bradyrhizobium* from *Rhizobium*, and the distinction is even more clear-cut at the genetic level. All bradyrhizobia form a closely-related cluster in terms of SSU ribosomal RNA sequence, very distinct from other rhizobia. The surprise is that this cluster also includes a number of bacteria that are not rhizobia and have been placed in other genera: *Rhodopseudomonas Nitrobacter, Blastobacter* and *Afipia*. The differences in their SSU sequences are not much greater than those that are often found between species classified in the same genus, but no one has yet made a careful comparative study of other properties, and the differences in certain aspects of their phenotype are so conspicuous that it would be premature at this stage to suggest that all these bacteria should be called *Rhodopseudomonas*.

Bradyrhizobium japonicum Jordan 1982 now covers all *Bradyrhizobium* strains that nodulate soybean effectively but do not fit the definition of *B. elkanii*. This is still a diverse assemblage from which more new species need to be disentangled in due course.

Bradyrhizobium elkanii Kuykendall et al. 1992. Phenotypic and genetic characters have been used to define a number of groups within the soybean-nodulating bradyrhizobia, and Kuykendall et al. proposed the name *B. elkanii* for group II strains. They did not provide a formal description of the new species, but this group does appear to have fairly clear natural boundaries.

Bradyhizobium liaoningense Xu et al. 1995. This species has recently been defined for extra slow growing soybean rhizobia that form a coherent DNA-DNA hybridization group. They also cluster together on the basis of growth tests, but this may largely reflect their low growth rate on all substrates. Only a partial SSU rRNA sequence is available.

Bradyrhizobium sp. There are many strains that clearly belong in *Bradrhizobium* but do not nodulate soybean. No species names have yet been defined for these, they are simply known as *Bradyrhizobium* sp., followed by the name of the legume host genus in parentheses. Some of these are closely related to particular soybean bradyrhizobia (*B. japonicum* or *B. elkanii*), and it will make more sense in the long run to drop soybean nodulation as an over-riding taxonomic character and to define species within *Bradyrhizobium* according to overall phenotypic and genotypic similarity. Some very interesting symbionts of *Aeschynomene indica* belong to this group (Young et al., 1991). In addition to forming stem nodules, these bacteria are photosynthetic, thus providing additional support for the grouping of bradyrhizobia with photosynthetic bacteria such as *Rhodopseudomonas palustris*. The name "*Photorhizobium*" has been applied to these bacteria (Eaglesham et al., 1990), but it has fortunately not been formally proposed; their SSU sequences (Wong et al., 1994) are close to those of other bacteria in this crowded cluster, which already includes at least five named genera.

Branch 6: the genus Azorhizobium

Azorhizobium caulinodans Dreyfus et al. 1988 nodulates both stems and roots of *Sesbania rostrata*. It is still the only named species in the genus, but a second species has been recognised by DNA-DNA hybridization assays (Rinaudo et al., 1991).

Should we split up the fast-growing rhizobia?

As I write this, proposals are in the air for the subdivision of the genus *Rhizobium*. *Sinorhizobium* has already been defined (de Lajudie et al., 1994), while "*Mesorhizobium*" still awaits formal description. Whenever nomenclature is changed there need to be very good reasons, because there are always negative effects as well as benefits. Any change will be confusing at first and will instil nostalgia and a sense of loss in those who find a new name has been slapped on "their" organism. More seriously, a name change creates a break with the earlier literature. Splitting a genus means that putting a name to a new isolate will be harder work: it will no longer be acceptable to record any fast-growing rhizobium as merely "*Rhizobium* sp.", so it is fortunate that the advent of the polymerase chain reaction (PCR) provides identification methods that are quicker, easier and more reliable than the traditional battery of growth tests. If narrow genus boundaries are adopted, further genera will almost certainly be needed as new rhizobia are brought into cultivation, and there is always a danger that so many intermediate species will be discovered that the boundaries between genera will become confused.

To set against all these problems, any change in the nomenclature needs to bring some powerful benefits. The benefits come from creating genera that are monophyletic and of an appropriate size. A monophyletic group is one that includes all the descendants, and only the descendants, of a particular ancestor; it can be snipped cleanly from the evolutionary tree. Clearly fast-growing rhizobia as a whole do not form such a group, they are interspersed with various other genera (Figure 1). The name *Sinorhizobium meliloti* is therefore much more informative than *Rhizobium meliloti* was: this organism is no more related to *R. leguminosarum* than it is to *Agrobacterium tumefaciens*, but it is a close relative of *S. fredii, S. teranga* and *S. saheli*. Furthermore, the fast-growing rhizobia are phylogenetically a very deep-branching group, more diverse than is customary within a single bacterial genus. As new species of rhizobia are being described at an increasing rate, it becomes helpful to classify them into smaller units so that we can remember what their relationships are. It is interesting, and encouraging, that most newly-described species fall quite unambiguously into one or another of the recognized genera. My view is that the groupings that I have discussed here are sufficiently well-supported and useful to merit recognition as separate genera, and that now is the appropriate time to do that.

There is an International Subcommittee on the Taxonomy of *Agrobacterium* and *Rhizobium* that offers advice and opinions on proposals for new nomenclature (e.g. Martinez-Romero and Jarvis, 1993). However, the Subcommittee is merely advisory, and ultimately the decision lies with journal editors and those who work with rhizobia: if they find the new names useful they will adopt them in their publications, if not, they will ignore them. Nomenclature, like the organisms it describes, evolves by survival of the fittest.

Acknowledgements

This field is moving very fast, and I am grateful to many people for sharing their data and ideas with me before publication. These especially include Philippe de Lajudie, Bert Eardly, Maria Fernandez, Kaisa Haukka, Brion Jarvis, Gisèle Laguerre, Kristina Lindström, Sarah Nour, Peter van Berkum, En-Tao Wang, and Ann Willems.

References

Allen O N and Allen E K 1981 The Leguminosae. University of Wisconsin Press, Madison, WI, USA.

Bouzar H 1994 Request for a judicial opinion concerning the type species of *Agrobacterium*. Int. J. Syst. Bacteriol. 44, 373–374.

Chen W X, Yan G H and Li, J L 1988 Numerical taxonomic study of fast-growing soybean rhizobia and a proposal that *Rhizobium fredii* be assigned to *Sinorhizobium* gen. nov. Int. J. Syst. Bact. 38, 392–397.

Chen W X, Li G S, Qi Y L, Wang E T, Yuan H L and Li J 1991 *Rhizobium huakuii* sp. nov. isolated from the root nodules of *Astragalus sinicus*. Int. J. Syst. Bacteriol. 41, 275–280.

Chen W X, Wang E T, Wang S Y, Li Y B, Chen X Q and Li J 1995 Characteristics of *Rhizobium tianshanense* sp nov., a moderately and slowly growing root nodule bacterium isolated from an arid saline environment in Xinjiang, People's Republic of China. Int. J. Syst. Bacteriol. 45, 153–159.

Dangeard P A 1926 Recherches sur les tubercles radicaux des Légumineuses. Le Botaniste, Series 16, Paris. 270 p.

De Lajudie P, Willems A, Pot B, Dewettinck D, Maestrojuan G, Neyra M, Collins M D, Dreyfus B, Kersters K and Gillis M 1994 Polyphasic taxonomy of rhizobia: emendation of the genus *Sinorhizobium* and description of *Sinorhizobium meliloti* comb. nov., *Sinorhizobium saheli* sp. nov., and *Sinorhizobium teranga* sp. nov. Int. J. Syst. Bacteriol. 44, 715–733.

Dreyfus B and Dommergues Y R 1981 Nodulation of *Acacia* species by fast- and slow-growing tropical strains of *Rhizobium*. Appl. Environ. Microbiol. 41, 97–99.

Dreyfus B, Garcia J L and Gillis M 1988 Characterization of *Azorhizobium caulinodans* gen. nov., sp. nov., a stem-nodulating nitrogen-fixing bacterium isolated from *Sesbania rostrata*. Int. J. Syst. Bacteriol. 38, 89–98.

Eaglesham A R J, Ellis J M, Evans W R, Fleischman D E, Hungria M and Hardy R W F 1990 The first photosynthetic N_2-fixing rhizobium: characteristics. *In* Nitrogen Fixation: Achievements and Objectives. Eds. P M Gresshoff, L E Roth, G Stacey and W E Newton. pp 805–811. Chapman and Hall, New York, USA.

Eardly B D, Materon L A, Smith N H, Johnson D A, Rumbaugh M D and Selander R K 1990 Genetic structure of natural populations of the nitrogen-fixing bacterium *Rhizobium meliloti*. Appl. Environ. Microbiol. 56, 187–194.

Eardly B D, Wang F S, Whittam T S and Selander R K 1995 Species limits in rhizobium populations that nodulate the common bean (*Phaseolus vulgaris*). Appl. Environ. Microbiol. 61, 507–512.

Elkan G H 1992 Taxonomy of the rhizobia. Can. J. Microbiol. 38, 446–450.

Frank B 1889 Über die Pilzsymbiose der Leguminosen. Ber. Deut. Bot. Gesell. 7, 332–346.

Fred E B, Baldwin I L and McCoy E 1932 Root nodule bacteria and leguminous plants. Univ. Wisconsin, Madison, WI, USA.

Geniaux E, Flores M, Palacios R and Martinez E 1995 Presence of megaplasmids in *Rhizobium tropici* and further evidence of differences between the two *R. tropici* subtypes. Int. J. Syst. Bacteriol. 45, 392–394.

George M L C, Young J P W and Borthakur D 1994. Genetic characterisation of *Rhizobium* sp. strain TAL 1145 that nodulates tree legumes. Can. J. Microbiol. 40, 208–215.

Graham P H, Sadowsky M J, Keyser H H, Barnet Y M, Bradley R S, Cooper J E, De Ley J, Jarvis B D W, Roslycky E B, Strijdom B W and Young J P W 1991 Proposed minimal standards for the description of new genera and species of root- and stem-nodulating bacteria. Int. J. Syst. Bacteriol. 41, 582–587.

Grimont P A D 1988 Use of DNA reassociation in bacterial classification. Can. J. Microbiol. 32, 541–546.

Jarvis B D W, Pankhurst C E and Patel J J 1982 *Rhizobium loti*, a new species of legume root nodule bacteria. Int. J. Syst. Bacteriol. 32, 378–380.

Jarvis B D W, Downer H L and Young J P W 1992 Phylogeny of fast-growing soybean-nodulating rhizobia supports synonymy of *Sinorhizobium* and *Rhizobium* and assignment to *Rhizobium fredii*. Int. J. Syst. Bacteriol. 42, 93–96.

Jenkins M B, Virginia R A and Jarrell W M 1987 Rhizobial ecology of the woody legume mesquite (*Prosopis glandulosa*) in the Sonoran desert. Appl. Environ. Microbiol. 53, 36–40.

Jordan D C 1982 Transfer of *Rhizobium japonicum* Buchanan 1980 to *Bradyrhizobium* gen. nov., a genus of slow-growing, root nodule bacteria from leguminous plants. Int. J. Syst. Bacteriol. 32, 136–139.

Jordan D C 1984 Rhizobiaceae. *In* Bergey's Manual of Systematic Bacteriology, Vol.1. Ed. N R Kreig. pp 234–256. Williams and Wilkins, Baltimore, USA.

Kuykendall L D, Saxena B, Devine T E and Udell S E 1992 Genetic diversity in *Bradyrhizobium japonicum* Jordan 1982 and a proposal for *Bradyrhizobiium elkanii* sp.nov. Can. J. Microbiol. 38, 501–505.

Laguerre G, Fernandez M P, Edel V, Normand P and Amarger N 1993 Genomic heterogeneity among French rhizobium strains isolated from *Phaseolus vulgaris*. Int. J. Syst. Bacteriol. 43, 761–767.

Lindström K 1989 *Rhizobium galegae*, a new species of root nodule bacteria. Int. J. Syst. Bacteriol. 39, 365–367.

Lindström K, van Berkum P, Gillis M, Martínez E, Novikova N and Jarvis B 1995 Report from the roundtable on *rhizobium* taxonomy. *In* Nitrogen Fixation: Fundamentals and Applications. Ed. I

A Tikhonovich, N A Provorov, V I Romanov and W E Newton. pp 807–810. Kluwer Acad. Publ., Dordrecht, the Netherlands.

Maidak B L, Larsen N, McCaughey M J, Overbeek R, Olsen G J, Fogel K, Blandy J and Woese C R 1994 The Ribosomal Database Project. Nucl. Acids Res. 22, 3485–3487.

Martinez-Romero E, Segovia L, Mercante F M, Franco A A, Graham P and Pardo M A 1991 *Rhizobium tropici*, a novel species nodulating *Phaseolus vulgaris* L. beans and *Leucaena* sp. trees. Int. J. Syst. Bacteriol. 41, 417–426.

Martinez-Romero E and Jarvis B D W 1993 International Committee on Systematic Bacteriology Subcommittee on the Taxonomy of *Agrobacterium* and *Rhizobium*. Int. J. Syst. Bacteriol. 43, 622.

Martinez-Romero E 1994 Recent developments in *Rhizobium* taxonomy. Plant and Soil 161, 11–20.

Norris D O 1965 Acid production by *Rhizobium*, a unifying concept. Plant and Soil 22, 143–166.

Nour S M, Fernandez M, Normand P and Cleyet-Marel J-C 1994 *Rhizobium ciceri* sp. nov., consisting of strains that nodulate chickpeas (*Cicer arietinum* L.). Int. J. Syst. Bacteriol. 44, 511–522.

Nour S M, Cleyet-Marel J-C, Normand P and Fernandez M P 1995 Genomic heterogeneity of strains nodulating chickpea (*Cicer arietinum* L) and description of *Rhizobium mediterraneum* sp. nov. Int. J. Syst. Bacteriol. 45.

Olsen G J, Woese C J and Overbeek R 1994 The winds of (evolutionary) change: breathing new life into microbiology. J. Bacteriol. 76, 1–6.

Oyaizu H and Sawada I 1994 Request for a judicial opinion concerning the type species of *Agrobacterium* - authors' reply. Int. J. Syst, Bacteriol. 44, 374.

Petersson B, Johansson K-E and Uhlén M 1994 Sequence analysis of 16S rRNA from mycoplasmas by direct solid-phase DNA sequencing. Appl. Environ. Microbiol. 62, 2456–2461.

Rinaudo G, Orenga S, Fernandez M P, Meugnier H and Bardin R 1991 DNA homologies among members of the genus *Azorhizobium* and other stem- and root-nodulating bacteria isolated from the tropical legume *Sesbania rostrata*. Int. J. Syst. Bacteriol. 41, 114–120.

Saitou N and Nei M 1987 The neighbor-joining method: a new method for reconstructing phylogenetic trees. Mol. Biol. Evol. 4, 406–425.

Sawada H, Ieki H, Oyaizu H and Matsumoto S 1993 Proposal for rejection of *Agrobacterium tumefaciens* and revised descriptions for the genus *Agrobacterium* and for *Agrobacterium radiobacter* and *Agrobacterium rhizogenes*. Int. J. Syst. Bacteriol. 43, 694–702.

Scholla M H and Elkan G H 1984 *Rhizobium fredii* sp. nov., a fast-growing species that effectively nodulates soybeans. Int. J. Syst. Bacteriol. 34, 484–486.

Segovia L, Young J P W and Martinez-Romero E 1993 Reclassification of American *Rhizobium leguminosarum* biovar. *phaseoli* type 1 strains in a new species, *Rhizobium etli* sp. nov. Int. J. Syst. Bacteriol. 43, 374–377.

Willems A and Collins M D 1993 Phylogenetic analysis of rhizobia and agrobacteria based on 16S ribosomal RNA gene sequences. Int. J. Syst. Bacteriol. 43, 305–313.

Wong F Y K Stackebrandt E, Ladha J K, Fleischman D E, Date R A and Fuerst J A 1994 Phylogenetic analysis of *Bradyrhizobium japonicum* and photosynthetic stem-nodulating bacteria from *Aeschynomene* species grown in separated geographical regions. Appl. Env. Microbiol. 60, 940–946.

Xu LM, Ge C, Cui Z, Li J and Fan H 1995 *Bradyrhizobium liaoningense* sp. nov., isolated from the root nodules of soybeans. Int. J. Syst. Bacteriol. 45, 706–711.

Yanagi M and Yamasato K 1993 Phylogenetic analysis of the family Rhizobiaceae and related bacteria by sequencing of 16S rRNA gene using PCR and DNA sequencer. FEMS. Microbiol. Lett. 107, 115–120.

Young J P W 1985 *Rhizobium* population genetics: enzyme polymorphism in isolates from peas, clover, beans and lucerne grown at the same site. J. Gen. Microbiol. 131, 2399–2408.

Young J P W 1991 Phylogenetic classification of nitrogen-fixing organisms. *In* Biological Nitrogen Fixation. Eds. G Stacey, R H Burris and H J Evans. pp 43–86. Chapman and Hall, New York, USA.

Young J P W, Downer H I and Eardly B D 1991 Phylogeny of the phototrophic rhizobium strain BTAi1 by polymerase chain reaction-based sequencing of a 16S rRNA gene segment. J. Bacteriol. 173, 2271–2277.

Section editor: H Lambers
Guest editors: G.H. Elkan and R.G. Upchurch

Plant and Soil **186:** 53–61, 1996.

Comparison of rhizobitoxine-induced inhibition of β-cystathionase from different bradyrhizobia and soybean genotypes*

K. Xiong and J.J. Fuhrmann
*Delaware Agricultural Experiment Station, Department of Plant and Soil Sciences, College of Agricultural Sciences, University of Delaware, Newark, DE 19717–1303, USA***

Key words: β-cystathionase, rhizobitoxine, secondary RT inhibition constant

Abstract

The enzyme β-cystathionase catalyzes the conversion of cystathionine to homocysteine in both plants and bacteria. Preparations of this enzyme taken from both *Salmonella* and spinach (*Spinacia oleracea* L.) have been shown to be irreversibly inhibited by low concentrations of rhizobitoxine (RT), a chlorosis-inducing phytotoxin produced by some strains of soybean bradyrhizobia. The sensitivities of β-cystathionase from bradyrhizobia and soybean are not well characterized. Therefore, we purified β-cystathionase from selected bradyrhizobia and soybean genotypes that have been shown to exhibit differences in RT production and apparent RT sensitivities, respectively. Enzyme purified from *E. coli* strain DH52 was used for comparison. The enzymes differed in their physiological properties and RT sensitivities. Overall, the β-cystathionase enzymes purified from bradyrhizobia were more sensitive to RT than were those from the soybean cultivars. Kinetic studies showed that the nature of the RT-induced inhibition also differed between the two sources. The enzymes from bradyrhizobia exhibited inhibition that was [RT]-dependent, whereas the enzymes from soybean showed a time-dependent inhibition. These contrasting characteristics may in part reflect differences in active site accessibility, amino acid components, and associated RT diffusion rates. However, in all cases the inhibition caused by RT showed a typical substrate-competitive inhibition pattern.

Abbreviations: RT – rhizobitoxine, PLP – pyridoxal phosphate, DTT – dithioerythritol, PMSF – phenylmethylsulfonyl fluoride, HTP – hydroxyapatite.

Introduction

Rhizobitoxine (RT) is an enol-ether amino acid that is produced by certain strains of bradyrhizobia. It induces distinct chlorosis on young leaves of soybeans. Owens et al. (1968) reported that RT is a potent inhibitor of β-cystathionase (E.C. 4.4.1.8) in *Salmonella*, and it was found subsequently that low concentrations of RT inhibit β-cystathionase in both plants and bacteria (Giovanelli et al., 1971; Owens et al., 1968). β-cystathionase catalyzes α,β-elimination of L-cystathionine to produce L-homocysteine. The enzyme also supports α, β-elimination of L-djenkolate with the formation of pyruvate, ammonia and S-thiomethylcysteine [*S*-(1-methylene-2-thiol) L-cysteine] (Giovanelli and Mudd, 1971). This last compound decomposes to give L-cysteine, formaldehyde, and H_2S. The inhibition therefore blocks the sulfur metabolic pathway (Datko and Mudd, 1982; Giovanelli et al., 1971, 1972; Negrutiu et al., 1985; Owens et al., 1968), inducing a methionine deficiency in cells which may contribute to the production of RT symptoms in the host plant. Kinetic studies of β-cystathionase from *Salmonella* have shown that inhibition may result from both noncompetitive and partially competitive inhibition, and the apparent enzyme affinity for the toxin is over 1,000 times that for its substrate (Owens et al., 1968). Inactivation of spinach (*Spinacia oleracea* L.) β-cystathionase by RT is an active-site-directed irreversible inhibition (Giovanelli et al., 1971). The holoenzyme form of β-cystathionase combines with RT to form a disassociable enzyme-RT complex, producing the inactivat-

* Published as Paper No. 1571 in the Journal Series of the Delaware Agricultural Experiment Station.
** FAX No: 302 831 3651. E-mail: kun@udel.edu

ed enzyme. However, the nature of the interaction(s) between RT and β-cystathionase from bradyrhizobia and soybeans remains unknown.

Previous studies have suggested that inhibition of β-cystathionase by RT in planta may be important to RT symptom development in the soybean plant (Giovanelli et al., 1971; Owens et al., 1968). In order to explore this possible causative relationship, we investigated RT-induced inhibition of β-cystathionase from different bradyrhizobia strains and soybean cultivars. The specific objectives of our study were to partially purify β-cystathionase from different bradyrhizobia strains and soybean cultivars, test for RT sensitivities of these enzymes under the defined conditions, and compare RT inhibitory characteristics of these enzymes by kinetic studies.

Materials and methods

Bacteria strains and soybean cultivars

Bradyrhizobium japonicum: USDA 110 and 123, non-RT-producing strains; *Bradyrhizobium elkanii*: USDA 61, 76, and 94, RT-producing strains; *Glycine max* [L.] Merr: SS FFR 565 (RT resistant), Williams (moderately RT resistant), and Forrest and Hutcheson (RT sensitive); *Escherichia coli*: DH52.

Enzyme purification from bradyrhizobia

Bacterial cells were suspended in extraction buffer (1:3 w/v): 10 m*M* Na_2HPO_4-NaH_2PO_4 buffer (pH 7.0) containing 0.1 m*M* dithioerythritol (DTT) and 0.01 m*M* pyridoxal phosphate (PLP). Cells were lysed using a Bead Beater (BioSpec Products Inc., Bartlesville, OK). β-cystathionase in the supernatant of cell lysate from a known mass of bradyrhizobia was purified by precipitation with ammonium sulfate (60% saturation), DE-52 anion exchange, ultrafiltration (MW cut-off: 30,000 daltons, PM 30 membrane, Amicon Inc.), Bio-Gel hydroxyapatite (HTP) chromatography, and Sephadex G100 chromatography. Resulting enzyme preparations having a specific activity above 1×10^4 U mg^{-1} protein were used (Deutscher, 1990; Dwivedi et al., 1982; Uren, 1987). All the purification processes were carried out at 4 °C.

Enzyme purification from soybeans

Cotyledons of dark-grown soybean seedlings were homogenized in 50 m*M* K_2HPO_4-KH_2PO_4 buffer (pH 7.0) containing DTT, 0.1 m*M*; PLP, 0.01 m*M*; EDTA, 0.1 m*M* and phenylmethylsulfonyl fluoride (PMSF), 0.1 m*M*. Attempts to recover enzyme using sodium phosphate buffer were unsuccessful. The resulting slurry was filtered through eight layers of cheese cloth, centrifuged (47,800 $\times$ *g*, 1 hr), and ultracentrifuged (104,200 $\times$ *g*, 6 hr). β-cystathionase in the resulting supernatant was then purified by precipitation with ammonium sulfate (40% saturation), DE-52 anion exchange, ultrafiltration (MW cut-off: 30,000 daltons, PM 30 membrane, Amicon Inc.), and Bio-Gel hydroxyapatite chromatography. All of the purification processes were carried out at 4 °C. Resulting enzyme preparations having a specific activity above 2×10^2 U mg^{-1} protein were used in all experiments.

Activity measurement

Definition of unit
A unit of enzyme was defined as that which produced 1 nmole of pyruvate per hour at pH 8.3 and 37 °C.

Method. A
This method is based on the reaction of pyruvate with 2,4-dinitrophenylhydrazine to produce 2,4-dinitrophenylhydrazon. It was used to measure residual activity of β-cystathionase in the enzyme system (Ruan and Peters, 1991). The reaction system consisted of: Tris-HCl buffer (pH 8.3), 50 m*M*; L-cystathionine, 2.5 m*M*; PLP, 0.01 m*M*; and a known quantity of enzyme preparation. After incubating at 37 °C for 1 hour, 2,4-dinitrophenylhydrazine (1 m*M* in 1 *M* HCl) was added to the system to a final concentration of 0.33 m*M*. After incubation at room temperature for 15 minutes, NaOH was added to a final concentration of 1.25 *M*, and color development was measured spectrophotometrically at 450 nm against a blank lacking substrate or enzyme.

Method B
Enzyme activity was determined by coupling β-cystathionase with an excess of lactate dehydrogenase in the presence of NADH (Giovanelli, 1987; Owens et al.,1968). The rate of pyruvate formation was followed by measuring the change in optical density at 340 nm. The reaction system consisted of Tris-HCl

(pH 8.3), 100 mM; L-cystathionine, 1 mM; PLP, 0.01 mM; NADH, 0.1 mM; 10 units mL^{-1} of L-lactate dehydrogenase, and enzyme solution.

pH effect on initial velocity of β-cystathionase

The initial velocity of β-cystathionase from bradyrhizobia was determined by using Method B to monitor the residual activity in the reaction system. The reaction was carried out in a 1-mL spectrophotometer cell at room temperature. The initial velocity of β-cystathionase from soybeans was determined by using Method A to monitor the residual activity in the reaction system. Tris-HCl (100 mM, pH 6.0 to 9.0) and glycine-NaOH (100 mM, pH 9.5 to 13.0) buffers were used. The reactions were carried out in a microtitration plate at 24 °C. The initial velocity of the enzyme was calculated from the time course of the enzyme reaction.

RT sensitivity of β-cystathionase

Initial inactivation velocity of β-cystathionase was calculated from residual activities obtained using Method A. The reactions for bacteria and soybean enzymes were carried out in microtitration plates at 24 °C using phosphate and Tris-HCl buffer systems (pH 8.3), respectively. Different concentrations of RT were established in the system for various durations to allow the RT to inactivate the enzyme. Residual enzyme activities were then measured, and initial velocity of RT inactivation of enzyme was calculated and plotted as a function of RT concentration. The secondary RT inhibition constant V_{RT} was then calculated.

Inhibition kinetics

The substrate L-cystathionine and inhibitor RT were added to a reaction system consisting of: Tris-HCl buffer (for soybean enzymes) or phosphate buffer (for bacteria enzymes), pH 8.3, 50 mM; L-cystathionine, 0.25 to 2.5 mM; PLP, 0.01 mM; and enzymes. The reaction was then initiated by adding the target β-cystathionase. Residual activity was measured by Method A at 24 °C. Plots of 1/V vs. 1/S were constructed, and the RT inhibition pattern of β-cystationase was defined. An RT dissociation constant (K_I) was calculated for each enzyme.

Native (anionic) PAGE: in situ staining for β-cystathionase activity

Non-denaturing PAGE of β-cystathionase was performed in a BioRad Protean™ II apparatus at 4 °C. The native gel contained: 1) stacking gel: 4% acrylamide with 2.7% bis, 125 mM Tris-HCl buffer (pH 6.8); 2) separating gel: 8% acrylamide with 2.7% bis, 45 mM Tris-borate and 1 mM EDTA buffer (TBE, pH 8.0). After electrophoresis (16 cm × 20 cm × 1.5 mm gel, 100 V, 5 hr,) in 45 mM TBE, the gel was washed and equilibrated (20 minutes at room temperature) with Tris-HCl buffer (pH 8.2), 100 mM; and PLP, 0.01 mM. The gel containing bradyrhizobia β-cystathionase was transferred to fresh Tris-HCl buffer (pH 9.5) with PLP, 0.01 mM; L-djenkolate, 1 mM; and bismuth trichloride, 0.5 mM; the phosphate buffer used previously with the bradyrhizobia enzymes was replaced by Tris-HCl buffer to avoid precipitation of bismuth phosphate. The gel containing soybean β-cystathionase was transferred to fresh Tris-HCl buffer (pH 8.2) with PLP, 0.01 mM; L-djenkolate, 2 mM; and bismuth trichloride, 0.5 mM. After 1 to 2 hour of incubation, areas of β-cystathionase activity were revealed by the formation of brown/black bands of bismuth sulfide (Gentry-Weeks et al., 1993).

Results

Purification of β-cystathionase from bradyrhizobia and soybean cultivars

Table 1 summarizes an example purification of bradyrhizobia β-cystathionase from USDA 110. The enzyme was purified 247-fold with 0.8% final recovery. Typically, β-cystahionase from bradyrhizobia precipitated at 60% ammonium sulfate saturation. The enzyme eluted from DE- 52 cellulose column at 150 to 250 mM NaCl, using a 0 to 500 mM NaCl gradient in 10 mM phosphate buffer (pH 7.0). In a Bio-Gel HTP column, it eluted at 30 to 50 mM Na_2HPO_4-NaH_2PO_4, using a 10 to 200 mM Na_2HPO_4-NaH_2PO_4 gradient. Differences in elution profiles were observed among the enzymes from bradyrhizobia using both systems indicating differences in amino acid composition (data not shown). Additionally, similar differences were observed using a Sephadex G-100 column, indicating that the enzymes from bradyrhizobia differed in structure or molecular weight in the native state (data not shown).

Table 1. Summary of purification of β-cystathionase from *Bradyrhizobium japonicum* USDA 110

Step	Total protein (mg)	Total activity (Units $\times 10^3$)	Specific activity (U mg^{-1} protein)	Purification (fold)	Yield (%)
Crude extract	5,856	1,830	313	1	100
60% $(NH_4)_2SO_4$ precipitate	1,558	796	511	1.6	43
DE-52 cellulose + ultrafiltration	127.5	690	5,412	17.3	37.7
Bio-Gel HTP	6.8	189	27,800	88.8	10.3
Sephadex G100	0.19	14.4	77,500	247.6	0.8

Purification of β-cystathionase was more difficult from soybean than from bradyrhizobia. The enzyme activity was found to be higher in cotyledons than other tissues but could not be detected directly in the crude extract. Following ultracentrifugation (104,200 $\times g$ for 6 hrs), β-cystathionase in the supernatant was precipitated at 40% ammonium sulfate saturation. The enzyme eluted from DE-52 cellulose column at about 200 to 300 m*M* KCl, using a 0 to 500 m*M* KCl gradient in 10 m*M* phosphate buffer (pH 7.0). On a Bio-Gel HTP column, it came off at 20 to 40 m*M* K_2HPO_4-KH_2PO_4, using a 10 to 200 m*M* K_2HPO_4-KH_2PO_4 gradient. As with the bradyrhizobia, enzyme from different soybean cultivars exhibited different elution profiles using both DE-52 and Bio-Gel HTP columns (data not shown). For instance, the enzyme from cultivar Williams was not retained by the DE-52 column.

The preparations of partially purified β-cystathionase from the various sources contained specific activities ranging from 1×10^4 to 5×10^4 U mg^{-1} protein for bradyrhizobia enzymes, 2×10^2 to 4×10^2 U mg^{-1} protein for soybean enzymes, and 7×10^3 U mg^{-1} protein for *E. coli* enzyme. These preparations produced 2 to 8 bands on 10% SDS-PAGE gel when stained with coumassie blue (data not shown). The purity of the enzyme preparations used in given experiment was constant. Because the specific activity varied among enzymes, the quantity of enzymes used in experiments was standardized based on number of units.

Rhizobitoxine sensitivity of β-cystathionase

The sensitivities of various β-cystathionase to RT were estimated using the secondary RT inhibition constant V_{RT}, which reflects the change of the initial RT inhibition velocity of the enzyme as RT concentration increases. The relative RT sensitivities of the enzymes were: *E. coli* > bradyrhizobia > soybean (Figure 1 and 2, Table 2). β-cystathionase from bradyrhizobia was about 10 fold more sensitive to RT than the enzyme from soybean, and about 10 fold more resistant to RT than the enzyme from *E. coli*. The RT sensitivity of bradyrhizobia β-cystathionase differed among the strains examined, and values of V_{RT} ranged from -0.34 to -0.67% activity min^{-1} μM^{-1} RT. The strain yielding the most sensitive β-cystathionase was USDA 94 which also has been shown to produce the greatest amount of RT among the strains investigated both in culture and as a nodule symbiont (Teaney and Fuhrmann, 1992).

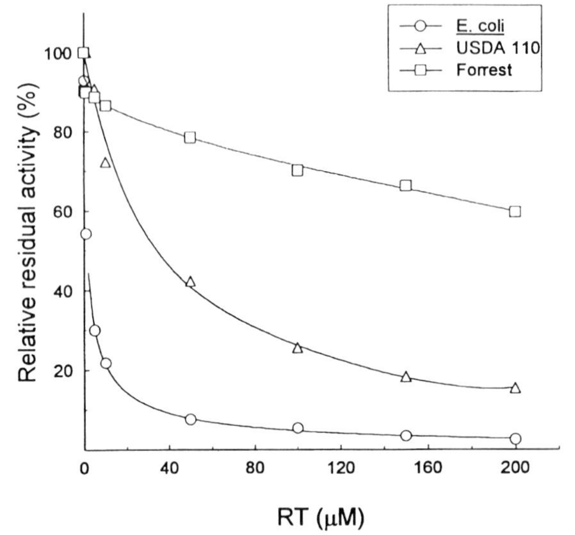

Figure 1. Residual activity of β-cystathionase purifed from selected sources as a function of RT concentration. The partially purified β-cystathionase (25 units) was preincubated with increasing concentrations of RT in 50 m*M* Tris-HC1 buffer (pH 8.3 for 15 minutes at room temperature). The residual activity was measured as described under Materials and methods. The relative residual activity was then calculated relative to the original activity.

β-cystathionase obtained from bradyrhizobia, soybean, *E. coli*, and spinach exhibited different inhibition kinetics (Figure 3). The inactivation of β-cystathionase from bradyrhizobia proceeded in two phases in a man-

Table 2. Selected parameters of β-cystathionase enzymes obtained from different sources[a]

Species	Strain/cultivar	V_{RT} (% Act. min^{-1} μM^{-1} RT)	K_M (m*M*)	K_I (μM)	pH optimum
E. coli	DH52	-4.25 ± 0.55	0.05	3.16	9.0
B. elkanii	USDA 61	-0.34 ± 0.03	0.29	32.7	9.0
	USDA 76	-0.38 ± 0.04	0.23	30.1	9.2
	USDA 94	-0.67 ± 0.09	0.22	22.0	9.5
B. japonicum	USDA 110	-0.47 ± 0.06	0.31	41.2	10.0
	USDA 123	-0.37 ± 0.04	1.07	77.8	9.2
G. max	Forrest	-0.031 ± 0.005	0.12	1048	8.2
	Hutcheson	-0.034 ± 0.004	0.12	702	8.2
	Williams	-0.050 ± 0.008	0.24	719	8.0
	SS FFR 565	-0.046 ± 0.004	0.21	625	8.2

[a] The sources of β-cystathionase include different strains/cultivars in different species listed in this table and Materials and methods. V_{RT} - secondary RT inhibition constant, which indicates the relationship between velocity of activity loss and RT concentration; K_m - Michaelis constant for L-cysatathionase; K_I - RT dissociation constant which is assumed to be at equilibrium and reflects the enzyme-RT affinity (K_I=[E][RT]/[E-RT]); pH optimum-the pH value at which the enzyme has the highest initial velocity with L-cystathionase as a substrate. All parameters were obtained at 24 °C using buffer systems described in Materials and methods

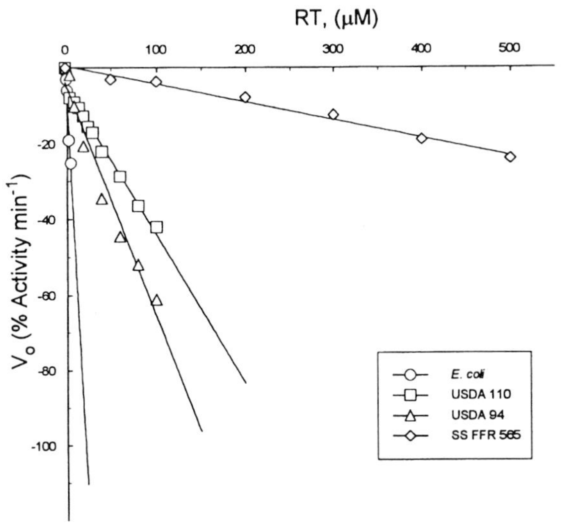

Figure 2. Initial velocity of RT-induced inactivation of β-cystathionase from *E. coli, B. elkanii* USDA 94, *B. japonicum* USDA 110, and *G. max* SS FFR 565 as the function of RT concentration. The initial velocity of RT inhibition of β-cystathionase was calculated from the RT inactivation time course. The reaction was carried out in microtitration plates at 24 °C using different RT concentrations and buffer systems (pH 8.3). Residual enzyme activities were measured at 24 °C as described under Materials and methods.

ner similar to that reported for spinach (Giovanelli et al., 1971), i.e. an initial rapid loss of activity followed by a phase in which the residual activity decreased at a very low rate. The amount of activity lost during the first phase was highly dependent on RT concentration and reflected the relative RT sensitivities of the enzymes (Figure 2 and 3, Table 2). Inactivation during this initial phase was also correlated with the RT dissociation constant K_I which reflected the affinities of the enzymes for RT (Table 2). In contrast, β-cystathionase from *E. coli* reacted with RT more slowly and in a more linear manner at lower RT concentrations, and the magnitude of its secondary RT inhibition constant V_{RT} was 6.3 to 12.3 times lower than that of β-cystathionase from bradyrhizobia and 85 to 137 times lower than that of β-cystathionase from soybeans (Figure 2, Table 2). β-cystathionase from soybeans reacted with RT in a manner similar to the enzyme from *E. coli* but at higher RT concentrations (Figure 3).

K_m for L-cystathionine and RT dissociation constant K_I

β-cystathionase from bacteria and plants showed a typical competitive inhibition pattern when using L-cystathionine as a substrate but differed in their K_m for L-cystathionine and RT dissociation constants K_I (Figure 4 and Table 2). β-cystathionase from *E. coli* had the lowest K_m for L-cystathionine (0.05 m*M*) which is consistent with the reported value 0.04 m*M* (Dwivedi et al., 1982; Uren, 1987). The K_m values for L-cystathionine for the bradyrhizobia enzymes varied from 0.22 m*M* to 1.07 m*M*. The enzymes obtained from USDA, 94 and USDA 110 had the lowest and highest K_m, respective-

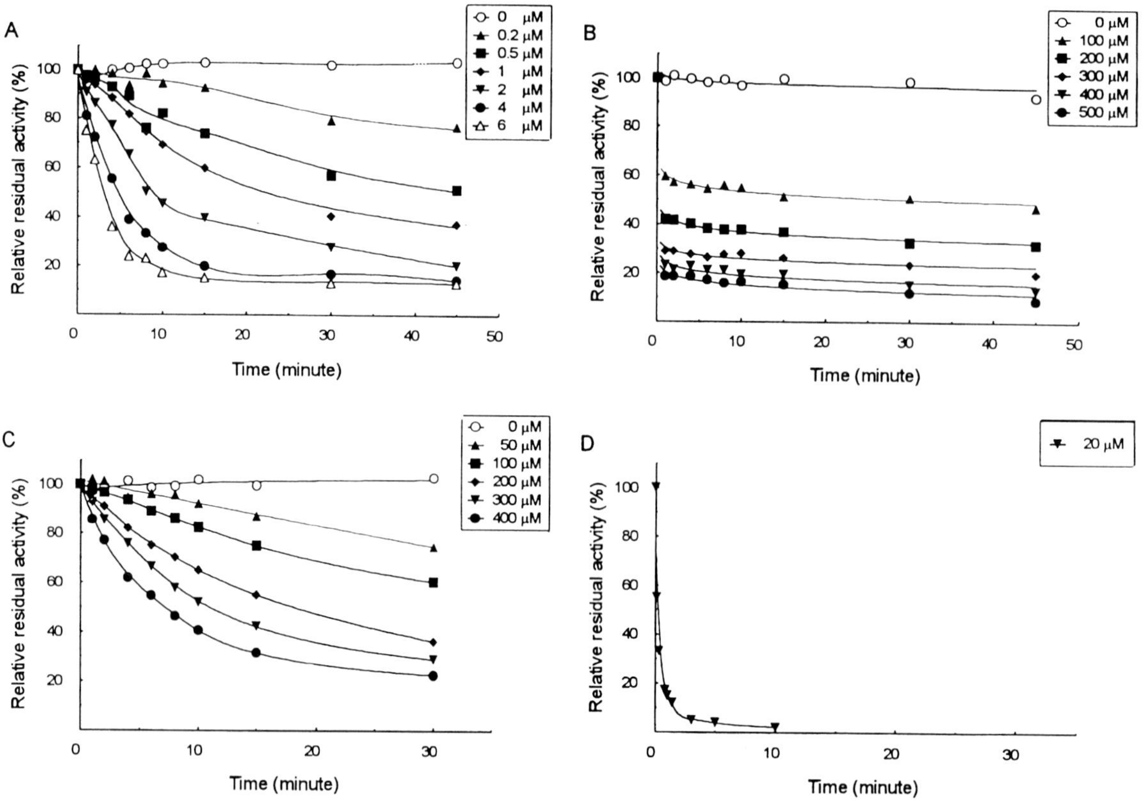

Figure 3. Time-course of inactivation of several β-cystathionase enzymes as influenced by RT concentration. β-cystathionase from: **A.** *E. coli*, **B.** *Bradyrhizobium* USDA 110; **C.** Soybean cultivar Forrest; and **D.** Spinach (replot, Giovanelli et al., 1971).

ly, among the bradyrhizobia enzymes tested. Among the soybean cultivars examined, the enzymes from the highly sensitive cultivars, Forrest and Hutcheson, had relatively low K_m values for L-cystathionine (0.12 m*M*) when compared to those from the more resistant cultivars, Williams and SS FFR 565 (0.24 and 0.21 m*M*, respectively).

β-cystathionase from *E. coli* and Forrest (soybean) had the lowest and highest RT dissociation constants K_I (3.16 μM, and 1048 μM respectively). Generally, the K_I values were negatively correlated with the secondary RT inhibition constants V_{RT} (Table 2). Among the bradyrhizobia examined, the value of K_I ranged from 22.0 μM to 77.8 μM with the lowest K_I value being observed for USDA 94.

pH optima and charge properties

The effect of pH on initial velocity of β-cystathionase was determined using L-cystathionine as a substrate. The optimal pH for the bacterial enzymes ranged from 9 to 10, whereas values for soybean were less variable and averaged approximately 8.2 (Table 2). At pH 7, the bradyrhizobial enzymes showed 0 to 5% of their maximum initial velocity, but corresponding values for the soybean enzymes ranged from 41.7 to 74.5% (Table 3). As the pH increased towards optimal pH, the initial velocity of the enzymes increased at different rates.

Results of the native PAGE indicated that β-cystathionase from bacteria and soybeans differed primarily in their charge characteristics, although molecular weight may have also had a minor effect. In particular, the enzyme from *E. coli* apparently had more net negative charge than any other enzyme tested at pH 8.2, and at least four different migration rates were observed among the bradyrhizobial enzymes (Figure 5).

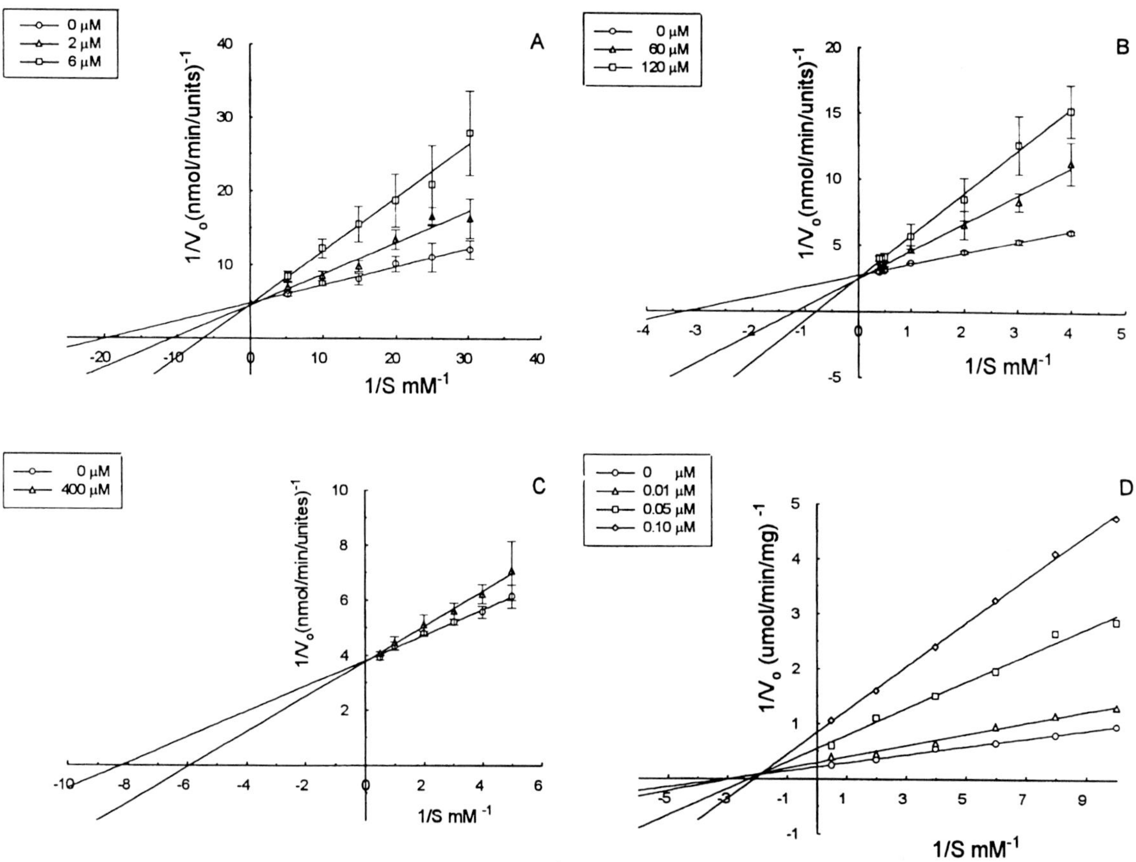

Figure 4. Lineweaver-Burk plots of the competitively and mixed inhibited Michaelis-Menten enzymes from: **A.** *E. coli*; **B.** *Bradyrhizobium* USDA 110; **C.** Soybean cultivar Forrest; and **D.** *Salmonella* (replot, data from Owens et al., 1968). Legends indicate RT concentrations.

Table 3. Effect of pH on the initial velocity of selected β-cystathionase enzymes

Source species	Strain/cultivar	Relative initial velocity at		
		pH 7.0	pH 7.5	pH 8.0
			(%)	
E. coli	DH52	23.9	50.0	78.3
B. elkanii	USDA 61	5.0	17.3	50.0
	USDA 76	4.1	nd[a]	51.4
	USDA 94	0.0	8.7	35.0
B. japonicum	USDA 110	1.0	7.1	27.6
	USDA 123	1.3	5.5	24.9
G. max	Forrest	50.8	78.8	97.0
	Hutcheson	45.3	71.7	99.1
	Williams	74.5	88.6	100.0
	SS FFR 565	41.7	65.4	91.9

[a] nd - Not determined.

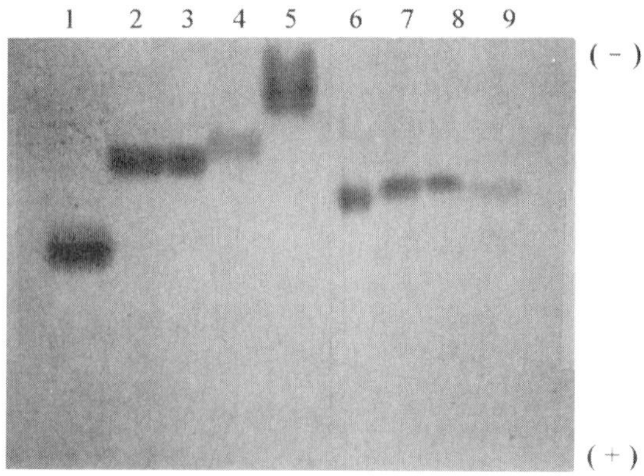

Figure 5. Demonstration of β-cystathionase activity by native (anionic) PAGE 8% acrylamide, 2.7% bis and *in situ* activity stain for the selected genotypes of bacteria and soybeans. Lane 1, *E. coli* DH 52, 840 units; lane 2, *B. elkanii* USDA 61, 610 units; lane 3, *B. elkanii* USDA 76, 615 units; lane 4, *B. elkanii* USDA 94, 2400 units; lane 5, *B. japonicum* USDA 110, 2020 units; lane 6, *G. max* Forrest, 420 units; lane 7, *G. max* Hutcheson, 240 units; lane 8, *G. max* SS FFR 565, 130 units; lane 9, *G. max* Williams, 80 units.

Discussion

During the purification procedure, it was noted that β-cystathionase activity could not be detected in the crude extracts (1:1 w/v) of dark-grown soybean cotyledons by the analysis method used in this study. Low enzyme expression and high protein contents in soybean tissues may have hindered activity detection. For the bradyrhizobia, the specific activity of β-cystathionase in crude extracts differed among strains (data not shown). This observation suggested different levels of β-cystathionase expression among bradyrhizobia.

The pH optima reported here for β-cystathionase from *E. coli*, bradyrhizobia, and soybeans (Table 2) differ from those reported previously for *Salmonella* (8.5; Guggenheim, 1971) and spinach (8.8; Giovanelli et al., 1971; Giovanelli, 1987). The differences in pH and apparent charge properties observed among the various enzyme sources in this study suggests that the corresponding amino acid compositions also differ (Figure 5; Table 2 and 3). Both intra- and inter-species differences were observed. It is possible that such differences in amino acid composition may be related to the observed range in RT sensitivities. Although the mobility of β-cystathionase on the native gel was not correlated with RT sensitivity when compared among all enzymes tested (Figure 5; Table 2), differences in enzyme mobilities on the native gel were possibly associated with changes in the RT sensitivity when compared among the RT-producing bradyrhizobia strains. Since these RT-producing strains belong to the same species, these changes may reflect amino acid substitutions in either the active center or an area affecting enzyme structure and regulation.

The RT dissociation constant K_I for β-cystathionase from spinach was determined to be 80 *μM* (Giovanelli et al., 1971). This value is similar to the values for bradyrhizobia enzymes (22.0 to 77.8 *μM*), but lower than for the soybean enzymes (625 to 1048 *μM*). The similarity between spinach and bradyrhizobia was also observed for the inactivation kinetic analyses (Figure 3). However, the reasons that soybean β-cystathionase had higher K_I values and different patterns of RT inactivation when compared to spinach and bradyrhizobial enzymes remain to be investigated.

Previous studies reported that RT-induced inhibition of β-cystathionase from *Salmonella* exhibited a mixed inhibition pattern (Owens et al., 1968) and that RT inhibition of β-cystathionase from spinach is an active-site-directed irreversible inhibition (Giovanelli et al., 1971). Conversely, the enzymes examined in this study all showed a typical substrate competitive pattern, i.e. RT binds to the substrate binding site. On the other hand, the time course studies of RT inactivation showed different inhibition types. Inhibition of β-cystathionase from *E. coli* and soybeans was found to be time dependent, whereas β-cystathionase from bradyrhizobia was RT concentration dependent. These differences may indicate that the enzymes differ in the accessibility of the RT binding site, regulation, and structure. Previously, other researchers have proposed two alternative hypotheses for two phase inactivation of spinach β-cystathionase by RT: (a) β-cystathionase exists in two forms (isoenzymes), one relatively susceptible to RT, the other relatively resistant; (b) β-cystathionase exists in one form only, which is completely converted by RT to a form with a small residual activity (Giovanelli et al., 1971). The results presented here support the second hypothesis, in that the time course analyses of RT inactivation of bradyrhizobial β-cystathionase revealed two phases of inactivations and no evidence of isoenzymes was found by native gel analysis (Figures 3 and 5) (Moss, 1979).

References

Datko A H and Mudd S H 1982 Methionine biosynthesis in *Lemna*: inhibitor studies. Plant Physiol. 69, 1070–1076.

Deutscher M P 1990 Guide to Protein Purification. Acad. Press. Inc., San Diego, USA.

Dwivedi C M, Ragin R C and Uren J R 1982 Cloning, purification, and charcterization of β-cystathionase from *Escherichia coli*. Biochemistry. 21, 3046–3069.

Gentry-Weeks C R, Keith J M and Thompson J 1993 Toxicity of *Bordetella avium* β-Cystathionase toward MC3T3-El Osteogenic Cells. J. Biol. Chem. 268, 7298–7314.

Giovanelli J and Mudd S H 1971 Transsulfuration in higher plants: Partial purification and properties of β-cystathionase of spinach. Biochim. Biophys. Acta 227, 654–670.

Giovanelli J, Owens L D and Mudd S H 1971 Mechanism of inhibition of spinach β-cystathionase by rhizobitoxine. Biochem. Biophys. Acta 227, 671–684.

Giovanelli J, Owens L D and Mudd S H 1972 β-Cystathionase - in vivo inactivation by rhizobitoxine and role of the enzyme in methionine biosynthesis in corn seedlings. Plant Physiol. 51, 492–503.

Giovanelli J 1987 Cystathionine β-lyase from spinach. Meth. Enzymol. 143, 443–449.

Guggenheim S 1971 β-Cystathionase (*Salmonella*). Meth. Enzymol. 17B, 439–442.

Moss D W 1979 Isoenzyme analysis. Spottiswoode Ballantyne Ltd., Colchester, UK.

Negrutiu I, De Brouwer D, Dirks R and Jacobs M 1985 Amino acid auxotrophs from protoplast cultures of *Nicotiana plumbaginifolia*, Viviani. Mol. Gen. Genet. 199, 330–337.

Owens L D, Guggenheim S and Hilton J L 1968 *Rhizobium*-synthesized photoxin: an inhibitor of β-cystathionase in *Salmonella typhimurium*. Biochem. Biophys. Acta 158, 219–225.

Ruan X and Peters N K 1991 Rapid and sensitive assay for the phytotoxin, rhizobitoxine. Appl. Environ. Microbiol. 57, 2097–2100.

Teaney G B III and J J Fuhrmann 1992 Soybean response to nodulation by bradyrhizobia differing in rhizobitoxine phenotype. Plant and Soil 145, 275–285.

Uren J R 1987 Cystathionase β-lyase from *Escherichia coli*. Meth. Enzymol. 143, 483–486.

Section editor: H Lambers
Guest editors: G.H. Elkan and R.G. Upchurch

Plant and Soil **186:** 63–67, 1996.

Assimilation of *nod* gene inducer ^{14}C-naringenin and the incorporation of labelled carbon atoms into the acyl side chain of a host-specific Nod factor produced by *Rhizobium leguminosarum* bv. viciae

J.R. Rao[1], J.E. Cooper[1], E.S.W. Everaert[2] and L. De Cooman[2]
[1]*Department of Applied Plant Science The Queen's University of Belfast, Belfast BT9 5PX, Northern Ireland, UK** and [2]*Laboratory of Plant Biochemistry, Rijksuniversiteit Gent, K.L. Ledeganckstraat 35, B-9000, Gent, Belgium*

Key words: ^{14}C-naringenin assimilation, Nod factor synthesis, *Rhizobium nod* gene-inducer

Abstract

The fate of ^{14}C-naringenin during its specific activation of *nod* genes in *Rhizobium leguminosarum* bv. viciae was examined. After incubation with either strain RBL5560 or its pSym-cured derivative in a medium supplemented with ^{14}C-naringenin at *nod* gene-inducing concentrations of 2 n*M* (ca. 12.5 kBq) plus cold acetate (0.5 μM), a radiocarbon inventory for the cells and supernatant extracts was obtained. The level of ^{14}C-label incorporation was also determined in the fractionated cellular components. Using ^{14}C-acetate at 0.5 μM (1036 kBq) and cold naringenin (2 n*M*) in incubations with strain RBL5560 as a separate treatment, the Nod metabolites were detected by thin layer and high performance liquid chromatographic methods and the data provided the basis for identification of the Nod factors from the supernatant obtained from ^{14}C-naringenin treatments. Subsequent radio-biochemical and chemical analyses revealed that RBL5560 cells assimilated ^{14}C-naringenin during the activation of *nod* genes. Our analysis also showed that labelled carbon atoms from the ^{14}C-naringenin were incorporated into the acyl moiety of a lipo-oligosaccharide Nod factor, NodRlv IV, present in the culture supernatants of RBL5560. The pSym-cured derivative failed to synthesize any Nod metabolites in a ^{14}C-naringenin supplemented medium. The tracing of flavonoid-derived carbon atoms to the acyl chain of a host-specific Nod factor, a moiety that defines host specificity for this *Rhizobium*, adds a new dimension to the signalling function of flavonoids in legume-*Rhizobium* interactions.

Abbreviations: Ac-acyl chain, ca-calculated approximately, dpm-disintegrations per minute, HPLC-High Performance Liquid Chromatography, pSym-symbiotic plasmid, *R.-Rhizobium*, TLC-Thin Layer Chromatography.

Introduction

Specific flavonoids present among those synthesized in and released from legume roots have been shown (Peters et al., 1986; Redmond et al., 1986) to induce the transcriptional activation of *nod* genes of various bacteria in the family Rhizobiaceae. This results in the synthesis of host specific lipo-oligosaccharide reverse signal molecules (Nod factors) which cause hair curling and cortical cell division in the plant root. It has been suggested that catabolism of flavonoids could be one of the earliest responses of rhizobia (Peters and Verma, 1990) in a complex *nod* gene regulatory process. Such speculation was in part prompted by the knowledge that these compounds can induce the expression of genes with unknown functions (Sadowsky et al., 1988). Recent investigations (Perret et al., 1994) indicated that such genes shared no homologies with nodulation gene promoters. Evidence for the rhizobial degradation of flavonoids is scarce, the first example being the utilization of a flavan-3-ol, catechin, by a *Rhizobium* sp., isolated from *Leucaena leucocephala* (Gajendiran and Mahadevan, 1988). Subsequently Rao et al. (1991) reported a novel form of C-ring cleavage in a flavone, quercetin, by *R loti*. Further studies with other *Rhizobium* species / biovars and their respective *nod* gene inducers confirmed that flavonoid degradation is a common metabolic feature

* FAX No: +441232668375

in members of this genus (Rao and Cooper, 1994). The fate of a flavonoid during and subsequent to *nod* gene induction is poorly understood. In this paper, we report on experiments to trace the fate of ^{14}C-naringenin, a *nod* gene inducing flavanone for *Rhizobium leguminosarum* bv. viciae, and to assess its contribution to the synthesis of a host specific reverse signal molecule (Nod factor) by the bacterium.

Materials and methods

Bacterial strains: *Rhizobium leguminosarum* bv. viciae strain RBL5560 (LPR5045 containing plasmid pJB5JI) and strain LPR5045 (a pSym-cured derivative lacking the plasmid pJB5JI) were obtained from the Department of Plant Molecular Biology, Botanical Laboratory, Leiden University, Leiden, The Netherlands.

Radioactive A-ring labelled (4a, 6, 8) ^{14}C-naringenin was synthesized enzymatically using the methods of De Cooman et al. (1993). ^{14}C-acetate (specific activity 40–60 mCi mmol^{-1}) was supplied by Amersham International, UK. HPLC-purified cold naringenin and cold acetate were obtained from Apin Chemicals, Abingdon, UK and Sigma Chemicals respectively.

The protocol followed for the growth of the bacteria in a defined medium was as described by Spaink et al. (1992). The labelling experiments were carried out as two separate trials, each employing the following treatments: the highly purified labelled (^{14}C-) naringenin (sp activity 6,259 kBq μmol^{-1}) was supplied to both strains at a concentration of 2n*M* (ca. 12.5 kBq) in the presence of cold acetate (0.5 μM). ^{14}C-acetate (sp. activity 2072 kBq μmol^{-1}) at a concentration of 0.5 μM (ca. 1036 kBq) together with cold naringenin (2 n*M*) was also supplied to the same two strains as a second separate treatment. Cells and cell-free supernatants were obtained after 2h incubation. RBL5560 cells treated with ^{14}C-naringenin were fractionated into various cellular components using the standard procedures described by Hanson and Phillips (1981). A radiocarbon inventory was obtained for the fractions corresponding to the cell components by monitoring the level of radioactivity in aliquots of the fractions spotted on a TLC and scanned in a Packard Instant Imager (Canberra Packard, UK). The supernatants were analysed for Nod metabolites by the TLC/HPLC methods described by Spaink et al. (1991), and the GC/MS methods of Roche et al. (1991). In our study we also used an Instant Imager for detecting radiolabelled Nod metabolites on the TLC plates. Nod factor(s) were then isolated from TLC/HPLC and characterized for their UV absorbance. Radioactivity was also measured in fractions of the Nod metbolite(s) obtained after an acid hydrolysis (boiling/6 *N* HCl/3 h) and reflux under nitrogen as well as an alkaline (cold/4 °C/2 *N* NaOH) digestion.

Results

The levels of radioactivity present in the various cellular components of RBL5560 after 2h incubation with ^{14}C-naringenin are depicted in Table 1. Thirty eight percent of the total radioactivity present in whole cells was found in proteins and cell wall components, while another 34% of the radiolabel was present in the fraction containing cellular lipids and alcohol soluble proteins. In contrast, only 9% and 5% of the radiocarbon in the cell fraction was found in the sub fractions presumed to contain lipopolysaccharide-protein complexes or alcohol soluble lipids and low molecular weight materials respectively; a further 4% of ^{14}C-label in the whole cell fraction was found in the nucleic acids pool.

An Instant Imager analysis (Figure 1) indicated the presence of a single strong clear signal on the silica TLC from the RBL5560 culture supernatant sample extracted with butanol from the ^{14}C-naringenin / cold acetate treatment. Radioactivity measurements after TLC and HPLC separation of the Nod metabolites from RBL5560 supplied with ^{14}C-naringenin (Table 2) indicated that the signal corresponding to the co-migrating Nod factors NodRlv IV (Ac, C18:4) and NodRlv V (Ac, C18:1) represented a mere 2189 dpm (3.4%) activity out of the total radioactivity of 62,714 dpm registered in 5 μL of the butanol extract spotted onto the silica TLC. The HPLC fraction corresponding to the Nod metabolite NodRlv IV (Ac, C18:4) had an activity of 1673 dpm. Upon acid and alkali hydrolysis, the residual lipophilic fractions from the chemically cleaved Nod metabolite retained ca. 1588 dpm radioactivity indicating that 95% of the radiolabel in the Nod factor resided in the fatty acid moiety of the molecule. An electron impact-mass spectrum of the HPLC peak eluting at 12.4 min (Figure 1B) showed that the labelled metabolite had molecular ion signals $(M+H)^+$ at 1088 and m/z at 204 and 395, which are characteristic of the Nod factor NodRlv IV (Ac, C18:4).

Table 1. Distribution of ^{14}C-label among the various cellular components of RBL5560 after incubation with ^{14}C-naringenin

Fraction	Corresponding cell component expected	Radioactivity[a] (dpm)	(%)
5% Trichloroacetic acid (TCA) soluble	Whole cell	15230	100
Cold - TCA soluble	Low molecular weight materials	750	4.9
75% Ethanol-soluble	Lipo-polysaccharide-protein complexes, alcohol soluble lipids	1458	9.5
7.5% Ethanol-diethyl ether (1:1) soluble	Cellular lipids, alcohol soluble proteins	5200	34.1
Hot - TCA soluble	Nucleic acids	620	4.0
Hot - TCA insoluble	Proteins, cell wall	5800	38.0

[a]dpm - radioactive disintegrations per minute, data presented are means from 2 trials consisting of three readings each, standard deviation in all cases < ± 2%.

Table 2. Radioactivity measurements after TLC and HPLC analysis of the Nod metabolites obtained from RBL5560 culture supernatants (^{14}C -naringenin treatment)

Sample	Activity[a] dpm	%
Butanol extract (5 μL)	62714	100
TLC signal corresponding to NodRlv IV (Ac, C18:4) and NodRlv IV (Ac, C18:1)	2189	3.4
HPLC fraction containing NodRlv IV (Ac, C18:4)	1673	2.6
Lipophilic residues obtained after acid/alkali hydroylsis of HPLC purified NodRlv IV (Ac, C18:4)	1588	2.5[b]

[a]dpm - radioactive disintegrations per minute was calculated from the mean of three counts per sample, per trial; standard deviation in all cases < ± 2%.
[b]The radioactivity (2.5%) present in the acyl moiety was obtained after chemical cleavage and this represented 95% of the radiolabel (= 1673 dpm) from the HPLC fraction containing NodRlv IV (Ac, C18:4).

Discussion

Our results clearly demonstrate that ^{14}C-naringenin supplied at a concentration of 2 n*M* is capable of inducing *nod* genes of *Rhizobium leguminosarum* bv. viciae and that *Rhizobium* cells assimilate the inducer during the synthesis of Nod factors. Data from the radiocarbon inventory experiments indicated that the radiolabel distribution within cell components was located principally in cellular proteins, lipids or cell wall components together with small amounts in nucleic acid and low molecular weight pools. The sources for the radioactivity in the various cell components may be unmetabolized portions of naringenin, its degrada-

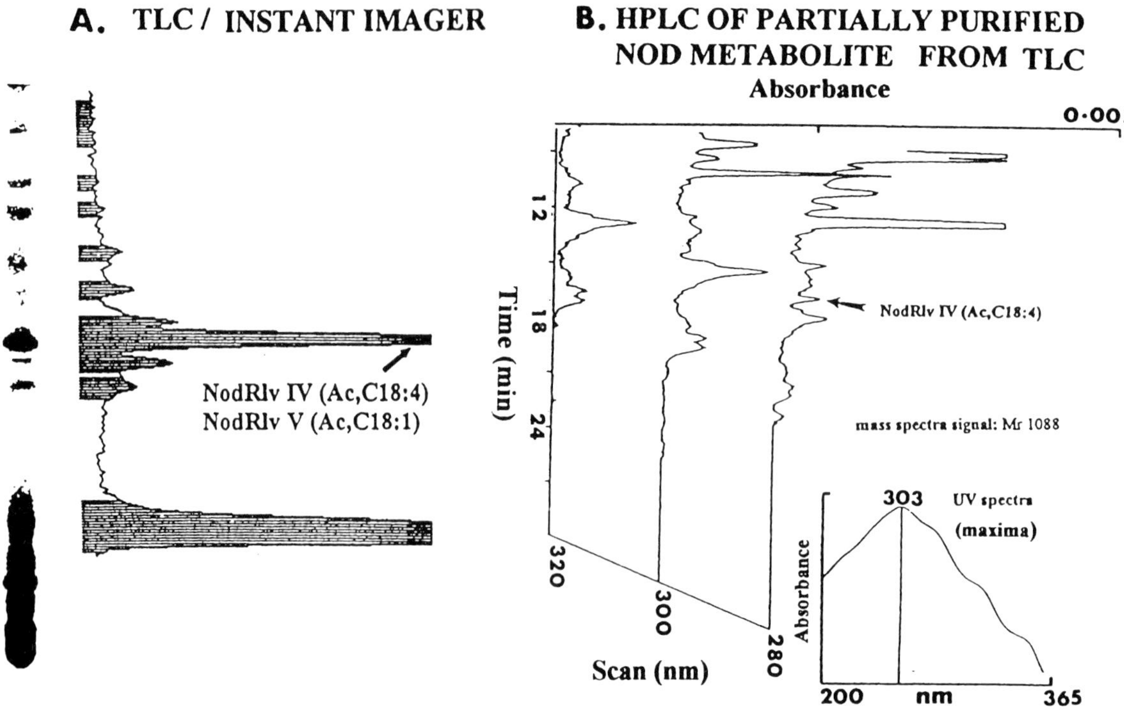

Figure 1. **A**. Instant Imager (Canberra Packard, UK) analyses of TLC of the sample from RBL5560 culture supernatant extracts from treatment supplied with ^{14}C-naringenin and cold acetate. A single strong signal on the electronic autoradiogram of the TLC at the position shown corresponded to the two co-migrating Nod metabolites Rlv IV (Ac, C18:4) and Rlv V (Ac, 18:1) identified from a separate treatment (lane not shown) in which cold naringenin and ^{14}C-acetate were supplied for the detection of Nod metabolites as described by Spaink et al. (1992). **B**. The crude Nod metabolite obtained from silica TLC of extracts of culture of RBL5560 induced with ^{14}C-naringenin was purified by HPLC and characterized for its UV absorbance and mass spectra by the methods described by Spaink et al. (1991). The HPLC fraction of the Nod factors from the ^{14}C-naringenin treatment identified a strongly labelled metabolite with a UV absorbance maximum of 303 nm and a relative molecular mass (Mr) of 1088, consistent with the presence of NodRlv IV (Ac, C18:4), a lipo-oligosaccharide which can induce root cortical cell division and nodule meristem formations on *Vicia sativa* roots. The symbiotic plasmid-cured derivative LPR5045, failed to synthesize any labelled Nod metabolites when supplied with ^{14}C-naringenin (data not shown).

tion products bound to cellular components (proteins, cell wall membrane lipids, nucleic acids), or radiocarbon assimilated into the particular macromolecular components themselves. Our recent studies (Rao and Cooper, 1995) with *Bradyrhizobium japonicum* indicated that its *nod* gene inducers and their degradation products were concomitantly present in the cells. These findings, when viewed together with our present data, point to a more complicated mechanism for the control of *nod* gene expression than the proposal by Hubac et al. (1994) involving enhanced retention of a single inducer molecule (luteolin) in the outer membrane of *R. meliloti*. Further studies using radiolabelled inducers are required to clarify the interaction with the *nodD* gene product or other receptor components that have been proposed in the current models (e.g. Spaink, 1994) to mediate the activation of transcription of the inducible *nod* genes.

Our data provide the first biochemical evidence of the fate of an inducer flavonoid during *nod* gene activation and establish a hitherto unsuspected contribution of carbon atoms from naringenin to the structure of a Nod metabolite. HPLC fractionation of Nod factors from the ^{14}C-naringenin supplemented RBL5560 culture filtrates yielded a labelled compound whose UV absorbance and mass spectral characteristics were consistent with the presence of the Nod metabolite Nod Rlv IV (Ac, C18:4), a host specific signal molecule that can induce cortical cell division on *Vicia sativa* roots (Spaink et al., 1991). Significant incorporation of the inducer label, supplied at a concentration that was 250 times lower than ^{14}C-acetate, into the Nod metabolite suggests that unique, synchronous intramolecular mechanisms at the subcellular level might be in operation at the time of assimilation of inducer flavonoid and *nod* gene induction in responding *Rhizobium* cells.

Furthermore, the tracing of flavonoid-derived carbon atoms into a component of the Nod factor molecule (the acyl side chain) which determines its host-specificity highlights the symbiotic importance of the mechanism.

Acknowledgements

This work was funded by the Leverhulme Trust and travel grants from The British Council and the Cyril Francis O'Reilly fund at the Queen's University of Belfast. We are grateful to RJH Okker and HP Spaink, Leiden, The Netherlands, for supplying *R. leguminosarum* bv. viciae strains and our sincere thanks go to F Fox, Canberra Packard, UK, for the Instant Imager analyses of our samples.

References

De Cooman L, Everaert E S W, Fache P, Vandecasteele K and Van Sumere C F 1993 Flavonoid biosynthesis in petals of *Rhododendron simsii*. Phytochemistry 33, 1419–1426.

Gajendiran N and Mahadevan A 1988 Utilization of catechin by *Rhizobium* sp. Plant and Soil 108, 263–266.

Hanson R S and Phillips J A 1981 Chemical composition: Chapter IV: *Metabolism*. Ed. W A Wood. *In* Manual of Methods for General Bacteriology. Ed. P Gerhardt. pp 328–364. American Society for Microbiology, Washington, DC, USA.

Hubac C, Ferran J, Tremolieres A and Kondorosi A 1994 Luteolin uptake by *Rhizobium meliloti*: evidence for several steps including an active extrusion process. Microbiology 140, 2769–2774.

Perret X, Fellay R, Bjourson A J, Cooper J E, Brenner S and Broughton WJ 1994 Subtraction hybridization and shot-gun sequencing: a novel approach to identify symbiotic loci. Nucleic Acids Res. 22, 1335–1341.

Peters N K, Frost J W and Long S R 1986 A plant flavone, luteolin, induces expression of *Rhizobium meliloti* nodulation genes. Science 233, 977–980.

Peters N K and Verma D P S 1990 Phenolic compounds as regulators of gene expression in plant-microbe interactions. Mol. Plant-Microbe Interactions 3, 4–8.

Rao J R, Sharma N D, Hamilton J T G, Boyd D R and Cooper J E 1991 Biotransformation of the pentahydroxyflavone quercetin by *Rhizobium loti* and *Bradyrhizobium* strains (*Lotus*). Appl. Environ. Microbiol. 57, 1563–1565.

Rao J R and Cooper J E 1994 Rhizobia catabolize *nod* gene inducing flavonoids via C-ring fission mechanisms. J. Bacteriol. 176, 5409–1565.

Rao J R and Cooper J E 1995 Soybean nodulating rhizobia modify *nod* gene inducers daidzein and genistein to yield aromatic products that can influence gene inducing activity. Mol. Plant-Microbe Interactions 8, 855–862.

Redmond J W, Batley M A, Djordjevic M A, Innes R W, Kuempel P L, and Rolfe B G 1986 Flavones induce expression of nodulation genes in *Rhizobium*. Nature 323, 632–635.

Roche P, Lerouge P, Ponthus C and Prome J-C 1991 Structural determination of bacterial nodulation factors involved in the *Rhizobium meliloti* - Alfalfa symbiosis. Biol. Chem. 266, 10933–10949.

Sadowsky M J, Olsen E R, Foster V E, Kosslak R M and Verma D P S 1988 Two host-inducible genes of *Rhizobium fredii* and characterization of the inducing compound. J. Bacteriol. 170, 171–178.

Spaink H P, Sheeley D M, Van Brussel A A N, Glushka J, York W S, Tak T, Geiger O, Kennedy E P, Reinhold V N and Lugtenberg B J J 1991 A novel highly unsaturated fatty acid moiety of lipo-oligosaccharide signals determines host specificity of *Rhizobium*. Nature 354, 125–130.

Spaink H P, Aarts A, Stacey G, Bloemberg G V, Lugtenberg B J J and Kennedy E P 1992 Detection and separation of *Rhizobium* and *Bradyrhizobium* Nod metabolites using thin layer chromatography. Mol. Plant-Microbe Interactions 5, 72–80.

Spaink H P 1994 The molecular basis for the host specificity of the *Rhizobium* bacteria. Antonie van Leeuwenhoek 65, 81–98.

Section editor: H Lambers
Guest editors: G.H. Elkan and R.G. Upchurch

Plant and Soil **186:** 69–74, 1996.

Corresponding 16S rRNA gene segments in Rhizobiaceae and *Aeromonas* yield discordant phylogenies

B.D. Eardly[1], F.-S. Wang[1] and P. van Berkum[2]
[1]*Department of Biology, The Pennsylvania State University, University Park, PA 19610-6009, USA* and*
[2]*Soybean and Alfalfa Research Laboratory USDA-ARS, Beltsville, Maryland, USA*

Key words: *Agrobacterium*, population genetics, *Rhizobium*, systematics, taxonomy

Abstract

Previous evidence has indicated that the 16S rRNA genes in certain species of *Aeromonas* may have a history of lateral transfer and recombination. A comparative analysis of patterns of 16S nucleotide sequence polymorphism among species of *Rhizobium* and *Agrobacterium* was conducted to determine if there is similar evidence for chimeric 16S genes in members of the Rhizobiaceae. Results from phylogenetic analyses and comparison of patterns of nucleotide sequence polymorphism in portions of rhizobial 16S genes revealed the same type of segment-dependent polymorphic site partitioning that was previously reported for *Aeromonas*. These results support the hypothesis that certain 16S segments in rhizobia may have a history of lateral transfer and recombination.

Abbreviations: 16S rRNA -16S ribosomal ribonucleic acid, 16S - the 16S rRNA gene.

Introduction

The two primary criteria by which taxonomic species of the Rhizobiaceae are defined are genomic DNA similarity (by DNA reassociation) and ribosomal RNA gene nucleotide sequence similarity (Graham et al., 1991). Occasional disagreements between the results of these two methods have been reported for certain species of *Aeromonas* (Martinez-Murcia et al., 1992) and other bacteria (Stackebrandt and Goebel, 1994). Sneath (1993) analyzed patterns of polymorphism in the nucleotide sequence data of Martinez-Murcia et al. (1992), and suggested that the 16S rRNA genes in certain species of *Aeromonas* may have a history of lateral transfer and recombination.

It is generally thought that ribosomal RNA genes are not subject to lateral transfer and recombination because of the probable adverse effect that changes in the nucleotide sequence of the 16S gene would have on translational efficiency (Niebel et al., 1987). Because bacterial phylogenies are often extrapolated from 16S rRNA gene sequence data, it is important to determine if these genes may have a history of recombination.

Chimeric genes are usually identified by the presence of abrupt transitions in linear patterns of nucleotide sequence polymorphism. Sneath (1993) reported that phylogenetic trees based on 16S sequences from certain species of *Aeromonas* were strongly dependent on the portion of the 16S gene included in the analysis. Similar segment-dependent estimates of 16S phylogeny have also been observed for species of Rhizobiaceae (K Lindstrom and J P W Young, pers. comm.). In *Rhizobium* populations, disagreements between estimates of genomic relatedness based on multilocus enzyme electrophoresis results and 16S restriction fragment length polymorphism analyses have also been reported (Eardly et al., 1995).

The present study was undertaken to compare patterns of 16S nucleotide sequence polymorphism in members of both the Aeromonadaceae and the Rhizobiaceae. The comparison revealed a common pattern of polymorphic site partitioning in certain species of both families, which may be explained by the lateral transfer and recombination of a 264 bp (promoter-proximal) segment of the 16S rRNA gene.

* FAX No: + 16103727974

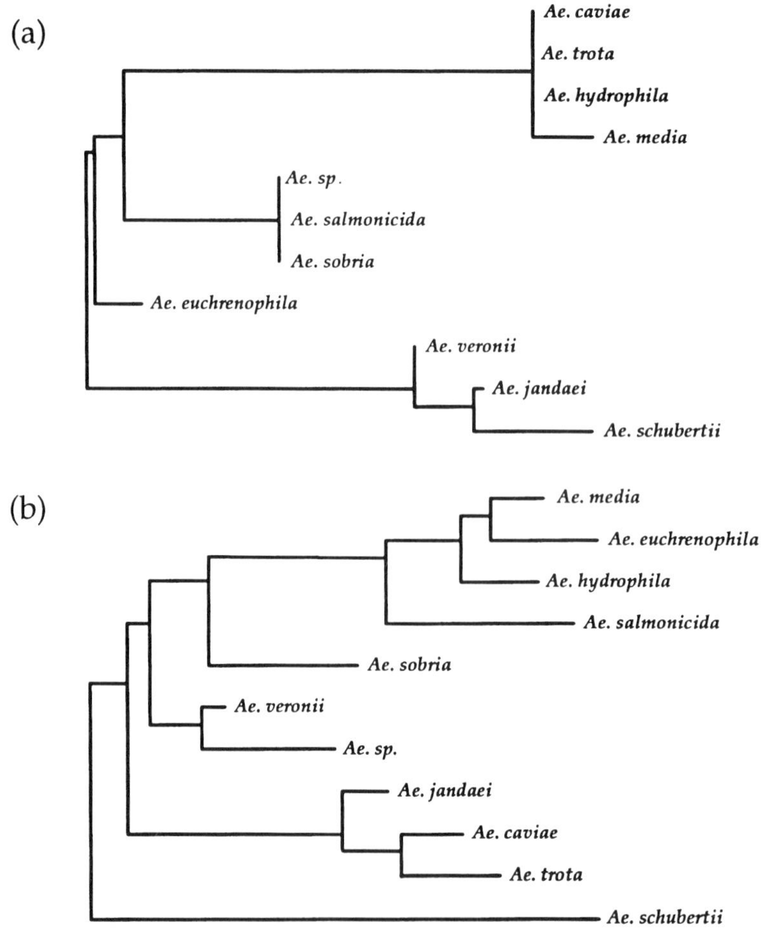

Figure 1. Neighbor-joining trees showing inferred phylogenetic relationships among eleven species of *Aeromonas* based on nucleotide sequence polymorphisms in the *(a)* 5' or *(b)* 3' portions of their 16S rRNA genes, divided at position 264 (Table 1). The species and their respective 16S nucleotide sequence (EMBL) accession numbers (in parentheses) are: *Ae. caviae* (X60409), *Ae. eucrenophila* (X60411), *Ae. hydrophila* (X60404), *Ae. jandaei* (X60413), *Ae. media* (X60410), *Ae. salmonicida* (X60407), *Ae. schubertii* (X60416), *Ae. sobria* (X60412), *Ae. sp.* (X60417), *Ae. trota* (X60415), and *Ae. veronii* (X60414).

Materials and methods

The strategy for the analysis of the rhizobial 16S genes in this study was based (in part) on the results of Sneath (1993) which indicated that patterns of 16S nucleotide sequence polymorphism in certain species of *Aeromonas* may be the result of lateral transfer and recombination. These results, together with unpublished observations of segment-dependent 16S phylogenetic tree topologies among species of rhizobia and agrobacteria (K Lindstrom and J P W Young, pers. comm.) were used as a guide in the selection of rhizobial gene segments for analysis.

The DNA sequences included in this study have been published elsewhere (Martinez-Murcia et al., 1992; Van Berkum et al., 1996 and Willems and Collins, 1993). The species included and their respective 16S sequence accession numbers are listed in the captions of Figures 1 and 2. The sequences were numbered and aligned (manually) according to published alignments for *Escherichia coli* and *Agrobacterium tumefaciens* (Dams et al., 1988). Phylogenetic tree reconstructions were generated by the neighbor-joining method of Saitou and Nei (1987) with Jukes-Cantor distances (Kumar et al., 1993) - alignment gaps were ignored in the analysis.

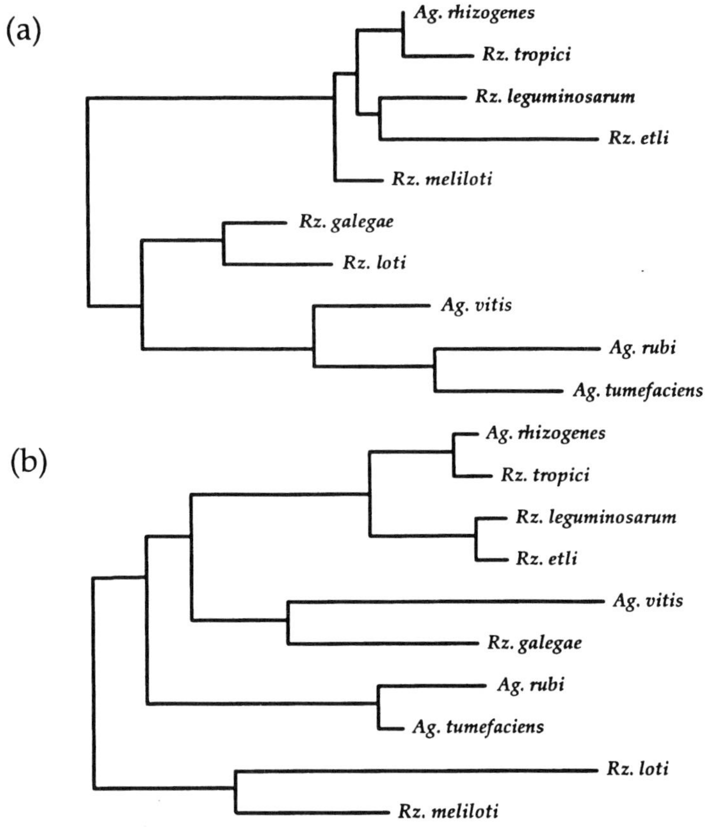

Figure 2. Neighbor-joining trees showing inferred phylogenetic relationships among ten species of Rhizobiaceae (*Agrobacterium* or *Rhizobium*) based on nucleotide sequence polymorphisms in the *(a)* 5' or *(b)* 3' portions of their 16S rRNA genes, divided at position 264 (Table 1). The species and their respective 16S nucleotide sequence (EMBL) accession numbers (in parenthesis) are: *Ag. rhizogenes* (X67224), *Ag. rubi* (X67228), *Ag. tumefaciens* (X67223), *Ag. vitis* (X67225), *Rz. etli* (GenBank no. U28916), *Rz. galegae* (X67226), *Rz. leguminosarum* (X67227), *Rz. loti* (X67229), *Rz. meliloti* (X67222), and *Rz. tropici* (X67234). For additional information on strains refer to Willems and Collins, (1993).

Results

As was reported by Sneath (1993), 16S phylogenetic tree topologies for the various species of *Aeromonas* depended on the portion of the gene included in the analysis. In his results, the 5' and 3' portions of the 16S gene (divided at nucleotide position 264) producing markedly different estimates of phylogenetic relatedness (also see Figure 1). Similar results were obtained for trees based on corresponding gene segments among species of Rhizobiaceae (Figure 2). In general, whole-gene trees for both groups (not shown) more nearly resembled those for the larger 3' gene segments.

Comparisons of individual polymorphic nucleotide sequence positions in the three species of Rhizobiaceae and *Aeromonas* are shown in Table 1. Only polymorphic sites through nucleotide positions 877 are shown because downstream (3') polymorphic sites revealed no clear patterns of nucleotide partitioning (data not shown). The three *Aeromonas* species included in the partitioning analysis were chosen on the basis of the results of Sneath (1993) who reported that estimates of 16S relatedness among species of *Aeromonas* can be influenced by the particular 16S gene segment being examined. Likewise, *R. loti, R. galegae*, and *A. tumefaciens* were also chosen because of their segment-dependent positions in the trees of Figure 2. The position of *R. meliloti* in these trees was also segment-dependent.

The alignment results in Table 1 show that in both families there is a distinct break or transition in the pattern of polymorphic nucleotide partitioning at the 16S nucleotide position numbered 264. A similar pattern was evident when the sequence for *R. meliloti* was compared to the other Rhizobiaceae species (results not shown). Position 264 precedes a long run of sequence homology (over 100 bps) in the 16S genes of both families.

Table 1. Polymorphic nucleotide positions in a portion of the 16S rRNA gene corresponding to the 5' and central domains of the 16S rRNA molecule in three species of *Rhizobiaceae* and in three species of *Aeromonadaceae*; position numbers correspond to those in *Escherichia coli* (Woese et al., 1983). Slashes denote within-family partitions for individual positions

16S rRNA nucleotide position number	Family					
	Rhizobiaceae			Aeromonadaceae		
	Rhizobium loti	*Rhizobium galegae*	*Agrobacterium tumefaciens*	*Aeromonas caviae*	*Aeromonas hydrophila*	*Aeromonas salmonicida*
66	G	G/	A			
73	T	T/	C			
98	A	A/	G			
103	C	C/	T			
129				A	A/	G
131				A	A/	T
132				T	T/	C
137	C	C/	G			
138	A	A/	T			
139	T	T/	G			
141	T/	C	C			
144	A	A/	G			
150	C	C/	T			
152	A	A/	G			
168	G	G/	A			
169	C	C/	T			
178	T	T/	C			
184	T/	C	C			
188	T	T/	A			
193	A/	G	G			
199				C	C/	G
218				G	G/	T
222	A/	G	G			
224	A	A/	T			
225	T	T/	A			
226	G	G/	T			
230				A	A/	G
231				T	T/	A
232				G	G/	A
258				A	A/	G
264	T/	A	A			
381	A/	C	C			
418	C	T	C			
420	C/	T	T			
425	G	A	G			
442	A/	C	C			
446	T/	A	A			
450	G	C	G			

Table 1. Continued

16S rRNA nucleotide position	Family					
	Rhizobiaceae			Aeromonadaceae		
	Rhizobium loti	*Rhizobium galegae*	*Agrobacterium tumefaciens*	*Aeromonas caviae*	*Aeromonas hydrophila*	*Aeromonas salmonicida*
457				C/	T	T
458				A/	G	G
459				G	A	G
460				T	T/	C
461				A/	G	G
462				G/	C	C
469				T/	C	C
470				C/	G	G
472				G	A	G
473				C/	T	T
474				T/	C	C
475				G/	A	A
476				G/	A	A
488	A/	T	T			
492	T/	G	G			
591	T/	A	A			
592	G/	T	T			
599	T/	C	C			
616	G/	A	A			
624	C/	T	T			
639	A/	G	G			
647	C/	G	G			
648	A/	T	T			
653	C/	T	T			
658	C/	A	A			
659	C/	T	T			
661	A/	G	G			
662	G/	A	A			
669	G/	A	A			
711	G	C	G			
737	C/	T	T			
743	C/	T	T			
744	T/	C	C			
746	G/	A	A			
747	G/	T	T			
818	T/	G	G			
823	G/	A	A			
824	A/	T	T			
826	C/	T	T			
837	C	C/	G			
838	A	A/	C			
843	T	T/	A			
848	T	T/	G			
849	G	G/	T			
874	G/	A	A			
876	T/	A	A			
877	C/	T	T			

Discussion

These results demonstrate that similar patterns of segment-dependent 16S nucleotide sequence polymorphism exist across different bacterial families, and support the contention that these genes may have a history of lateral transfer and recombination among species of the same genus (Sneath, 1993). The subject of genetic recombination in rRNA genes in bacteria is controversial because it is assumed that genetic recombination would likely result in changes in the 16S rRNA molecule that would adversely affect the translational efficiency (and hence Darwinian fitness) of the recombinant cell. It should also be considered however that this extreme requirement for sequence conservation, which makes the 16S gene so useful for phylogenetic studies, may also provide the long runs of sequence homology necessary for homologous recombination.

Because organismal phylogenies are commonly extrapolated from 16S phylogenies, it is important to recognize whether these genes may have a history of genetic recombination. However the existence of natural or authentic 16S chimeras among closely related species may be difficult to prove (statistically), because the resulting patterns of polymorphism may involve only a few nucleotides (Robison-Cox et al., 1995). Nevertheless the results of the current study, that suggest that there are corresponding chimeric segments across highly divergent bacterial families, provides an alternative line of evidence in support of the hypothesis that 16S genes may be subject to lateral transfer and recombination.

Acknowledgements

The authors thank K Lindstrom and J P W Young for helpful discussions and for sharing unpublished results.

References

Dams E, Hendriks L, Van de Peer Y, Neefs J-M, Smits G, Vandenbempt I and De Wachter R 1988 Compilation of small ribosomal subunit RNA sequences. Nucl. Acids Res. 16, r87–r173.

Eardly B D, Wang F-S, Whittam T S and Selander R K 1995 Species limits in *Rhizobium* populations that nodulate the common bean (*Phaseolus vulgaris*). Appl. Environ. Microbiol. 61, 507–512.

Graham P H, Sadowsky M J, Keyser H H, Barnet Y M, Bradley R S, Cooper J E, DeLey J, Jarvis B D W, Roslycky E B, Strijdom B W and Young J P W 1991 Proposed minimal standards for the description of new genera of species of root- and stem-nodulating bacteria. Int. J. Syst. Bacteriol. 41, 582–587.

Kumar S, Tamura K and Nei M 1993 MEGA: Molecular Evolutionary Genetic Analysis, version 1.01. The Pennsylvania State University, University Park, PA 16802, USA

Martinez-Murcia A J, Benlloch S and Collins M D 1992 Phylogenetic interrelationships of the genera *Aeromonas* and *Plesiomonas* as determined by 16S ribosomal DNA sequencing: lack of congruence with results of DNA-DNA hybridizations. Int. J. Syst. Bacteriol. 42, 412–421.

Niebel H, Dorsch M and Stackebrandt E 1987 Cloning and expression in *Escherichia coli* of *Proteus vulgaris* genes for 16S ribosomal RNA. J. Gen. Microbiol. 133, 2401–2409.

Robison-Cox J F, Bateson M M and Ward D M 1995 Evaluation of nearest neighbor methods for detection of chimeric small-subunit rRNA sequences. Appl. Environ. Microbiol. 61, 1240–1245.

Saitou N and Nei M 1987 The neighbor-joining method: a new method for reconstructing phylogenetic trees. Mol. Biol. Evol. 4, 406–425.

Sneath P H A 1993 Evidence from *Aeromonas* for genetic crossing-over in ribosomal sequences. Int. J. Syst. Bacteriol. 43, 626–629.

Stackebrandt E and Goebel E M 1994 Taxonomic note: A place for DNA-DNA reassociation and 16S rRNA sequence analysis in the present species definition in bacteriology. Int. J. Syst. Bacteriol. 44, 846–849.

Van Berkum P, Beyene D and Eardly B D 1996 Phylogenetic relationships among *Rhizobium* species nodulating the common bean (*Phaseolus vulgaris* L.) Int. J. Syst. Bacteriol. 46, 240–244.

Willems A and Collins M D 1993 Phylogenetic analysis of rhizobia and agrobacteria based on 16S rRNA gene sequences. Int. J. Syst. Bacteriol. 43, 305–313.

Woese C R, Gutell R, Gupta R and Noller H F 1983 Detailed analysis of the higher-order structure of 16S-like ribosomal ribonucleic acids. Microbiol. Rev. 47, 621–669.

Section editor: H Lambers
Guest editors: G.H. Elkan and R.G. Upchurch

Plant and Soil **186**: 75–79, 1996.

Rapid evaluation of peat-base legume inoculant using immunomagnetic beads for cell retrieval and fluorescent nucleic acid probes for viability analysis

P.E. Olsen and W.A. Rice
Northern Agriculture Research Centre, Agriculture and Agri - Food Canada, Beaverlodge, Alberta, Canada TOH OCO *

Key words: antibody, immunomagnetic, legume inoculant, magnet, *Rhizobium*

Abstract

Problems associated with traditional methods of evaluating legume inoculants using the most-probable-number plant nodule grow-out test include technician time, expense, growth chamber space, and the length of time (30 days) required to conduct the tests. Through the use of specific monoclonal antibodies, a biotin-labelled intermediate antibody, and streptavidin-labelled magnetic micro-beads, we were able to rapidly remove and purify rhizobial cells from peat inoculants used as quality control samples in the Canadian Legume Inoculant and Pre-Inoculated Seed Testing Program. Viability of the recovered cells was evaluated with a commercial kit employing differential fluorescent nucleic acid probes allowing viability to be rapidly determined via visual examination with a fluorescence microscope. Total elapsed time for complete evaluation of an inoculant was about 90 minutes.

Introduction

Agriculture and Agri-Food Canada conducts the only program in North America which regulates rhizobial legume inoculant products and pre-inoculated seed. This program monitors commercial inoculants in an attempt to identify and eliminate poor quality products. Standards have been established in terms of the number of viable *Rhizobium* and *Bradyrhizobium* bacteria which must be present in legùme inoculants to ensure adequate nodulation (Olsen et al., 1994a). The most-probable-number (MPN) technique often used for evaluation of inoculants involves growing *Rhizobium*-free legume plants in growth chambers, applying dilutions of the inoculant product to the plants, and waiting for nodules to develop on the plants (Anonymous, 1991). This is an expensive process which normally takes 30 days to complete. Moreover, by the time that a faulty inoculant has been identified it is often too late to detain it and prevent it from being used by farmers. Faster means of evaluating legume inoculants are needed.

Polystyrene micro-beads covalently labelled with streptavidin (SA) and containing sufficient iron oxide to make them magnetisable are commercially available (Dynal, Inc.; Spherotech, Inc.). These beads can be rapidly pelleted from a liquid suspension into a small volume with a magnet. We used a strain-specific anti-*Rhizobium meliloti* strain NRG34 mouse monoclonal antibody (MAb) to specifically agglutinate the rhizobia in liquid suspensions of peat inoculant and then labelled the agglutinated MAbs and rhizobia with a sheep anti-mouse Ig-biotin (SAMBi) conjugate. Upon addition of streptavidin-labelled beads, the agglutinated *Rhizobium* cells became bound to the beads via the combination of the specific MAb, the biotin-labelled anti-mouse antibodies, and the streptavidin on the beads. The beads and bound rhizobia were then separated from the rest of the inoculant with a magnet (see schematic, Figure 1).

Two differential nucleic acid fluorescent stains (Molecular Probes, Inc.) were added to the *Rhizobium* cell suspension and then examined under an epifluorescence microscope equipped with the proper fluorescence filter set. Cells with intact membranes (viable) fluoresce bright green, whereas cells with damaged cell membranes (non-viable) fluoresce bright red. By establishing the ratio of green to red cells, the numbers of live and dead rhizobia in the immunobead puri-

* FAX No: + 14033548171. E. mail: RICEW@EM.AGR.CA

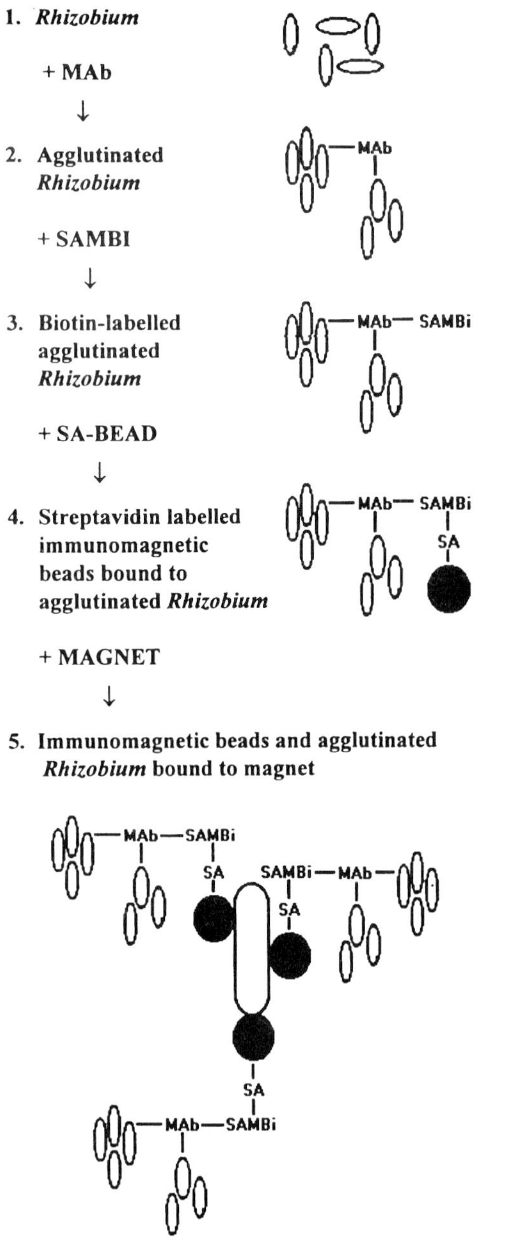

Figure 1. Schematic diagram of procedure used to retrieve *Rhizobium* cells from peat inoculant.

fied cell extract were determined. Counting the total number of purified cells in a Petroff-Hauser counting chamber and then applying the "live/dead" ratio resulted in a minimum number of live rhizobia, of the strain for which the MAb was specific, which must have been in the original inoculant sample. The whole procedure was accomplished in 90 minutes.

Materials and methods

Monoclonal antibody

The MAb involved in this work (produced by Northern Agriculture Research Centre hybridoma cell line no. F12P92F) was used in the form of diluted (1: 500) ascites fluid, and was highly specific for *R. meliloti* strain NRG34. The MAb was of IgM class and was effective in the agglutination of NRG34 cells. The production, characterization, and specificity of this antibody have been previously described (Olsen et al., 1994b). While the procedure described in this paper used a specific mouse monoclonal antibody, there is no technical requirement that the antibody be monoclonal. A specific rabbit antisera capable of agglutinating the rhizobia, in conjunction with a sheep or goat anti-rabbit Ig-biotin conjugate, should work as well.

Peat-base inoculant

The inoculants used in this study were prepared as controls for use in the regulatory Canadian Legume Inoculant and Pre-Inoculated Seed Testing Program (Olsen et al., 1994a). Approximately 250 bags (45 g final weight) were prepared in Dec., 1994 for the 1995 testing season. The inoculants were made by the addition of 17 mL of dilute (1: 50), 48 h yeast extract-mannitol broth culture of *R. meliloti* strain NRG34 to 28 g bags of gamma-irradiated peat followed by 14 days maturation at 26 °C. Viable *Rhizobium* cell counts in these inoculants ranged from 4 - 6 $\times$ 10^9 per g and they were virtually contaminant free.

Immunomagnetic beads

Polystyrene micro-beads (2.8 μm in diameter) which contained sufficient iron oxide to make them magnetisable and which were covalently labelled with streptavidin were obtained from Dynal, Inc., cat. no. 112.05. When suspended in liquid, these beads can be rapidly pelleted into a small volume with a neodynium-iron boron magnet (magnetic particle concentrator) also available from Dynal.

Differential fluorescent probes

Two differential nucleic acid stains were obtained from Molecular Probes, Inc., (LIVE/DEAD BacLight Viability Kit, Molecular Probes, cat. no. L-7007). The basis of this kit is nucleic acid staining via differential stain entry through the cell membrane. Under epifluorescence exposure with the correct light wavelengths, cells with intact membranes (viable) fluoresce bright green and cells with damaged cell membranes (non-viable) fluoresce bright red. Manufacturer's instructions were followed, but experience led us to apply about twice the recommended concentration of stains to achieve best results.

Analytical procedure

Peat inoculant in the bag was mixed and 1.0 g removed and placed in a test tube. Four mL of phosphate-buffered saline containing 0.05% Tween 20 (PBST) was added and mixed with the inoculant by vortexing. After settling for 5 min, 2 mL of suspension was removed from the top and divided into 0.5 mL samples in 1.5 mL micro-centrifuge tubes. One μL of ascitic fluid containing MAb specific to *R. meliloti* strain NRG34 was added per 0.5 mL inoculant suspension, mixed, and 1 μL of sheep anti-mouse Ig-biotin conjugate (Boehringer Mannheim, cat. no. 1089 285) was then added and mixed. Fifty μL of streptavidin-labelled immunomagnetic beads (10 mg beads per mL) was washed twice with PBST, resuspended in 50 μL PBST and added to the inoculant-antibody mix. The tube containing the sample was then rotated slowly (0.5 rpm) using an adapted LKB liquid chromatography pump for 60 min at room temperature. The beads and bound rhizobia were pelleted with the magnet and then washed twice, without resuspension, with PBST. The mix was then resuspended and the beads detached from the bacteria by vortexing and then separated from the bacteria by magnetization. The purified rhizobia were removed with the supernatant liquid and deagglutinated by 2 washes in purified H_2O and resuspended to 0.75 mL in H_2O. Viable rhizobia were quantified by standard plate-count techniques using yeast-extract mannitol agar (Vincent, 1970).

Three μL of each of the LIVE/DEAD BacLight kit stains were mixed and 4 μL of the mix added to 0.5 mL of the purified deagglutinated rhizobia in H_2O and examined with a Zeiss Photomicroscope III using an epifluorescence RS III condenser and a 100 watt mercury lamp. An Omega optical fluorescence filter set (cat. no. XF52) was fitted to allow cells stained red and cells stained green to be seen simultaneously. Total cell counts, and red and green fluorescing cell counts were then made using a Petroff-Hauser counting chamber.

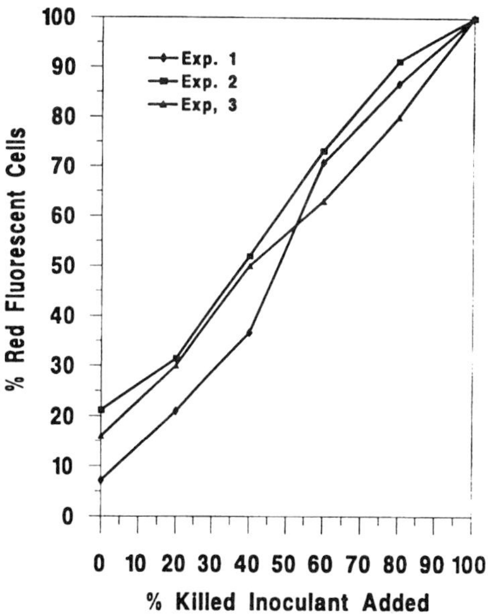

Figure 2. Relationship between the number of rhizobial cells exhibiting red fluorescence (indicating non-viability) and increasing levels of deliberately killed rhizobial inoculant added to normal live inoculant. ♦ = unpurified cells in inoculant suspension (1 g inoculant in 6 mL phosphate-peptone diluent). ■ = purified cells from inoculant suspension (1 g inoculant in 6 mL diluent. ▲ = purified cells from inoculant suspension (1 g inoculant in 12 mL diluent).

Results

On the basis of increasing proportions of killed inoculant added to live inoculant and concomitant measurement of the live/dead cell ratio by means of the fluorescent red and green nuclear stains, the fluorescence determination of rhizobial viability worked well (Figure 2). Three separate experiments showed that between 7% and 21% of the rhizobial cells present in different quality control packages (plate counts = 4.9 – 5.1 × 10^9 per g) were non-viable (showed strong red fluorescence). Experiment no. 1 was conducted using unpurified cells in a peat inoculant suspension in PBST. Experiments 2 and 3 used immunomagnetic bead purified rhizobial cells from differing levels of initial peat dilution in PBST. With increasing levels of added killed inoculant, the inoculant mix showed a direct relationship between the amount of killed inoculant added and the ratio of red to green cells counted. The samples containing 100% killed cells showed 100% red fluorescent cells.

Repeated washing of peat inoculant with extraction buffer did not completely remove all viable cells from the peat. Approximately 20% of the total initial viable

Table 1. Recovery of *Rhizobium meliloti* strain NRG34 viable cells from peat inoculants using immunomagnetic beads

Sample no.	Inoculant plate count[b]	IMMBRC[a] plate count[b]	Percent viable rhizobia recovery
1	4.6×10^9	1.9×10^9	41.3
2	4.1×10^9	9.8×10^8	23.9
3	5.6×10^9	2.3×10^9	41.1
4	4.2×10^9	2.0×10^9	47.6
5	4.9×10^9	1.6×10^9	32.7
6	4.5×10^9	2.3×10^9	51.1
7	5.1×10^9	2.2×10^9	43.1
8	4.4×10^9	1.7×10^9	38.6

[a] Immunomagnetic bead recovered cells.
[b] All values are relative to 1.0 g of inoculant.

cells were still attached to the peat and detectable by plating even after ten washes (suspension, vortexing and centrifugation) in PBST (data not shown). The level of Tween 20 in the buffer used to extract cells from peat inoculants had little effect in enhancing cell extraction over a wide range of concentration ranging from zero to 1.25% (data not shown).

Recovery of rhizobial cells from inoculant suspensions using immunomagnetic beads was not complete. Estimates of viable cell recovery ranged from 23.9% to 51.1% in tests run with eight different control inoculants (Table 1). Washing of the magnetized bead-cell pellet by removal from the magnet, addition of buffer, vortexing, and repelleting of the beads resulted in physical loss of a large proportion of cells in the pellet (data not shown). Vortexing of the bead-cell mass after resuspension in purified H_2O was, in fact, an effective method for near 100% separation of beads and simultaneous deagglutination of the bacteria (needed to obtain meaningful plate counts).

Discussion

The Legume Inoculant and Pre-Inoculated Seed Testing Program conducted by Agriculture and Agri-Food Canada requires that all samples collected for testing be accompanied at all times by a control inoculant of known rhizobial count. Both the sample and the control inoculant are always tested. The purpose of the control is to establish that sample inoculants were not damaged by any conditions of transport or storage between the time of sampling and testing. Control inoculants constitute up to 50% of the samples received and tested. These inoculants, which are contaminant free, can also be effectively analysed by plate count, so that the savings in time needed to obtain a result using immunomagnetic beads is only the few days necessary for colony growth. If, however, the technique is applicable to the analysis of heavily contaminated inoculants (which cannot be analysed by plate count and require MPN testing), then the savings in time to a result would be on the order of 30 days. This amount of time saved would provide time for regulatory action, if required. Work is anticipated to evaluate the new technique with other rhizobial species and with contaminated inoculant products. The technique described has the potential for analysis of official test samples because official results are simply "pass or fail" based upon the demonstration of the presence of a defined minimum of viable rhizobial cells. The recovery levels obtained with control inoculants are approximately 10 fold those required to "pass" an ordinary peat inoculant.

Dye (1994) has reported enrichment of *Rhizobium* in a model system using immunomagnetic beads. His model used antibody-labelled beads to selectively bind rhizobia from aqueous media and pioneered the use of immunomagnetic beads with rhizobia. Immunomagnetic bead capture of other bacteria from aqueous suspension has been reported, with recovery effectiveness ranging from 40 to 100% (Molloy et al., 1995). Our technique demonstrates the recovery of rhizobia from a complex and dense particulate environment. The evaluation of these inoculants using immunomagnetic beads for cell recovery and fluorescent nucleic acid probes for viability analysis rapidly provided a minimum number of viable rhizobial cells of known species (and strain) which must have been present per unit weight in the inoculant.

References

Anonymous 1991 The methods of testing legume inoculant and pre-inoculated seed products. Fertilizers Act, Section 23, Regulations. Feed and Fertilizer Division, Government of Canada, Ottawa. 15 p.

Dye M 1994 The enrichment of *Rhizobium* from a model system using immunomagnetic separation. J. Microbiol. Meth. 19, 235–245.

Molloy P, Brydon L, Porter A J and Harris W J 1995 Separation and concentration of bacteria with immobilized antibody fragments. J. Appl. Bacteriol. 78, 359–365.

Olsen P E, Rice W A, Bordeleau L M and Biederbeck V O 1994a Analysis and regulation of legume inoculants in Canada: The need for an increase in standards. Plant and Soil 161, 127–134.

Olsen P E, Wright S, Collins M and Rice W 1994b Patterns of reactivity between a panel of monoclonal antibodies and forage *Rhizobium* strains. Appl. Environ. Microbiol. 60, 654–661.

Vincent J M 1970 A manual for the practical study of the root-nodule bacteria. International Biological Programme Handbook No. 15. Blackwell Scientific Publications, Ltd., Oxford, UK.

Section editor: H Lambers
Guest editors: G.H. Elkan and R.G. Upchurch

Plant and Soil **186:** 81–85, 1996.

Detection of genetic variation in *Bradyrhizobium japonicum* USDA 110 variants using DNA fingerprints generated with GC rich arbitrary PCR primers

James N. Mathis and David E. McMillin
Department of Biology, State University of West Georgia, Carrollton, GA 30118, USA and Department of Biological Sciences, Clark-Atlanta University, Atlanta, GA 30314, USA*

Key words: Bradyrhizobium japonicum colony morphology variants, DNA fingerprints, GC rich arbitrary primers RAPD-PCR

Abstract

Bradyrhizobium japonicum USDA 110 has been shown to contain several genetically similar naturally occurring colony morphology variants. These variants differ in symbiotic nitrogen fixation ability and in the utilization of various carbon substrates. They have been shown to share extensive DNA homology and appear to be derived from a common ancestor. Despite these similarities certain *B. japonicum* USDA 110 variants have been shown to be devoid of symbiotic nitrogen fixation. One of these variants (L2-110), however, was recently shown to possess significant levels of ex planta nitrogen fixation and to synthesize functional dinitrogenase enzyme within bacteroids. In an effort to identify genetic markers which could explain differences in symbiotic nitrogen fixation between *B. japonicum* variants, DNA fingerprints were generated by PCR using arbitrary primers. Two of these primers with GC rich sequences were able to differentiate between *B. japonicum* USDA 110 variants I-110, L2-110, and MN-110. Unique markers have now been identified which could be examined further to determine if they explain the differences in symbiotic nitrogen fixation between USDA 110 variants.

Introduction

Bradyrhizobium japonicum strain USDA 110 has been shown to contain several colony morphology variants which differ in their symbiotic nitrogen fixation abilities and/or their abilities to utilize carbohydrates such as D-mannitol (Kuykendall and Elkan, 1976; Mathis et al., 1986a, b, c). One of these variants, L2-110, which lacks symbiotic nitrogen fixation ability was recently shown to possess significant levels of ex planta nitrogen fixation (relative to other symbiotically efficient variants, I-110 and MN-110) and to synthesize functional dinitrogenase enzyme within nodules (Bradburne et al., 1994). Furthermore, DNA sequence analyses of the *nif* structural genes have revealed the same sequence in both L2-110 and MN-110, thus indicating identical nitrogenase structure. In further attempts to demonstrate specific genetic differences between USDA 110 colony morphology variants several analyses have been used: DNA-DNA homology (Mathis et al., 1986b), restriction enzyme analysis of genomic DNA (Mathis et al., 1986a, c; Mielenz et al., 1979), Southern blot analysis of *nif* structural genes (Mathis et al., 1986a, c), two dimensional polyacrylamide gel analysis of total proteins (Mathis and Bradburne, personal communication) and restriction fragment length polymorphism analysis using random probes (Mathis, personal communication). None of these techniques, however, has demonstrated a specific genetic difference between the various USDA 110 variants.

Recently several different PCR (polymerase chain reaction) based techniques have been developed which are useful in the detection of variations between bacterial strains. Among these techniques are REP-PCR (repetitive extragenic palindromic, Woods et al., 1993), ERIC-PCR (enterobacterial repetitive intergenic consensus, Judd et al., 1993) and PCR amplification using arbitrary primers (RAPD, Williams et al., 1990). This report describes how one of these techniques, RAPD, has been used to identify variations

* FAX No: + 14048366633.

in *B. japonicum* USDA 110 variants which differ in carbohydrate utilization and nitrogen fixation ability.

Materials and methods

Bacterial strains used and conditions for growth

Bradyrhizobium japonicum USDA 110 variants I-110, L2-110 (Kuykendall and Elkan., 1976) and MN-110 (Mathis et al., 1986a, b) were used in this study, since they represent the various phenotypes observed for *B. japonicum* USDA 110 variants (Table 1). Strains USDA 122, USDA 466, USDA 468 and USDA 110 were obtained from P Van-Berkum, USDA-ARS, Beltsville, MD, USA. USDA 122 was selected as a reference strain for comparison with USDA 110 due to its relatively low DNA-DNA homology with USDA, 110 (2% homology under stringent reassociation conditions, Hollis et al., 1981). All USDA 110 variants and control strains were grown to stationary phase in yeast-extract mannitol medium (Mathis et al., 1986b). Following growth of cells in yeast-extract mannitol medium, cells were harvested twice by centrifugation (5000 × *g*) and then resuspended each time in sterile 0.95%, sodium chloride. Treatment with 0.95% sodium chloride, prior to DNA isolation, removed much of the extracellular material produced during growth which would otherwise interfere with DNA isolation.

Table 1. Characteristics of *Bradyrhizobium japonicum* USDA 110 colony morphology variants used in this study

Variant	N_2-fixation		Mannitol utilization
	Symbiotic	Ex planta	
I-110[a]	+[b]	+[c]	-[a]
L2-110[a]	-[b]	+[c]	+[a]
MN-110[d]	+[d]	+[c]	+[d]

[a] Kuykendall and Elkan (1976).
[b] Mathis et al. (1986c).
[c] Bradburne et al. (1994).
[d] Mathis et al. (1986a).

DNA extraction and PCR amplification of DNA

High molecular weight genomic DNA was isolated from *B. japonicum* cells using a CTAB (hexadecyltrimethyl ammonium bromide) procedure (Doyle and Doyle, 1990). Cells were harvested at 3000 × *g* in a Kontes microfuge tube and crushed in liquid nitrogen with a Kontes pellet pestle. CTAB extraction buffer [0.5 mL, (4% CTAB, 2 *M* Tris-hydrochloride (pH 8.0), 5 *M* sodium chloride, 0.5 *M* EDTA)] was added, vortexed and incubated at 65 °C for 45 minutes. Following incubation, the tube was centrifuged and the supernatant collected. The supernatant was then extracted twice with an equal volume of phenol / chloroform / isoamyl alcohol (25:24:1, v/v). One additional extraction of chloroform / isoamyl alcohol (24:1, v/v) was also performed. DNA was then precipitated by addition of one tenth volume of 3 *M* sodium acetate and three volumes of isopropanol. Following centrifugation at 6,000 × *g*, the pellet was dried and resuspended in TE [10 m*M* Tris-hydrochloride (pH 7.5), 1 m*M* EDTA]. DNA was then quantitated using a TKO-100 mini-fluorometer (Hoeffer Scientific, San Francisco, CA, USA).

ERIC- and REP-PCR analyses were performed as described previously (Judd et al., 1993). RAPD-PCR was carried out in a 50 μL reaction volume containing 25 ng of genomic DNA, 1.75 m*M* magnesium chloride, 100 μM each of dATP, dCTP, dGTP, dTTP, 200 μg mL^{-1} bovine serum albumin, 10 m*M* Tris-hydrochloride (pH 8.3), 50 m*M* potassium chloride, 30 p moles primer and 1.0 unit of *Taq* polymerase (Perkin Elmer, Norwalk, CT, USA). All tubes were initially incubated at 94 °C for 3 minutes and then run for a cycle consisting of denaturation at 94 °C for 1 minute, annealing at 35 °C for 2 minutes and extension at 72 °C for 2 minutes. This cycle was followed by 40 cycles of denaturation for 1 minute at 94 °C, annealing at 37 °C for 1.5 minutes, and extension for 2 minutes at 72 °C using a MJ thermocycler (MJ Research Inc., Watertown, MA, USA). At the conclusion of the PCR cycles, all tubes were incubated for 7 minutes at 72 °C to allow the amplification process to go to completion. The resulting PCR products were then separated on 2% agarose gels and visualized using ethidium bromide staining (Sambrook et al., 1989). In order to determine which primers might be useful in detecting polymorphisms between USDA 110 colony morphology variants, 28 primers were screened using representative DNA samples from USDA 110 and USDA 122. The following primers were screened, OPE-01 - OPE-10 and OPO-1 - OPO-10 (Operon Company, Alameda California, USA), ARP-2, ARP-5, ARP-6 (McMillin and Muldrow, 1992), ARP-7 (GTACGTGGCG), SPH-1 (Harrison et al., 1992), MICROSAT, [(T or A)(T or A)GAGGGTGG], CRL-7 and CRL-13 (Kubelik and Szabo, 1995). Each of the primers which amplified and detected polymorphisms between USDA 110 and

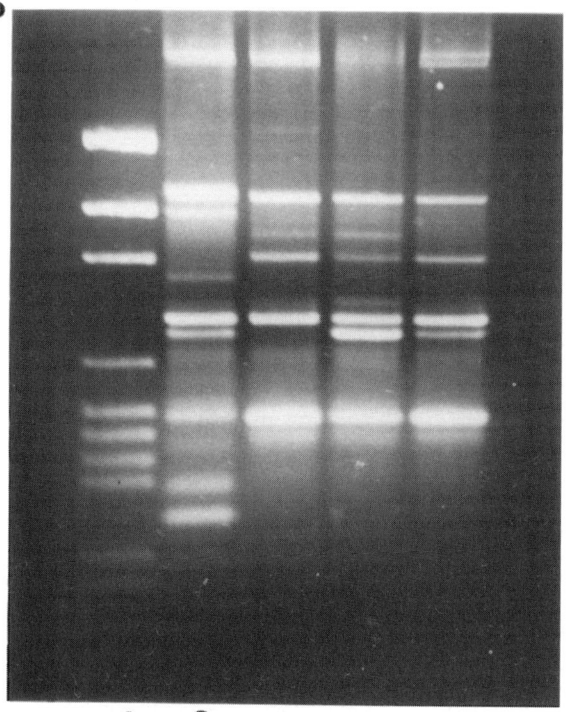

Figure 1. DNA fingerprints generated by RAPD-PCR of *B. japonicum* USDA 110 colony morphology variants using arbitrary primer CRL-7 (GCCCGCCGCC). DNA samples were separated on a 2% agarose gel; Lane 1, molecular weight marker, pGEM, fragment sizes in nucleotides are denoted; Lane 2, USDA 122; Lane 3, L2-110; Lane 4, I-110; Lane 5, MN-110.

USDA 122 were used to amplify DNA from *B. japonicum* USDA 110 variants.

Results and discussion

DNAs from the two reference strains of *B. japonicum* (USDA 110 and USDA 122) showed differences in RAPD-PCR amplification profiles for 17 of the 28 primers screened (61%, data not shown). No amplification was noted for six of the 28 primers (21%) and five primers gave the same pattern of amplification for each control strain (18%, data not shown). Primers which were able to differentiate between the reference strains were used for amplification of each of the *B. japonicum* USDA 110 variants.

In the analysis of *B. japonicum* USDA 110 colony morphology variants by RAPD-PCR two primers were found to give clear and repeatable variations in pattern between I-110, L2-110 and MN-110; CRL-7 (Figure 1) and ARP-7 (data not shown). In *B. japonicum* strain USDA 122 amplification products of 1750, 1700, 1600, 1100, 850, 800, 480, 350 and 300 nucleotides were observed using primer CRL-7 (Figure 1). Each USDA 110 variant shared common bands of 1700, 850 and 480 nucleotides with strain USDA 122. A band at 1200 nucleotides and a diffuse band at 450 nucleotides were observed in each USDA 110 variant but were not found in strain USDA 122 (Figure 1). Differences in PCR pattern were also observed between USDA 110 variants, for example a faint band at 1300 nucleotides was observed in L2-110 and I-110 but not in MN-110 (Figure 1). Interestingly, I-110 and L2-110 were derived from the same USDA 110 culture, whereas MN-110 was derived separately (Kuykendall and Elkan, 1976; Mathis et al., 1986b). A band of 800 nucleotides was observed in Fix^+ USDA 110 variants I-110 and MN-110 and Fix^+ strain USDA 122 but was absent in Fix^- variant L2-110 (Figure 1). Future efforts will focus on cloning and sequencing this 800 nucleotide fragment present in all the Fix^+ isolates but absent in L2-110. It may then be determined if absence of this DNA sequence can be further associated with the Fix^- phenotype of this variant. Previous studies indicate that L2-110 may possess an unusual defect in symbiotic nitrogen fixation not previously characterized (Bradburne et al., 1994). PCR primer ARP-7 also provided distinct DNA fingerprints for each of the variants (data not shown). No clear correlations in banding pattern with respect to mannitol utilization ability were noted for variant DNAs amplified with either primer CRL-7 (Figure 1) or ARP-7, despite clear differences in DNA-fingerprint for each of the variants with each primer.

Both of the primers which detected genetic polymorphisms between each of the *B. japonicum* USDA 110 variants are high in GC content relative to other primers which failed to separate these closely related variants. One of these, CRL-7, was made up exclusively of GC residues. Harrison et al. (1992) observed a similar result when analyzing *Rhizobium leguminosarum* biovar. *trifolii* strains using RAPD, zero of eight 7-10 nucleotide primers with approximately 50% GC were able to amplify *R. leguminosarum* DNA. However, several other primers of similar size with high GC content were able to amplify these same DNAs, producing numerous bands in each of the amplification patterns obtained, and differentiating between the various strains of *R. leguminosarum*. Our data would indicate a similar enhancement in the number of bands and ability to differentiate *B. japonicum* strains using GC rich primers.

Other investigators have used REP-PCR and ERIC-PCR to differentiate very closely related *B. japonicum* serocluster 123 strains (Judd et al., 1993), we were not able to differentiate between I-110, L2-110 and MN-110 with these techniques (data not shown). However, various other strains of *B. japonicum* such as USDA 122, USDA 110 and 110- serogroup strains USDA 466 and 468 were differentiated using these methods (data not shown). These data lend further evidence to the genetic relatedness of *B. japonicum* USDA 110 variants and their apparent derivation from a common ancestor.

In this report clear differences have been demonstrated in DNA structure between each of the *B. japonicum* USDA 110 colony morphology variants using RAPD-PCR with GC rich arbitrary primers. These differences may demonstrate genetic variations which could account for the phenotypic differences between these variants. Numerous other genetic analyses have failed to detect such differences. It may be possible to apply these techniques to field or laboratory situations where genetic variations between closely related *B. japonicum* strains or isolates have not been elucidated. Alternatively the fate of very closely related organisms in the environment could be followed using this methodology. Finally this report adds further evidence to the idea that GC rich bacteria such as *R. leguminosarum* and *B. japonicum* require GC rich primers for RAPD-PCR to be effective.

Acknowledgements

We thank Tony Griffin and Julie Baggett for excellent technical assistance, and Tom Beggs for help with photography. This investigation was partially supported by an RCMI award, G12RR03062, from the Division of Research Resources, National Institutes of Health and by West Georgia College Faculty Research Support Grants awarded by the Learning Resources Committee.

References

Bradburne J A, Mathis J N and Israel D W 1994 *nif* Gene expression in a Nif$^+$, Fix$^-$ *Bradyrhizobium japonicum* variant. FEMS Microbiol. Lett. 123, 91–98.

Doyle J J and Doyle J L 1990 Isolation of plant DNA from fresh tissue. Focus 12, 13–15.

Harrison S P, Mytton L R, Skot L, Malcolm D and Cresswell A 1992 Characterization of *Rhizobium* isolates by amplification of DNA polymorphisms using random primers. Can. J. Microbiol. 38, 1009–1015.

Hollis A B, Kloos W E and Elkan G H 1981 DNA:DNA hybridization studies of *Rhizobium japonicum* and related Rhizobiaceae. J. Gen. Microbiol. 123, 215–222.

Judd A K, Schneider M, Sadowski M J and De Bruijn F 1993 Use of repetitive sequences and the polymerase chain reaction technique to classify genetically related *Bradyrhizobium japonicum* serocluster 123 strains. Appl. Environ. Microbiol. 59, 1702–1708.

Kubelik A R and Szabo L J 1995 High-GC primers are useful in RAPD analysis of fungi. Curr. Genet. 28, 384–389.

Kuykendall L D and Elkan G H 1976 *Rhizobium japonicum* derivatives differing in nitrogen-fixing efficiency and carbohydrate utilization. Appl. Environ. Microbiol. 32, 511–519.

Mathis J N, Barbour W M, Miller T B, Israel D W and Elkan G H 1986a Characterization of a mannitol utilizing, nitrogen-fixing *Bradyrhizobium japonicum* USDA 110 derivative. Appl. Environ. Microbiol. 52, 81–85.

Mathis J N, Israel D W, Barbour W M, Jarvis B D W and Elkan G H 1986b Analysis of the symbiotic performance of *Bradyrhizobium japonicum* USDA 110 and I-110 and discovery of a new mannitol-utilizing, nitrogen-fixing USDA 110 derivative. Appl. Environ. Microbiol. 52, 75–80.

Mathis J N, Kuykendall L D and Elkan G H 1986c Restriction endonuclease and *nif* homology patterns of *Bradyrhizobium japonicum* USDA 110 derivatives with and without nitrogen fixation competence. Appl. Environ. Microbiol. 51, 477–480.

McMillin D E and Muldrow L L 1992 Typing of toxic strains of *Clostridium difficile* using DNA fingerprints generated with arbitrary polymerase chain reaction primers. FEMS Microbiol. Lett. 92, 5–10.

Mielenz J R, Jackson L E, O'Gara F and Shanmugam K T 1979 Fingerprinting bacterial chromosomal DNA with restriction endonuclease *Eco*RI; comparison of *Rhizobium* spp. and identification of mutants. Can. J. Microbiol. 25, 803–807.

Sambrook J, Fritsch E F and Maniatis T 1989 Molecular Cloning: A Laboratory Manual. Cold Spring Harbor Laboratory Press, Cold Spring Harbor, NY, USA.

Williams J G K, Kubelik A R, Livak K J, Rafalski J A and Tingey S V 1990 DNA polymorphisms amplified by arbitrary primers are useful as genetic markers. Nucleic Acids Res. 18, 6531–6535.

Woods C R, Versalovic J, Koeuth T and Lupski J R 1993 Whole-cell repetitive element sequence-based polymerase chain reaction allows rapid assessment of clonal relationships of bacterial isolates. J. Clin. Mirobiol. 31, 1927–1931.

Section editor: H Lambers
Guest editors: G.H. Elkan and R.G. Upchurch

Plant and Soil **186:** 87–92, 1996.

Mimosine produced by the tree-legume *Leucaena* provides growth advantages to some *Rhizobium* strains that utilize it as a source of carbon and nitrogen

Muchdar Soedarjo[1,2] and Dulal Borthakur[1,3]
[1]*Department of Plant Molecular Physiology and* [2]*Department of Microbiology, University of Hawaii, Honolulu, HI 96822, USA.* [3]*Corresponding author*.*

Key words: degradation, *Leucaena*, mimosine, *Rhizobium*, toxicity

Abstract

Growth of most *Rhizobium* strains is inhibited by mimosine, a toxin found in large quantities in the seeds, foliage and roots of plants of the genera *Leucaena* and *Mimosa*. Some *Leucaena*-nodulating strains of *Rhizobium* can degrade mimosine (Mid^+) and are less inhibited by mimosine in the growth medium than the mimosine-non-degrading (Mid^-) strains. Ten Mid^+ strains were identified that did not degrade 3-hydroxy-4-pyridone (HP), a toxic intermediate of mimosine degradation. However, mimosine was completely degraded by these strains and HP was not accumulated in the cells when these strains were grown in a medium containing mimosine as the sole source of carbon and nitrogen. The mimosine-degrading ability of rhizobia is not essential for nodulation of *Leucaena* species, but it provides growth advantages to *Rhizobium* strains that can utilize mimosine, and it suppresses the growth of other strains that are sensitive to this toxin.

Introduction

Leucaena spp. are very common leguminous trees in many tropical and subtropical countries. In recent years the potential for economic use of *Leucaena* spp. for various commercial purposes has been recognized. Its wood and twigs can be used for a paper industry. Because of its high tolerance to drought, resistance to pests and diseases and the ability to grow in a wide variety of soil types, it has become an important tree in agroforestry. However, *Leucaena* plants contain a toxin, mimosine, which is known to be harmful to animals (Brewbaker and Hylin, 1965; Hegarty et al., 1964). Structurally, mimosine [β-*N*(3-hydroxy-4-pyridone)-a-aminopropanoic acid] is an analog of dihydroxyphenylalanine with a 3-hydroxy-4-pyridone ring instead of a 3,4-dihydroxyphenyl ring.

Mimosine has general antimitotic activity that blocks the cell cycle at late G1 (Boehme and Lenardo, 1993; Khanna and Lavin, 1993). It was found to arrest cell division in cultured human cells (Hoffman et al., 1991; Telfer and Green, 1993; Watters et al., 1994) and in Chinese hamster cells (Mosca et al., 1992). It also inhibited replication in yeast (Levenson and Hamlin, 1993) and cultured petunia leaf cells (Perennes et al., 1993). Recently, Feldman and Schonthal (1994) have shown that mimosine prevents the serum-stimulated synthesis and activation of histone H1 kinase, a crucial regulator of cell cycle progression. It is also known to chelate metals, bind pyridoxal phosphate and inhibit the enzymes tyrosine decarboxylase, tyrosinase (Thomson et al., 1969) and ribonucleotide reductase (Dai et al., 1994).

We found that young leaves and pods of *Leucaena* contain as high as 8–10% mimosine on a dry weight basis (Soedarjo and Borthakur, 1996). The toxicity of mimosine has been shown particularly in connection with animal husbandry. Because of the toxic effect of mimosine, sheep unaccustomed to *Leucaena* shed their wool approximately 7 to 14 days after feeding on it (Hegarty et al., 1964; Reis et al., 1975). Similarly, mice having 10% *Leucaena* seed or 1% mimosine in their diet showed loss of hair (Crounse et al., 1962). In spite of the high toxic effects of mimosine on animals, *Leucaena* leaves are used as a fodder for sheep and cattle in some countries since ruminants in some parts

* FAX No: + 18089563542. E-mail: dulal@hawaii.edu

of the world contain a mimosine-degrading bacterium that degrades mimosine to nontoxic products (Allison et al., 1992).

Enzymes in the macerated leaves of *Leucaena* can rapidly degrade mimosine to 3-hydroxy-4-pyridone (HP) (Lowry et al., 1983). Previous studies of the metabolism of mimosine established that in the rumen of some *Leucaena*-fed ruminants mimosine is converted to HP before being completely degraded (Allison et al., 1992). Recently, we showed that mimosine induces a mimosine-degrading enzyme activity in some strains of *Rhizobium* that nodulate *Leucaena* (Soedarjo et al., 1994). In the present work, it is shown that mimosine provides growth advantages to some *Leucaena*-nodulating *Rhizobium* strains that utilize it as a source of carbon and nitrogen, and suppresses the growth of other strains that cannot catabolize it.

Materials and methods

Media and growth conditions

Rhizobium strains were grown in YEM, TY, RM and RP media as described previously (Soedarjo et al., 1994). In order to determine the toxic effects of mimosine, *Rhizobium* strains were grown on TY broth containing 3 m*M* mimosine. The cell densities were measured every 10 to 12 h for 2 days. Toxic effects of mimosine on *Rhizobium* was also determined by spotting 10 μL aliquots from serial dilutions, 10^{-1} to 10^{-6}, of a fresh culture grown in TY broth onto YEM agar containing 1 m*M* mimosine.

Rhizobium strains were screened for their ability to utilize mimosine or HP as the sole source of carbon and nitrogen by streaking them on RM agar plates and incubating them at 28°C for 4 days. The growth rate of *Rhizobium* strains in liquid RM medium with different mimosine concentrations was determined by inoculating 50 mL RM medium in 250 mL screw-cap bottles with 0.5 mL of fresh stationary phase *Rhizobium* culture. Cultures were grown at 28°C with shaking and growth was determined by measuring the cell density as absorbance at 600 nm in a Spectronic 20 colorimeter (Bausch and Lomb, Rochester, NY). Aliquots were also plated from each culture to check for possible contamination. Each treatment was replicated three times. Cultures were maintained until the stationary phase was achieved.

Mimosine and HP determination

Mimosine was purchased from Sigma Chemical Co., St. Louis, Mo. HP was a gift from Dr Thomas K Hemscheidt. The mimosine or HP concentration in a sample was determined by means of HPLC (Beckman model 110A) using a C_{18} column (Ranin Instrument Co., Woburn, MA) and an absorbance detector at 280 nm (Tangendjaja and Wills, 1980). Elution of mimosine or HP was obtained using a solvent system of 0.2% orthophosphoric acid in deionized water at a flow rate of 1 mL/min. In this system mimosine and HP peaks appeared approximately at 2.7 min and 4.8 min, respectively.

Results

Mimosine is toxic to most strains of Rhizobium

Strains belonging to various *Rhizobium* spp. were streaked on YEM agar containing 1 m*M* mimosine. Although almost all strains grew, most of them did not form single colonies and grew much slower than on YEM without mimosine, suggesting that mimosine is toxic to most strains. When 10 to 10^4 freshly-grown cells in 10 μL of these strains were spotted on YEM agar containing 1 m*M* mimosine, growth was not visible in most spots where fewer than 10^3 bacteria were used. When more than 10^3 cells of a sensitive strain were plated on the mimosine-containing medium, aggregates of many cells together could grow overcoming the toxic effects of mimosine. Such inhibitory effects of mimosine were also seen on TY agar. However, some *Leucaena*-nodulating strains of *Rhizobium* did not show inhibitory effects due to mimosine when they were spotted on YEM agar containing mimosine. Later, it was found that these strains degraded the toxin and used it as a source of carbon and nitrogen (Mid$^+$). TAL1145 is such a *Leucaena*-nodulating Mid$^+$ strain that was originally isolated from the nodules of *Leucaena diversifolia* in Australia. We isolated a mutant MS1246 that cannot degrade mimosine (Mid$^-$) by Tn3Ho*gus*-insertion in the *midD* gene of TAL1145. Strains TAL1145 and MS1246 were grown in TY broth containing 3 m*M* mimosine. Compared to the growth in TY broth without mimosine, the growth of TAL1145 in TY with mimosine was inhibited by 30% (Figure 1). However, an 85% inhibition was observed in the growth of the Mid$^-$ strain MS1246 when its growth was compared in TY medium with and without mimo-

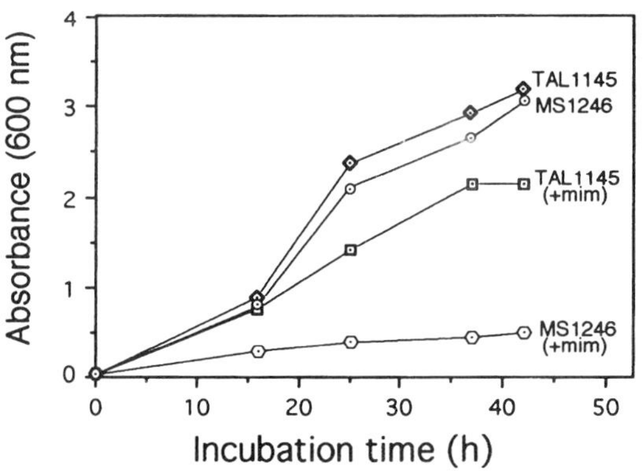

Figure 1. The effects of mimosine on the growth of the strain TAL1145 and MS1246. These strains were grown in TY broth with 3 m*M* mimosine (+ mim) and the growth of these cultures was compared with cultures grown in TY broth without mimosine.

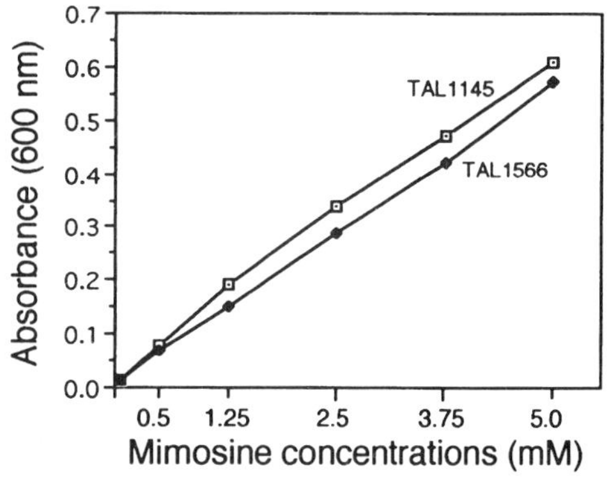

Figure 2. Growth of *Rhizobium* strains TAL1145 and TAL1566 in RM medium containing 0–5 m*M* mimosine. The cells were grown at 28°C for 48 h with shaking. It took 30–36 h for the cultures to attain stationary phase when 5 m*M* mimosine was present in the RM broth.

sine. Analysis of the culture supernatants by HPLC showed that strain MS1246 did not utilize any mimosine while TAL1145 utilized it completely. In another experiment, TY broth containing 3 m*M* mimosine was co-inoculated with an equal amount of cells of MS1246 and TAL1145 and grown for 72 h. The mixed culture was then diluted and plated on YEM agar containing different antibiotics to distinguish the colonies of the two strains. More than 90% of the colonies were of TAL1145, suggesting that mimosine in the medium provided a growth advantage to this strain by suppressing the growth of MS1246.

Some Leucaena-*nodulating* Rhizobium *strains can degrade mimosine*

Strains of different *Rhizobium* spp. were streaked on RM medium that contains mimosine as the sole source of carbon and nitrogen. Strains that can degrade mimosine (Mid$^+$) grew on the RM agar plates in 3–4 days. These strains were also tested in RM broth. Of the 92 strains originally isolated from the nodules of various *Leucaena* species, only 37 degraded mimosine (Table 1). When mimosine was completely utilized in the RM medium, the cultures turned from light yellow to colourless. This was verified by HPLC analyses of the culture media. *Rhizobium* NGR234 and *R. tropicii* CIAT899 that nodulate *Leucaena* did not utilize mimosine. Similarly, none of the *Rhizobium* strains isolated from the nodules of other mimosine-containing plants such as *Mimosa invisa* and *M. pigra* degraded mimosine. More than 200 strains belonging to different *Rhizobium* spp. were tested for growth in RM medium. None of these strains utilized mimosine.

When *Rhizobium* strains TAL1145 and TAL1566 were grown in RM medium, the final cell densities, determined by absorbancy at 600 nm, depended on the mimosine concentrations in the medium (Figure 2). The disappearance of mimosine from RM medium inoculated with these strains was monitored by HPLC. The time taken to utilize mimosine completely in the RM broth depended on the concentrations. For example, 1.25 m*M* mimosine disappeared in 15–18 h while 2.5 m*M* mimosine was completely utilized in 18–24 h. In the RM broth containing the highest concentrations of mimosine (5 m*M*), the entire amount disappeared completely in less than 36 hours whereas the mimosine concentrations of the RM broth containing the least concentration of mimosine (0.5 m*M*) inoculated with the Mid$^-$ strains remained unchanged even when the incubation was extended to 48 hours (data not shown).

The Mid$^+$ strains were screened for their ability to degrade HP which is known to be a degradation product of mimosine. Ten strains were identified that cannot utilize HP as a source of carbon and nitrogen (Pid$^-$) (Table 1). In this way, the *Leucaena*-nodulating rhizobia were classified into three groups on the basis of mimosine and HP degrading abilities: (i) Mid$^+$Pid$^+$, (ii) Mid$^+$Pid$^-$, and (iii) Mid$^-$Pid$^-$.

Table 1. Degradation of mimosine and HP by *Leucaena*-nodulating rhizobia collected from different geographic origin

Strains	Legume host	Origin	Mimosine	HP
TAL82	*Leucaena leucocephala*	Hawaii, USA	+	+
TAL582	*L. leucocephala*	Australia	+	+
TAL589	*L. leucocephala*	Hawaii, USA	+	+
TAL595	*L. leucocephala*	Hawaii, USA	+	+
TAL598	*L. leucocephala*	Hawaii, USA	+	+
TAL599	*L. leucocephala*	Hawaii, USA	+	+
TAL721	*L. leucocephala*	Colombia	+	+
TAL995	*L. leucocephala*	Philippines	+	+
TAL996	*L. leucocephala*	Philippines	+	+
TAL1005	*L. leucocephala*	Philippines	+	-
TAL1143	*L. leucocephala*	Hawaii, USA	+	-
TAL1145	*L. diversifolia*	Australia	+	+
TAL1564	*L. diversifolia*	Hawaii, USA	+	-
TAL1566	*L. diversifolia*	Hawaii, USA	+	+
TAL1571	*L. shannoni*	Hawaii, USA	+	-
TAL1861	*L. retusa*	Puerto Rico	+	+
TAL1862	*L. retusa*	Texas, USA	+	+
TAL1863	*L. retusa*	Texas, USA	+	+
TAL1864	*L. retusa*	Texas, USA	+	+
TAL1943	*L. macrophylla*	Honduras	+	-
TAL1951	*L. macrophylla*	Honduras	+	-
MS1	*L. leucocephala*	Hawaii, USA	+	+
MS2	*L. leucocephala*	Hawaii, USA	+	-
MS3	*L. leucocephala*	Hawaii, USA	+	-
MS4	*L. leucocephala*	Hawaii, USA	+	+
MS5	*L. leucocephala*	Hawaii, USA	+	-
MS8	*L. leucocephala*	Hawaii, USA	+	+
MS11	*L. leucocephala*	Hawaii, USA	+	+
MS12	*L. leucocephala*	Hawaii, USA	+	+
MS14	*L. leucocephala*	Hawaii, USA	+	+
MS15	*L. leucocephala*	Hawaii, USA	+	+
MS16	*L. leucocephala*	Hawaii, USA	+	+
MS17	*L. leucocephala*	Hawaii, USA	+	+
MS18	*L. leucocephala*	Hawaii, USA	+	+
MS19	*L. leucocephala*	Hawaii, USA	+	+
MS22	*L. leucocephala*	Hawaii, USA	+	-
MS23	*L. leucocephala*	Hawaii, USA	+	+

+, degrading; -, nondegrading. Degradation of mimosine and HP was assayed in RM medium RP medium containing 2 m*M* mimosine and 2 m*M* HP, respectively.

Physiological differences between Mid$^+$Pid$^+$ and Mid$^+$Pid$^-$ strains

MS22, a strain that could not utilize HP (Mid$^+$Pid$^-$) was studied in more detail. When grown in RM medium, this strain grew to the same cell density as TAL1145, indicating that MS22 utilized mimosine as efficiently as TAL1145 (Figure 3A). Strain 8002 which does not degrade either mimosine or HP (Mid$^-$Pid$^-$) was used as a negative control. To determine whether HP accumulates in MS22 cells, the cells grown in RM medium were harvested and lysed. HP could not be detected in the lysate of either MS22 or TAL1145, suggesting that mimosine was completely degraded by both strains (data not shown). When these two strains were grown in medium to which HP was added instead

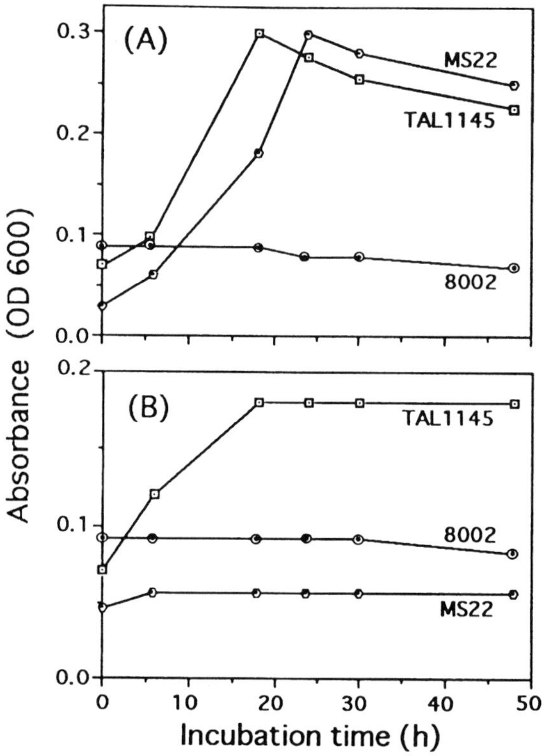

Figure 3. Utilization of mimosine and HP by Mid^+Pid^+ strain TAL1145 and Mid^+Pid^- strain MS22. Mid^-Pid^- strain 8002 was used as a control. Strains were grown in RM medium containing 2 m*M* mimosine (**A**) and RP medium containing 2 m*M* HP (**B**) as described in the Materials and methods section.

of mimosine, strain TAL1145 (Mid^+Pid^+) grew to a cell density of 0.18 within 18 hours whereas the cell density for the Mid^+Pid^- strain MS22 did not increase (Figure 3B). These cultures reached the stationary phase at a relatively low cell density because the amount of HP was exhausted.

Discussion

Mimosine is a naturally-occurring toxic compound present in the shoots and roots of *Leucaena*. We also found mimosine in the nodules of *Leucaena*. Mimosine or a mimosine-like compound is present in the root exudate of *Leucaena* (M Soedarjo and D Borthakur, unpublished results). However, only a limited number of *Rhizobium* strains can utilize mimosine as a selective growth substance. The ability to catabolize mimosine is not required for nodulation or nitrogen fixation since many strains that cannot catabolize mimosine, such as CIAT899 and MS13, can form nitrogen-fixing nodules on *Leucaena*. Mimosine utilization may be a specialized mechanism that some rhizobia and other microorganisms living in the rhizosphere of *Leucaena* have developed to survive. It is conceivable that the ability to catabolize mimosine provides a competitive advantage to Mid^+ strains of *Rhizobium* in the rhizosphere. We have shown that mimosine provides a growth advantage to the Mid^+ strain TAL1145 over the Mid^- strain MS1246 by suppressing the growth of the latter. Ability to degrade mimosine may provide another advantage to the Mid^+ strains by providing a source of carbon and nitrogen for growth. However, the Mid^+ strains do not utilize mimosine when they are growing in a medium containing tryptone and yeast extract. The mimosine-degrading enzyme activity in *Rhizobium* may be suppressed in the presence of other preferred sources of carbon and nitrogen. Recently, we did competition experiments using the Mid^+ strain TAL1145 and its Mid^- mutants and found that the Mid^- mutants are competition-defective (M Soedarjo and D Borthakur, unpublished results). This suggests that the ability to utilize mimosine provides a competitive advantage to strain TAL1145 in the rhizosphere. Mimosine may provide a source of carbon and nitrogen to the Mid^+ rhizobia and suppress the growth of other mimosine-sensitive microorganisms in the rhizosphere of *Leucaena*. Thus the *Leucaena* plants may provide a selective advantage to the Mid^+ strains by releasing mimosine in the root exudate.

We found that most Mid^+ strains also degrade HP which is a degradation product of mimosine. In our growth studies, strains such as MS22 that degrade mimosine but not HP were expected to accumulate HP when grown in RM medium. However, we did not detect HP in cells of MS22 growing on mimosine. On the other hand, strain MS22 grew to the same cell density as the Mid^+Pid^+ strain TAL1145 in RM medium. Since exogenous HP is not utilized by the strain MS22, we postulate that mimosine but not HP may be transported by this strain. Recently, we isolated several Mid^- strains of TAL1145 that cannot degrade mimosine but degrade HP, suggesting that mimosine degradation by *Rhizobium* is completed in at least two steps (D Borthakur and M Soedarjo, manuscript in preparation). In the first step, mimosine is converted to HP which may be then degraded to pyruvate. In the mutants that we have recently isolated, the first step of mimosine degradation is blocked. We used strain TAL1145 in most experiments because this is an effective and competitive strain for nodulation of *Leucae-*

na and it is genetically well characterized (George et al., 1994; Parveen and Borthakur, 1994; Pooyan et al., 1994). We are currently doing genetic analysis of mimosine and HP degradations in TAL1145.

Acknowledgements

This work was supported by the USDA Grant # 92-37305-7875 and a grant from the University of Hawaii Research Council. We would like to thank Thomas K Hemscheidt for providing HP and Leslie R Berger for critically reading the manuscript. This is paper number 4161 of the Hawaii Institute of Tropical Agriculture and Human Resources Journal Series, University of Hawaii, Honolulu.

References

Allison M J, Mayberry W R, Mcweeney C S and Stahl D A 1992 *Synergistes jonesii*, gen. nov., sp. nov: a rumen bacterium that degrades toxic pyridinediols. Syst. Appl. Microbiol. 15, 522–529.

Boehme S A and Lenardo M J 1993 Ligand-induced apoptosis of mature T lymphocytes (propriocidal regulation) occurs at distinct stages of the cell cycle. Leukemia 7, 45–49.

Brewbaker J L and Hylin J W 1965 Variations in mimosine content among *Leucaena* species and related mimosaceae. Crop. Sci. 5, 348–349.

Crounse R G, Maxwell R D and Blank H 1962 Inhibition of growth of hair by mimosine. Nature 194, 694–695.

Dai Y, Gold B, Vishwanatha J K and Rhode S L 1994 Mimosine inhibits viral synthesis through ribonucleotide reductase. Virology 205, 210–216.

Feldman S T and Schonthal A 1994 Negative regulation of Histone H1 kinase expression by mimosine, a plant amino acid. Cancer Res. 54, 494–498.

George M L C, Young J P W and Borthakur D 1994 Genetic characterization of *Rhizobium* sp. strain TAL1145 that nodulates tree legumes. Can. J. Microbiol. 40, 208–215.

Hegarty M P, Schinckel P G and Court R D 1964 Reaction of sheep to the consumption of *Leucaena glauca* Benth. and to its toxic principle mimosine. Aust. J. Agric. Res. 15, 153–167.

Hoffman B D, Hanauske-Abel H M, Flint A and Lalande M 1991 A new class of reversible inhibitors. Cytometry 12, 26–32.

Khanna K K and Lavin M F 1993 Ionizing radiation and UV induction of p53 protein by different pathways in ataxia-telangiectasia cells. Oncogene 8, 3307–3312.

Levenson V and Hamlin J L 1993 A general protocol for evaluating the specific effects of DNA replication inhibitors. Nucleic Acids Res. 21, 3997–4004.

Lowry J B, Maryanto and Tangendjaja B 1983 Optimising autolysis of mimosine to 3-hydroxy-4(1H)pyridone in green tissue of *Leucaena leucocephala*. J. Sci. Food Agric. 34, 529–533.

Mosca P J, Dijkwel P A and Hamlin J A 1992 The plant amino acid mimosine may inhibit initiation at origins of replication in Chinese hamster cells. Mol. Cell Biol. 12, 4375–4383.

Parveen N and Borthakur D 1994 Construction of a single-transposon-insertion mutant in *Rhizobium* sp. strain TAL1145 from a double-insertion mutant. Lett. Appl. Microbiol. 19, 142–145.

Perennes C, Qin L X, Glab N and Bergounioux C 1993 Petunia p34cdc2 protein kinase activity in G2/m cells obtained with a reversible cell cycle inhibitor, mimosine. FEBS Lett. 333, 141–145.

Pooyan S, George M L C and Borthakur D 1994 Isolation and characterization of a gene for nodule development linked to the *ndvA* and *ndvB* genes in *Rhizobium* sp. strain TAL1145. Symbiosis 17, 201–215.

Reis P J, Tunks D A and Hegarty M P 1975 Fate of mimosine administered orally to sheep and its effectiveness as a defleecing agent. Aust. J. Biol. Sci. 28, 495–501.

Soedarjo M, Hemscheidt T K and Borthakur D 1994 Mimosine, a toxin present in leguminous trees (*Leucaena* spp.) induces a mimosine-degrading enzyme activity in some strains. Appl. Environ. Microbiol. 12, 4268–4272.

Soedarjo M and Borthakur D 1996 Simple procedures to remove mimosine from young leaves, seeds and pods of *Leucaena leucocephala* used as food. Int. J. Food Sci. Technol. 36, 103–109.

Tangendjaja B and Wills R B H 1980 Analysis of mimosine and 3-hydroxy-4-(1H)-pyridone by high-performance liquid chromatography. J. Chromatogr. 202, 317–318.

Telfer J F and Green C D 1993 Placental alkaline phosphatase activity is inversely related to cell growth rate in HeLaS3 cervical cancer cells. FEBS Lett. 329, 238–2344.

Thompson J F, Morris C J and Smith I K 1969 New naturally occurring amino acids. Annu. Rev. Biochem. 38, 137–158.

Watters D J, Beamish H J, Marshall K A, Gardiner R A, Seymour G J and Lavin M F 1994 Accumulation of HL60 leukemia cells in G2/M and inhibition of cytokinesis caused by two marine compounds, bistratene A and cyclohexazoline. Cancer Chemo. Pharmacol. 33, 399–409.

Section editor: H Lambers
Guest editors: G.H. Elkan and R.G. Upchurch

Plant and Soil **186:** 93–98, 1996.

Subnanomolar concentrations of membrane chitolipooligosaccharides from *Rhizobium leguminosarum* biovar *trifolii* are fully capable of eliciting symbiosis-related responses on white clover

Guy G. Orgambide[1], Saleela Philip-Hollingsworth[1], Pedro F. Mateos[2], Rawle I. Hollingsworth[3] and Frank B. Dazzo[1]
[1]*Department of Microbiology, Michigan State University, East Lansing, MI 48824, USA**, [2]*Departamento de Microbiologia y Genetica, Facultad de Farmacia, Universidad de Salamanca, 37007 Salamanca, Spain and* [3]*Departments of Biochemistry and Chemistry, Michigan State University, East Lansing, MI 48824 USA*

Key words: chitolipooligosaccharide, *Rhizobium-Trifolium repens* symbiosis, root hair

Abstract

Axenic seedling bioassays were performed on white clover, vetch, and alfalfa to assess the variety and dose responses of biological activities exhibited by membrane chitolipooligosaccharides (CLOSs) from wild type *Rhizobium leguminosarum* bv. *trifolii* ANU843. Subnanomolar concentrations of CLOSs induced deformation of root hairs (Had) and increased the number of foci of cortical cell divisions (Ccd) in white clover, some of which developed into nodule meristems. In contrast, ANU843 CLOSs were unable to induce Had in alfalfa and required a 10^4-fold higher threshold concentration to induce this response in vetch. Also, ANU843 CLOSs were not mitogenic on either of these non-host legumes. In addition, CLOS action also increased chitinase activity in white clover root exudate. Thus, the membrane CLOSs from wild type *R. leguminosarum* bv. *trifolii* are fully capable of eliciting various symbiosis-related responses in white clover in the same concentration range as extracellular CLOSs of other rhizobia on their respective legume hosts. These results and our earlier studies indicate that membrane CLOSs represent one of many different classes of bioactive metabolites made by *R. leguminosarum* bv. *trifolii* which elicit more intense symbiosis-related responses in white clover than in other legumes. Therefore, CLOSs evidently play an important role in symbiotic development, but they may not be the sole determinant of host-range in the *Rhizobium*-clover symbiosis.

Abbreviations: Ccd-cortical cell division, CLOS-chitolipooligosaccharide, Had-root hair deformation.

Introduction

The most thoroughly examined class of bioactive Nod metabolites produced by *Rhizobium* consists of a family of chitolipooligosaccharides (CLOSs), first discovered in *R. meliloti* by Lerouge et al. (1990), and now documented for almost all known rhizobia (Carlson et al., 1994, and references therein). At subnanomolar concentrations, extracellular CLOSs can elicit root hair deformations (Had) and cortical cell divisions (Ccd) that may ultimately develop into nodule primordia (Relic et al., 1994; Truchet et al., 1991). Extracellular CLOSs have also been reported to increase the activity of chitinase enzymes in root exudate, which is potentially important because chitinases can selectively degrade certain CLOS molecules (Staehelin et al., 1995).

Although it is generally thought that CLOSs are excreted into the extracellular milieu, our studies have shown that CLOSs accumulate primarily in cell membranes of various wild type rhizobia, including *R. leguminosarum* bv. *trifolii* ANU843, *R. meliloti* 2011, and *R.* sp. NGR234 (Cedergren et al., 1995; Orgambide et al., 1995; Philip-Hollingsworth et al., 1995). In addition, wild type ANU843 accumulates bioactive diglycosyl diacylglycerols (DGDGs) in its membranes during flavone-activated; *nod*-related metabolism (Orgambide et al., 1994). These find-

* FAX No: +1517353 8953. E-mail: 23249mgr@msu.edu

ings raise the question of whether the membrane CLOSs exhibit the spectrum of activities and concentration dependence previously described for ANU843 DGDGs, and for extracellular CLOSs of other rhizobia. To address this question, we used quantitative microscopy to perform Had and Ccd bioassays of membrane CLOSs from ANU843 on axenic seedlings of white clover as the host, and on alfalfa and vetch as non-host legumes. We also assessed whether chitinase activity in root exudate is affected during growth of white clover plants in the presence of ANU843 CLOSs.

Materials and methods

Purification of membrane CLOSs

The extraction and purification of membrane CLOSs from *R. leguminosarum* bv. *trifolii* wild type strain ANU843 used in this study have been described elsewhere (Orgambide et al., 1995). A 10^{-5} *M* stock of CLOSs was prepared in NF medium, assuming a 1,000 D nominal molecular weight. The CLOS stock was steamed for 30 min and verified as sterile by plating aliquots on BIII medium and trypticase soy agar.

Axenic bioassays

ANU843 CLOSs were bioassayed for Had and Ccd activities on axenic seedlings of white clover (*Trifolium repens* L. cv Dutch), alfalfa (*Medicago sativa* cv. Gemini), and hairy vetch (*Vicia hirsuta*) under microbiologically controlled conditions previously described (Hollingsworth et al., 1989; Orgambide et al., 1994; Philip-Hollingsworth et al., 1991). Morphological responses were examined by conventional brightfield microscopy and laser scanning microscopy (Orgambide et al., 1994).

Influence of ANU843 CLOSs on chitinase activity in white clover root exudate

Roots of axenic white clover seedlings were grown through stainless steel wire platforms into sterile NF medium (Dazzo, 1982). Three days after planting, the growth medium was supplemented with sterile ANU843 CLOSs at a final concentration of 10^{-9} *M* and incubated for 1 additional day. Root exudates of CLOS-treated and untreated cultures were collected, concentrated by ultrafiltration using Amicon PM10 membranes, and clarified by centrifugation at 10,800 × *g* for 15 min at 4 °C. Chitinase activity was measured by a spectrofluorometric assay using 4-methylumbelliferyl-β-D-N,N′,N″-triacetylchitotriose substrate (Kuranda and Robbins, 1987). The presence of chitinase isozymes in root exudate was examined by native vertical gel electrophoresis followed by detection of enzyme activity using an agarose overlay technique with glycol-chitin as substrate and Calcofluor Fluorescent Brightner No. 28 (Sigma Chemical Co., St. Louis, MO) as a stain for residual substrate (Pan et al., 1991).

Results and discussion

Membrane CLOSs of R. leguminosarum *bv.* trifolii *ANU843 elicit Had and Ccd responses in white clover*

Figure 1 summarizes the large diversity of CLOS species isolated from membranes of wild type ANU843 (Orgambide et al., 1995; Philip-Hollingsworth et al., 1995). This family of CLOSs applied in the subnanomolar range elicited Had and Ccd responses on the host legume white clover, but not on the non-host legumes alfalfa or hairy vetch under our experimental conditions, which are optimized for white clover (Table 1). The threshold concentration of ANU843 CLOSs for induction of these responses on white clover was at or near 10^{-11} *M*, whereas the threshold concentration required for morphological responses on hairy vetch was approximately 10^4-fold higher. ANU843 CLOSs in the concentration range of 10^{-7} to 10^{-11} *M* did not elicit Had or Ccd responses on alfalfa above background levels (Table 1). These results indicate that Had and Ccd in the host white clover are elicited by ANU843 membrane CLOSs in the same concentration range as has been described for ANU843 DGDGs (Orgambide et al., 1994) and for extracellular CLOSs of other rhizobia on their respective legume hosts (Relic et al., 1994; Spaink et al., 1991; Truchet et al., 1991).

Various root hair deformations were elicited by ANU843 CLOSs applied to white clover roots at the optimum concentration of 5 × 10^{-9} *M* (Figure 2A-C), with an irregular distortion being most frequent (Figure 2B). Other types of CLOS-induced Had included corkscrews, swollen tips, branches, and combinations of these on the same root hairs (Figure 2B-C). This diversity of Had morphologies induced by CLOSs was very similar to the Had phenotypes induced on clover by a live inoculum of wild type *R. leguminosarum*

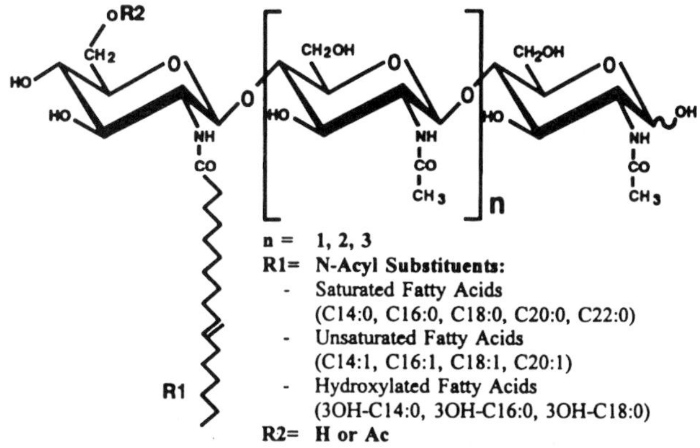

Figure 1. Summary of the structural diversity of CLOS species in membranes of wild type *R. leguminosarum* bv. *trifolii* ANU843 (from Orgambide et al., 1995, Philip-Hollingsworth et al., 1995).

Table 1. Host range and dose responses of root hair deformation (Had) and foci of cortical cell divisions (Ccd) induced by membrane chitolipooligosaccharides from *R. leguminosarum* bv. *trifolii* ANU843

Plant	ANU843 CLOS (*M*)	Response (X ± s.e.m.)	
		Had[a]	Ccd[b]
Dutch white clover	10^{-5}	1.3 ± 0.3	8.1± 0.7
	10^{-7}	2.2 ± 0.3	7.3 ± 0.9
	5×10^{-9}	2.9 ± 0.2	nd[c]
	10^{-9}	2.4 ± 0.2	5.7 ± 0.6
	10^{-11}	1.8 ± 0.2	4.9 ± 0.9
	10^{-13}	0.5 ± 0.2	1.6 ± 0.3
	0	0.2 ± 0.1	0.2 ± 0.1
Alfalfa	10^{-7}	0.1 ± 0.1	1.2 ± 0.3
	5×10^{-9}	0 ± 0	nd
	10^{-9}	0 ± 0	0.8 ± 0.3
	10^{-11}	0 ± 0	0.5 ± 0.3
	0	0 ± 0	1.0 ± 0.6
Vetch	10^{-7}	0.8 ± 0.2	1.5 ± 0.5
	10^{-9}	0 ± 0	1.0 ± 0.4
	10^{-11}	0.1 ± 0.1	0.8 ± 0.4
	0	0 ± 0	0.3 ± 0.3

[a]Had scores are % deformed root hairs: 0 = 0%, 1 = 1–25%, 2 = 26–50%, 3 = 51–75%, 4 = > 75%
[b]Ccd scores are number of foci per plant.
[c]nd, not determined.

bv. *trifolii* (Bhuvaneswari and Solheim, 1985; Li and Hubbell, 1969; Yao and Vincent, 1976). This general observation was confirmed by quantitative image analysis of the branching morphology as one Had phenotype (Figure 2C), which showed that the average geometric angle of the branch at the vertex connecting the two tips of the same root hair were very similar (71 ± 13 degrees [n = 51] for CLOS-induced; 67 ± 11 degrees for bacteria-induced (Dazzo and Petersen, 1989).

Figures 2D-I illustrate various developmental stages of foci of CLOS-induced Ccd in clover. The earliest Ccd responses occurred in well-isolated foci of the inner cortex (Figure 2E). Laser scanning microscopy

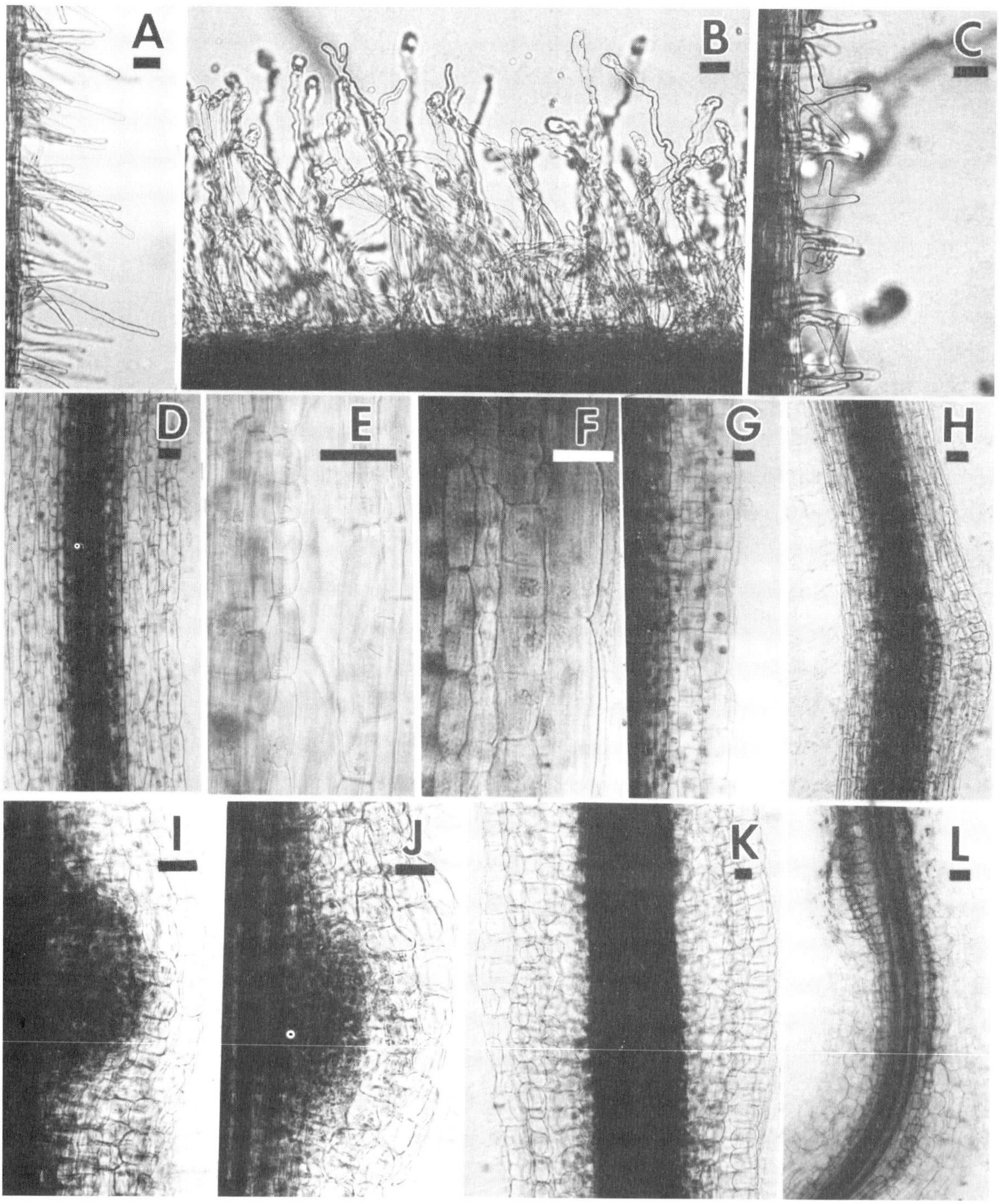

Figure 2. Induction of morphological responses on axenic seedling roots of white clover treated with ANU843 CLOSs. **A** and **D**. Untreated negative control for Had and Ccd bioassays, respectively. **B**. The most common type of root hair deformations. **C**. Examples of root hair branching used in image analysis. **E-I**. Development of foci of cortical cell divisions, showing a progression ranging from a few cells in the innermost cortical layer to multiple cortical cells on one side of the root, resembling a nodule-like meristem. **J**. Nodule primordium in the cortex of roots inoculated with live cells of ANU843. **K**. 'Collaroid' foci of cell divisions. **L**. Extensive cell divisions involving the pericycle, endodermis, and cortex. *Bar* scales are 50 μm.

showed the repeated anticlinal walls that have divided these separate cortical cells (Figure 2F). Figure 2G shows a more laterally extended zone of cell divisions involving multiple rows of the cortex, restricted to only one side of the root. Continued development of some of these foci of Ccd caused a progressive unilateral distension of the cortex (Figure 2H), occasionally resembling rudimentary nodule-like meristems (Figure 2I). However, under these axenic growth conditions, no fully developed nodules emerged from white clover roots treated with purified ANU843 CLOSs. For comparison, a true nodule primordium elicited by ANU843 bacteria is shown in Figure 2J.

ANU843 CLOSs also induced two other types of foci of cell divisions in axenic white clover roots. One structure was of the 'collaroid' type involving the entire ring of the cortex (Figure 2K). Although the collaroid type of Ccd was induced by CLOSs less frequently than the unilateral foci described above, it was the most common type of Ccd induced in white clover by ANU843 DGDGs (Orgambide et al., 1994). By minimizing the depth of field while controlling the light intensity, brightfield microscopy revealed another type of foci of cell divisions with extensive involvement of the pericycle, endodermis and cortex within the root (Figure 2L).

Membrane CLOSs of ANU843 elicit molecular responses in white clover

Extracellular CLOSs from *R. meliloti* have been shown to promote up to a 6-fold increase in level of chitinolytic activity in root exudate of its legume host, alfalfa (Staehelin et al., 1995). ANU843 CLOSs induced a 4.8-fold increase in chitinase activity in white clover root exudate (1.25 vs. 6.00 U mg^{-1} protein for untreated control and CLOS-treated plants, respectively). Thus, the ability of ANU843 membrane CLOSs to elicit this molecular response and its magnitude are similar to the response elicited by extracellular CLOSs of other rhizobia. Activity gel electrophoresis of concentrated root exudates from axenic seedlings grown with or without CLOSs revealed a single band of chitinase activity having the same electrophoretic mobility (figure not shown). In addition to these quantitative changes in symbiosis-related proteins described here, Bjourson and colleagues have recently found by an analysis of differentially displayed cDNA that ANU843 membrane CLOSs induce transcripts encoding various early nodulins in Dutch white clover roots (Bjourson et al., 1995). Considered collectively, these results indicate that ANU843 membrane CLOSs are able to elicit various molecular response in the white clover host.

Based on these findings, we conclude that the membrane chitolipooligosaccharides of wild type *R. leguminosarum* bv. *trifolii* are fully capable of eliciting Had, Ccd, and various molecular responses on white clover in the same concentration range as extracellular CLOSs of other rhizobia on their respective legume hosts. It is obvious from these and other data (Dazzo et al., 1991; Hollingsworth et al., 1989; Orgambide et al., 1994; Philip-Hollingsworth et al., 1991) that white clover displays symbiosis-related responses to many different classes of metabolites made by its compatible microsymbiont, *R. leguminosarum* bv. *trifolii*. Under our bioassay conditions, all of the above metabolites elicit more intense responses in white clover than in other legumes. In conclusion, *R. leguminosarum* bv. *trifolii* CLOSs evidently plays an important role in symbiotic development, but they may not be the sole determinant of host-range in the *Rhizobium*-clover symbiosis.

Acknowledgements

We thank A Squartini, K Nadler, J Salzwedel, K Ninke and A Bjourson for helpful suggestions, and J Palma and J Whallon for technical assistance. This work was supported by United States Department of Agriculture National Research Initiative Grant: No 94–03537 and by the Michigan Agricultural Experiment Station. This work is dedicated in memory of Professor David H Hubbell, who attended this 15th North American Symbiotic Nitrogen Fixation Conference in Raleigh, North Carolina (August 13–17, 1995) and died unexpectedly on August 26, 1995. His studies on the infection process in the *Rhizobium*-legume root nodule symbiosis were inspirational and continue to be a bright beacon light to a better understanding of plant-microbe interactions.

References

Bhuvaneswari T V and Solheim B 1985 Root hair deformation in the white clover/*Rhizobium trifolii* symbiosis. Physiol. Plant. 63, 25–34.

Bjourson A J, Cooper J E, Orgambide G and Dazzo F B 1995 Isolation of symbiosis-specific plant mRNA from *Trifolium repens* roots by DDRT-PCR and subtraction hybridization-PCR. *In* Nitrogen Fixation: Fundamentals and Applications. Eds. I

Tikhonovich, N Provorov, V Romanov and W Newton. p 493. Kluwer Academic Publishers, Dordrecht, the Netherlands.

Carlson R W, Price N P and Stacey G 1994 The biosynthesis of rhizobial lipo-oligosaccharide nodulation signal molecules. Mol. Plant-Microbe Interact. 7, 684–695.

Cedergren R, Lee J, Ross K and Hollingsworth R I 1995 Common links in the structure and cellular localization of *Rhizobium* chitolipooligosaccharides and general *Rhizobium* membrane phospholipid and glycolipid components. Biochemistry 34, 4467–4477.

Dazzo F B 1982 Leguminous root nodules. *In* Experimental Microbial Ecology. Eds. R Burns and J H Slater. pp 431–446. Blackwell Scientific Publications, Oxford, UK.

Dazzo F B and Petersen M A 1989 Applications of computer-assisted image analysis for microscopic studies of the *Rhizobium*-legume symbiosis. Symbiosis 7, 193–210.

Dazzo F B, Truchet G L, Hollingsworth R I, Hrabak E M, Pankratz H S, Philip-Hollingsworth S, Salzwedel J L, Chapman K, Appenzeller L, Squartini A, Gerhold D and Orgambide G G 1991 *Rhizobium*-lipopolysaccharide modulates infection thread development in white clover root hairs. J. Bacteriol. 173, 5371–5384.

Hollingsworth R, Squartini A, Philip-Hollingsworth S and Dazzo F B 1989 Root hair deforming and nodule initiating factors from *Rhizobium trifolii*. *In* Signal Molecules in Plants and Plant-Microbe Interactions. Ed. B J J Lugtenberg. pp 387–393. Springer-Verlag, Berlin, Germany.

Kuranda M J and Robbins P W 1987 Cloning and heterologous expression of glycosidase genes from *Saccharomyces cerevisiae*. Proc. Natl. Acad. Sci. USA 84, 2585–2589.

Lerouge P, Roche P, Faucher C, Maillet F, Truchet G, Promé J-C and Dénarié J 1990 Symbiotic host-specificity of *Rhizobium meliloti* is determined by a sulphated and acylated glucosamine oligosaccharide signal. Nature 344, 781–784.

Li D and Hubbell D H 1969 Infection thread formation as a basis of nodulation specificity in *Rhizobium*-strawberry clover association. Can. J. Microbiol. 15, 1133–1136.

Orgambide G G, Lee J, Hollingsworth R I and Dazzo F B 1995 Structurally diverse chitolipooligosaccharide Nod factors accumulate primarily in membranes of wild type *Rhizobium leguminosarum* biovar. *trifolii*. Biochemistry 34, 3832–3840.

Orgambide G G, Philip-Hollingsworth S, Hollingsworth R I and Dazzo F B 1994 Flavone-enhanced accumulation and symbiosis-related biological activity of a diglycosyl diacylglycerol membrane glycolipid from *Rhizobium leguminosarum* biovar *trifolii*. J. Bacteriol. 176, 4338–4347.

Pan S Q, Ye X S, and Kuc J 1991 A technique for detection of chitinase, β-1,3-glucanase, and protein patterns after a single separation using polyacrylamide gel electrophoresis or isoelectrofocusing. Phytopathology 81, 970–974.

Philip-Hollingsworth S, Hollingsworth R I and Dazzo F B 1991 N-acetylglutamic acid: an extracellular *nod* signal of *Rhizobium trifolii* ANU843 that induces root hair branching and nodule-like primordia in white clover roots. J. Biol. Chem. 266, 16854–16858.

Philip-Hollingsworth S, Orgambide G G, Bradford J J, Smith D K, Hollingsworth R I and Dazzo F B 1995 Mutation or increased copy number of *nodE* has no effect on the spectrum of chitolipooligosaccharide Nod factors made by *Rhizobium leguminosarum* bv. *trifolii*. J. Biol. Chem. 270, 20968–20977.

Relić B, Talmont F, Kopcinska J, Golinowski W, Promé J C and Broughton W J 1994 Biological activity of *Rhizobium* sp. NGR234 Nod-factors on *Macroptilium atropurpureum*. Mol. Plant-Microbe Interact. 6, 764–774.

Spaink H P, Sheeley D M, van Brussel A A N, Glushka J, York W S, Tak T, Geiger O, Kennedy E P, Reinhold V N and Lugtenberg B J J 1991 A novel highly unsaturated fatty acid moiety of lipo-oligosaccharide signals determines host specificity of *Rhizobium*. Nature 354, 125–130.

Staehelin C, Schultze M, Kondorosi E and Kondorosi A 1995 Lipo-chitooligosaccharide nodulation signals from *Rhizobium meliloti* induce their rapid degradation by the host plant alfalfa. Plant Physiol. 108, 1607–1614.

Truchet G, Roche P, Lerouge P, Vasse J, Camut S, De Billy F, Promé J C and Dénarié J 1991 Sulphated lipooligosaccharide signals of *Rhizobium meliloti* elicit root nodule organogenesis in alfalfa. Nature 351, 670–673.

Yao P Y and Vincent J M 1976 Factors responsible for curling and branching of clover root hairs by *Rhizobium*. Plant and Soil 45, 1–16.

Section editor: H Lambers
Guest editors: G.H. Elkan and R.G. Upchurch

Plant and Soil **186:** 99–106, 1996.

Analysis of the two nodulins, sucrose synthase and ENOD2, in transgenic *Lotus* plants

L. Skøt, F.R. Minchin, E. Timms, M.T. Fortune, K.J. Webb and A.J. Gordon
*Institute of Grassland and Environmental Research, Plas Gogerddan, Aberystwyth, Dyfed, SY23 3EB, Wales, UK**

Key words: antisense RNA, ENOD2, *Lotus*, nodulin, oxygen diffusion barrier, sucrose synthase

Abstract

A partial sucrose synthase clone was isolated from a nodule cDNA library of *Lotus japonicus*, which had strong homology to sucrose synthase genes from *Vicia faba* and other plant species. The transcript was 2.7 kb in size and expression was enhanced in nodules compared to roots of *L. japonicus*. Two overlapping genomic clones containing sucrose synthase sequences were isolated from *L. japonicus* with a view to mapping and analysing the promoter of this gene.

The early nodulin ENOD2 is a (hydroxy)proline-rich cell wall protein, which is expressed in the inner cortex surrounding the infected region of root nodules. In order to see if this nodulin is involved in the operation of the oxygen diffusion barrier, *L. corniculatus* was transformed with antisense ENOD2 gene constructs derived from the promoter and coding sequence of soybean. Two of the 20 antisense plant lines had significantly lower dry weight and nitrogenase activity as well as a different O_2 profile through the nodules when compared to control plant lines, but there was no difference in ENOD2 transcriptional activity between the control and antisense plants. Possible reasons for this observation are discussed.

Abbreviations: kDa - kilodalton, kb - kilobases, CAT - chloramphenicol acetyl transferase, CaMV - Cauliflower Mosaic Virus; Nos - Nopaline Synthase.

Introduction

Symbiotic nitrogen fixation in legume root nodules is a very energy-intensive process that relies on a continuous supply of photosynthates in the form of sucrose to provide a source of ATP and reductants, as well as carbon skeletons to assimilate fixed nitrogen. In root nodules, sucrose may be cleaved by either sucrose synthase (UDP-glucose: D-fructose-2-glucosyl-transferase, EC 2.4.1.13) to UDP-glucose and fructose, or by alkaline invertase (β-D-fructofuranoside fructohydrolase, EC 3.2.1.26) to glucose and fructose, the former being energetically more favourable (Gordon,1992). Sucrose synthase (SS) has been characterized from soybean root nodules as a homotetrameric enzyme with a subunit molecular weight of 90 kDa (Morell and Copeland, 1985). Thummler and Verma (1987) showed that the nodulin NGm-100 from soybean nodules was identical to the SS subunit, and that it was 20 times more abundant in nodules than in roots. A partial cDNA clone encoding this nodulin had high homology with maize SS (Thummler and Verma, 1987). The first full length cDNA encoding SS was isolated from a nodule specific cDNA library of *Vicia faba* (Küster et al., 1993). This cDNA contained a reading frame encoding a protein with a predicted molecular weight of 92 kDa.

The suggested importance of this enzyme for nodules is underlined by the fact that transcript levels, protein concentration and enzyme activity of SS decrease more drastically than other nodulins such as aspartate aminotransferase, glutamine synthetase and leghaemoglobin in response to environmental stresses such as salt, NO_3, drought and defoliation (Gordon et al., 1993; Gonzalez et al., 1995; Gordon and Skøt, *in preparation*). We describe here the isolation of a partial cDNA clone, with strong homology to other plant sucrose synthase genes, from a nodule cDNA library of *Lotus japonicus*. The aim of this work is to analyse how this gene is regulated in nodules using the newly

* FAX No: +44 1970 828357; E-mail: leif.skot@bbsrc.ac.uk

developed *Agrobacterium tumefaciens* mediated transformation system of *Lotus japonicus* (Handberg and Stougaard, 1992; Webb et al., 1994).

The need for aerobic metabolism in root nodules necessitates the operation of mechanisms to protect the oxygen sensitive nitrogenase enzyme from irreversible damage by O_2. One of these is the oxygen carrying protein leghaemoglobin which facilitates the transfer of O_2 within the infected cells to the site of oxidative phosphorylation in the bacteroids (Hunt and Layzell, 1993). In addition to this, a large body of evidence based on physiological experiments and mathematical analyses strongly suggest that a variable gaseous diffusion barrier exists to regulate flux of O_2 to the infected cells (Hunt and Layzell, 1993; Witty et al., 1986). How the barrier operates has not been established, but several hypotheses, none of them mutually exclusive, have been proposed. A model incorporating the various hypotheses was described by Minchin (1996). This "rolling wave" hypothesis involves infected region components consisting of changes in affinity of Lb to O_2; changes in mitochondrial respiration rates, and glycoprotein occlusions in intercellular spaces between the infected cells, perhaps involving cross-linking to lectins and flavonoids. If these responses are insufficient to deal with a change in the internal O_2 concentration (O_i), then components in the nodule cortex would come into effect. They probably include an osmocontractile response of cells in the inner cortex, and intercellular space occlusions formed by cross-linking involving glycoproteins, lectins, isoflavanoids or (hydroxy)proline-rich proteins in the mid-cortex.

The early nodulin gene ENOD2 encodes a (hydroxy)proline rich cell wall protein whose expression is induced during nodule morphogenesis (Franssen et al., 1987). In situ hybridisation studies and promoter-reporter gene analysis in transgenic *Lotus corniculatus* and white clover root nodules showed that this gene is expressed in the inner and mid-cortex where the barrier is located (Lauridsen et al., 1993; Van de Wiel et al., 1990). This has led to the suggestion that the ENOD2 gene product is involved in the establishment and operation of the barrier (Van de Wiel et al., 1990). To test this hypothesis, *Lotus corniculatus* was transformed with antisense ENOD2 DNA constructs under the control of its own promoter. Here we report some results of the analysis of the transgenic antisense plant lines.

Materials and methods

Plant transformations and growth conditions

Agrobacterium tumefaciens strain LBA4404 containing a binary vector with the antisense sucrose syrithase construct was used to transform *Lotus japonicus* using the procedure developed by Handberg and Stougaard (1992) and adapted in our laboratory (Webb et al., 1994). *Lotus corniculatus* was transformed with ENOD2 constructs using *A. rhizogenes* and the hairy root system first described by Stougaard et al. (1986) with modifications as described (Webb et al., 1990). The control plants were transformed with the β-glucuronidase reporter gene (*uidA*) under the control of the ENOD2 promoter from soybean (Lauridsen et al., 1993). The antisense plants contained the reversed soybean ENOD2 coding sequence instead of the *uidA* gene. Regenerated transformants were planted in pots with vermiculite and inoculated with *Rhizobium loti*. The plants were grown in a growth room with the following regime: 16/8 hour day/night and 20 °C/20 °C with fluorescent and incandescent light sources. Soybean growth conditions and stress treatments were as described by Gonzalez et al. (1995).

Nucleic acid manipulations and CAT assay

Extraction of DNA from plant tissue was as described by Handberg and Stougaard (1992). RNA extractions and Northern blot hybridisation analysis was carried out as described by Gordon et al. (1993) using cDNA probes labelled with ^{32}P by the random prime method. Nucleotide sequencing was carried out using an automated sequencer and a reaction kit (Applied Biosystems 373A). Nucleotide sequence analysis was carried out using the manual for the EGCG (VMS) 8.0 program package (Peter Rice, The Sanger Centre, Hinxton Hall Cambridge, UK). A CAT assay was performed on extracts from hairy root cultures using a ^{14}C labelled chloroamphenicol precursor as described (Stougaard et al., 1986).

Physiological analyses

Measurements of oxygen profiles through root nodules were performed with oxygen microelectrodes as described (Witty et al., 1987). Nitrogenase activity of whole plants was measured using the flow-through system described by Minchin et al. (1992).

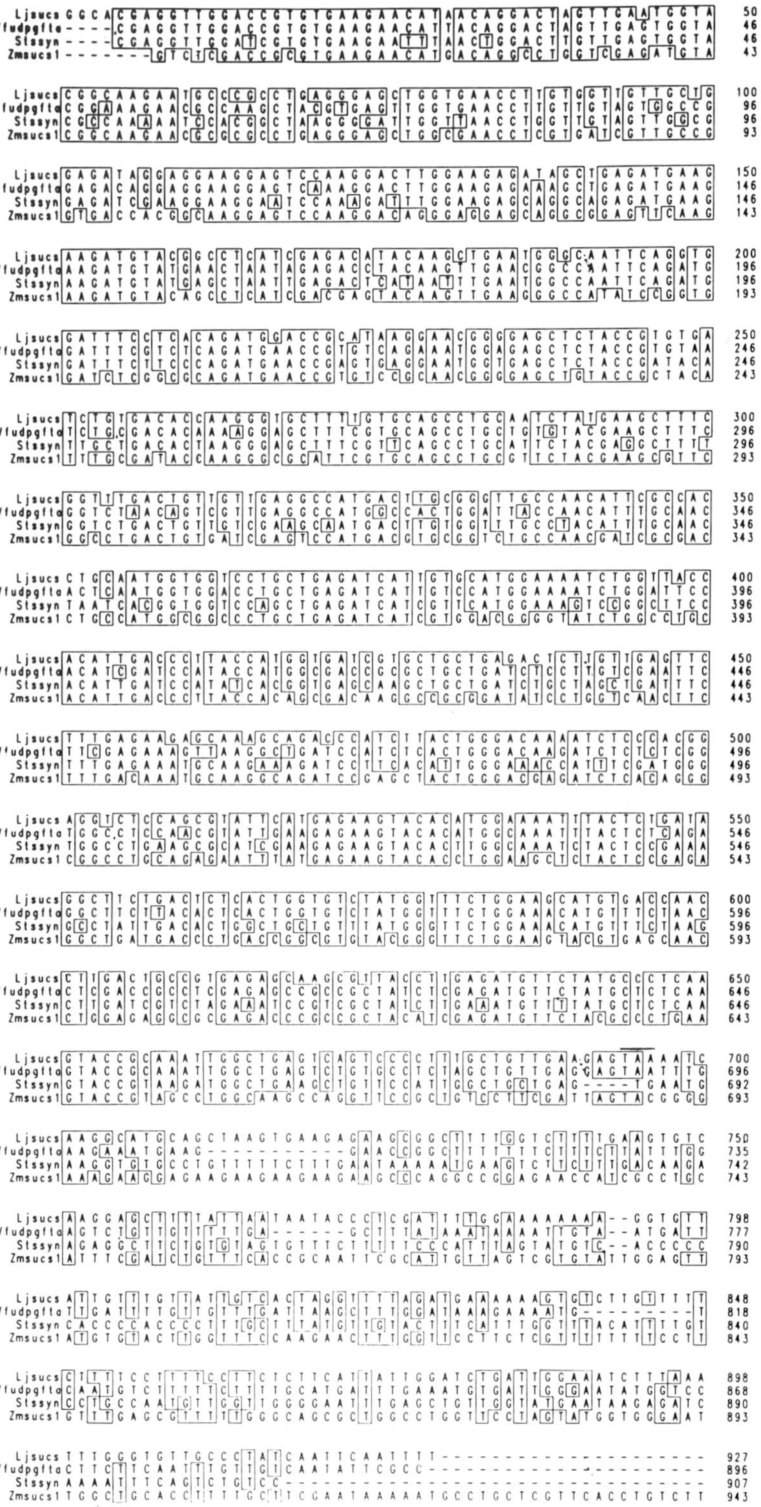

Figure 1. Nucleotide sequence of the partial cDNA clone Ljsucs from *L. japonicus*, and comparison to three other sucrose synthase cDNAs. Vfudpgfta - broadbean (Küster et al., 1993); Stssyn - potato (Salanoubat and Belliard, 1987); Zmsucs1 - maize (Werr et al., 1985). A putative termination codon has been overlined.

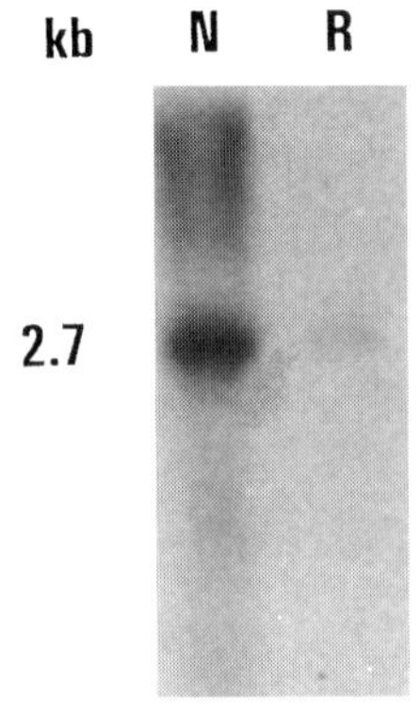

Figure 2. Northern blot hybridisation analysis of 10 μg of total RNA from nodules and roots of *Lotus japonicus* using the sucrose synthase cDNA clone as a probe.

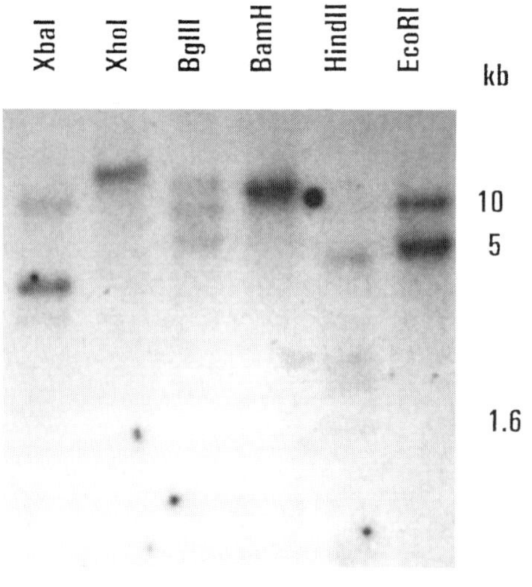

Figure 3. Southern blot hybridisation analysis of total DNA (10 μg per lane) from *Lotus japonicus* digested with various restriction endonucleases using the Ljsucs cDNA as a probe.

Results

Sucrose synthase

Using a partial cDNA clone of the sucrose synthase nodulin gene from soybean (Thummler and Verma, 1987) as a probe, a 1 kb cDNA clone was isolated from a custom made λ-Uni-ZAP^{TM}XR (Stratagene, La Jolla, California, USA) cDNA library of *Lotus japoncicus* root nodules. The nucleotide sequence of this cDNA, Ljsucs, had strong homology to several sucrose synthase sequences in the database. Figure 1 shows the sequence of Ljsucs in comparison to sucrose synthase sequences from broad bean (76.9% homology), potato (74.6% homology) and maize (73.0% homology). Northern blot hybridisation analysis with Ljsucs as a probe revealed a strong signal of 2.7 kb in RNA from nodules of *L. japonicus* and a weaker one from roots (Figure 2). This result, together with the homology data, suggest strongly that Ljsucs encodes the 3′ end of a nodule enhanced sucrose synthase mRNA. Southern blot hybridisation analysis with total DNA from *L. japonicus* cut with various restriction endonucleases suggests the presence of a small gene family (Figure 3).

The 1 kb cDNA clone was used to make an antisense plant transformation construct in the binary vector pROK2, between the CaMV 35S promoter and the Nos terminator. *Lotus japonicus* has been transformed with this construct, and plants are currently regenerating. Using the same clone as a probe, two overlapping genomic clones of 14.4 and 15 kb in size were isolated from a custom made genomic library in λFIX^{TM} II (Stratagene, La Jolla California, USA). At present these clones are being mapped with a view to identify and analyse the 5′ promoter region of the sucrose synthase gene.

ENOD2

This transformation was carried out using a cointegration vector which also contained the CAT gene driven by the CaMV 35S promoter in the T-DNA (Hansen et al., 1989). A CAT assay was used to screen the hairy root lines for the presence of the transgenes. Twenty CAT-positive antisense lines were regenerated and analysed further. The dry weight and nitrogenase activity of six of the antisense lines 8 weeks after planting is shown in Figure 4. The two lines A2-5 and A3-5 had significantly lower dry weight and nitrogenase activity compared to the control line. Antisense plants with a defective oxygen barrier would be expected to increase the nitrogenase-linked respiration linearly in response to elevated O_2 concentrations until inhibitory levels were reached in the infected tissue whereas normal plants would be able to keep the nitrogenase-linked respiration more or less constant by closing the barrier. However, there were no consistent differences in this characteristic between the antisense plants and the control. The oxygen profiles of nodules from the two antisense lines with lower dry weight and nitrogenase activiity were consistently different from those of the control line. The zone in which the rapid decline in O_2 concentration could be seen was wider in the two antisense lines (Figure 5). This phenotype indi-

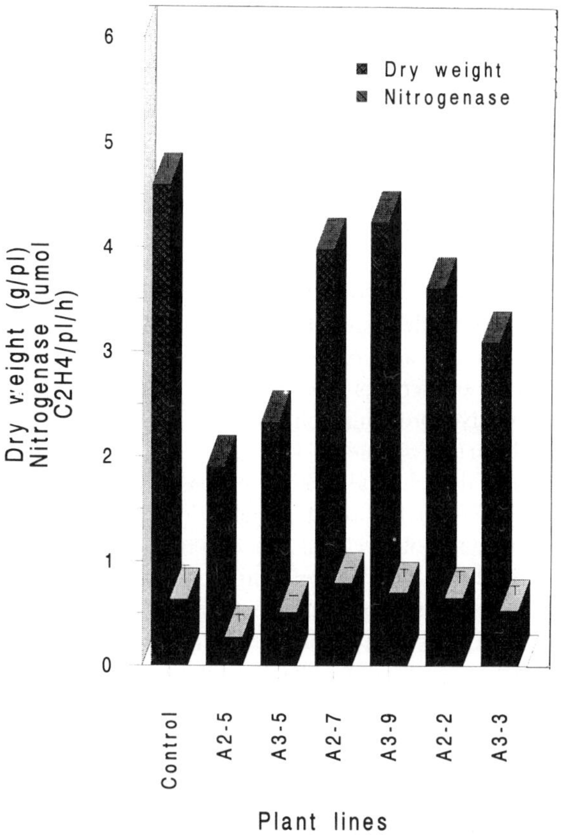

Figure 4. Dry weight and nitrogenase activity of transgenic control and antisense ENOD2 *Lotus corniculatus* plant lines 8 weeks after planting. Values of the control, A2-5 and A3-5 samples represent the mean and SE of 6 replicate plants from two experiments. The other samples represent the mean and SE of 3 replicate plants.

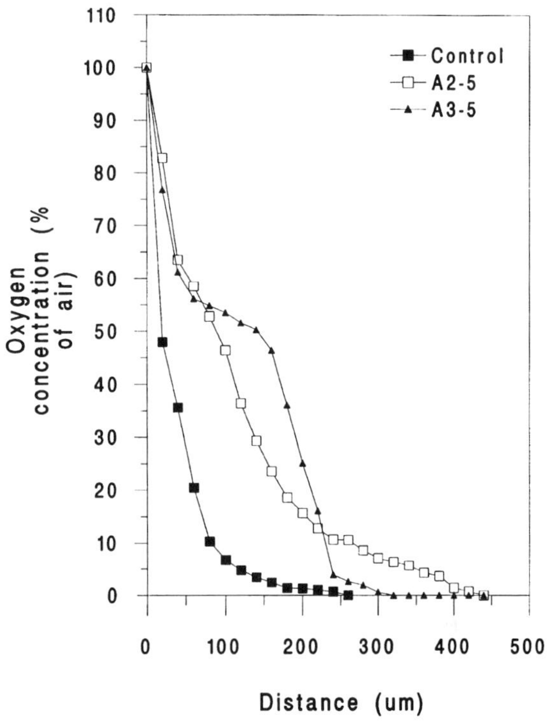

Figure 5. Profile of O_2 concentration through nodules from transgenic ENOD2 control and antisense plant lines. The profiles are representative of at least 5 tracings through different nodules from each plant line.

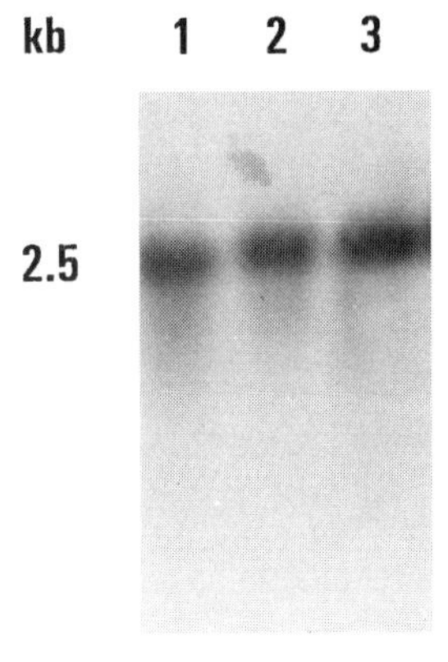

Figure 6. Northern blot hybridisation analysis of 10 μg of total RNA from nodules of transgenic control and antisense ENOD2 *L. corniculatus* plant lines using the cNA clone pGmENOD2 as a probe. Lane 1 - Control; Lane 2 - A2-5; Lane 3 - A3-5.

cated that the gaseous diffusion barrier was defective, and that measurable O_2 levels were encountered in the infected zone.

Northern blot hybridisation analysis of RNA isolated from root nodules of the transgenic plants did not indicate a significant difference in transcript levels of the indigenous *Lotus* ENOD2 gene (Figure 6). The size of this transcript was 2.5 kb, whereas the antisense transcript has a predicted size of approximately 1.2 kb. Using a cDNA, clone pGmENOD2 of the soybean ENOD2 gene as a probe (Van de Wiel et al., 1990), allows the detection of the antisense as well as the sense transcript, but there was no consistent signal of that size detected in RNA from any of the antisense lines (Figure 6).

In order to analyse further the possible involvement of the ENOD2 gene product in the operation of the barrier, wild type soybean plants were exposed to various environmental stresses which are known to cause a change in the gaseous diffusion resistance into root nodules. Exposure of the root system to 50% O_2

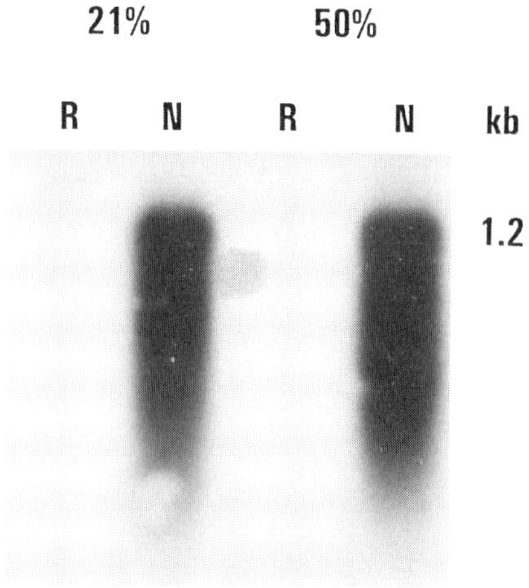

Figure 7. Northern blot hybridisation analysis of ENOD2 transcript level, in soybean nodules (N) and roots (R) before and after exposure to 50% O_2 for 24 hours.

for 24 hours did not result in any significant change in ENOD2 transcript levels (Figure 7). However, nodules from plants exposed to drought did have higher levels of ENOD2 transcript than control plants (Figure 8). The same trend was evident after salt stress, but less pronounced, whereas NO_3 resulted in an increase in one of the 3 replicate samples (Figure 8).

Discussion

Sucrose synthase

There are three reasons to suggest that the Ljsucs cDNA described here represents a partial nodule enhanced sucrose synthase mRNA from *L. japonicus*. Firstly, it has strong homology to sucrose synthase sequences from many other plant species, in particular the nodule enhanced cDNA from broadbean (Küster et al., 1993). Secondly, the size of the transcript is similar to that of other sucrose synthase mRNAs. Thirdly, transcript levels were higher in nodules than in roots. It is, possible, however, that relative levels of expression in nodules, roots and other tissue may vary with plant age.

The Southern blot hybridisation data suggest the presence of a small gene family in *L. japonicus*. How the expression of this gene is regulated, must await the analysis of the promoter region of the genomic sucrose synthase DNA, which we have isolated.

ENOD2

This work shows that ENOD2 is expressed in functional root nodules, and not just during nodule morphogenesis. This is a prerequisite for a gene product to be involved in the operation of the gaseous diffusion barrier. The two antisense plant lines A2-5 and A3-5 were different from the coritrol lines. Both the dry weight and nitrogenase activity was lower and the oxygen profiles through the nodules from these lines were consistently different. This phenotype suggests that the barrier was affected.

A glycoprotein identified in the cortex of pea and soybean root nodules and in the infection thread matrix (VandenBosch et al., 1989) was the first to be implicated in the operation of the barrier. This was based on microscopic observations of soybean root nodules harvested from plants grown at different oxygen levels for 28 days. Immunological labelling demonstrated that more glycoprotein occluded the intercellular airspaces in the inner cortex of nodules from plants grown at 40% O_2 than at 21% and 10% (James et al., 1991). Similar results were obtained with lupin nodules, where an increase in the diffusion resistance, induced by lowering the root temperature, removal of shoots, darkening or NO_3^-, was correlated with an increase in glycoprotein occlusion of the intercellular airspaces of the inner cortex (Iannetta et al., 1993). One hypothesis to explain how this occlusion occurs, involves cross-linking of the glycoprotein or ENOD2. This phenomenon was observed in soybean cell cultures with the glycoprotein and a (hydroxy)proline rich cell wall protein similar to ENOD2, 2-5 minutes after being elicited with glutathione or fungal elicitors (Bradley et al., 1992). Such cross-linking could result in the formation of a water-filled "gel" in the intercellular air-spaces, leading to higher resistance to gas diffusion (Minchin, 1996). If ENOD2 is involved in the operation of the barrier, a decrease in the expression of this gene would be expected to lead to less cross-linking and therefore to a less efficient barrier, where the fall in free O_2 is more gradual, which was what we observed in nodules from plant lines A2-5 and A3-5 (Figure 5). However, these two lines exhibited similar levels of ENOD2 mRNA to the controls (Figure 6). This suggests that the observed phenotype is not caused by a decrease in ENOD2 transcription. It is, however, impossible to draw any conclusions regarding the levels of the ENOD2 gene product. This would require the use of ENOD2-specific antibodies.

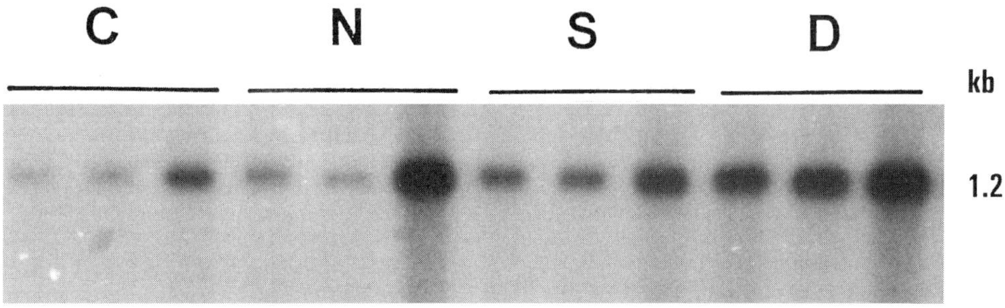

Figure 8. Northern blot hybridisation analysis of ENOD2 transcript level, in soybean nodules isolated from plants after exposure to various environmental stresses for 6 days. C - control; N - 10 m*M* NO_3^- ; S - 150 m*M* NaCl; D - drought.

The apparent lack of effect of antisense RNA on the levels of sense RNA could be due to the heterologous origin of the transgenic ENOD2 DNA, although there was a good degree of homology to the Lotus DNA as judged by the hybridisation analyses. However, the size of the *Lotus* transcript was 2.5 kb, whereas the soybean transcript was 1.2 kb. Whether this is due to a difference in the number of pentapeptide repeat units, which these genes mainly consist of, or a qualitative difference, could influence the homology and consequently the antisense strategy.

Although it is impossible to conclude that the phenotypes described here are caused by an antisense effect, it is also difficult to dismiss them as artefacts of the transformation. Firstly, the effect was observed in two independent plant lines, ruling out a positional effect of T-DNA insertion. Secondly, previous transformations of *L. corniculatus* with wild type *Agrobacterium rhizogenes* strains have demonstrated that the root nodules were similar to the control root nodules in relevant physiological properties (Webb et al., 1990). Furthermore, the control plants used in this work were transformed with the ENOD-prornoter-*uidA* gene construct.

The fact that expression of the ENOD2 gene in soybean was not affected by 24 hours exposure to 50% O_2 does not rule out its involvement in the operation of the barrier. The response could be post-transcriptional involving no changes in transcriptional activity. Drought-stressed soybean plants displayed increased levels of ENOD2 mRNA, suggesting that this gene is involved in responses to certain environmental stresses. Further work is obviously needed to elucidate the role of ENOD2. Experiments with the transgenic ENOD2-promoter-*uidA*-fusion plants are in progress to answer this question.

Acknowledgements

IGER is grant aided by the Biotechnology and Biological Sciences Research Council. We wish to thank Peter Lauridsen and Henk Franssen for ENOD2 plant transformation constructs and cDNA clone, Jens Stougaard for cDNA and genomic libraries of *Lotus japonicus*, and Sue Mizen for technical assistance.

References

Bradley D J, Kjellbom P and Lamb C J 1992 Elicitor- and wound-induced oxidative cross-linking of a proline-roch plant cell wall protein: a novel, rapid defense response. Cell 70, 21–30.

Franssen H, Thomson D V, Idler K, Kormelink R, van Kammen A and Bisseling T 1987 Nucleotide sequence of two soybean ENOD2 early nodulin genes encoding Ngm-75. Plant Mol. Biol. 14, 103–106.

Gonzalez E M, Gordon A J, James C L and Arrese-Igor C 1995 The role of sucrose synthase in the response of soybean nodules to drought. J. Exp. Bot. 46, 1515–1523.

Gordon A J 1992 Carbon metabolism in the legume nodule. *In* Carbon Partitioning within and between Organism. Eds. C J Pollock, J F Farrar and A J Gordon. pp 133–162. BIOS Scientific Publishers, Oxford, UK.

Gordon A J, Ougham H J and James C L 1993 Changes in levels of gene transcripts and their corresponding proteins in nodules of soybean plants subjected to dark-induced stress. J. Exp. Bot. 44, 1453–1460.

Handberg K and Stougaard J 1992 *Lotus japonicus*, an autoganous, diploid legume species for classical and molecular genetics. Plant J. 2, 487–496.

Hansen J, Jørgensen J-E, Stougaard J and Marcker K A 1989 Hairy roots - a short cut to transgenic root nodules. Plant Cell Rep. 8, 12–15.

Hunt S and Layzell D R 1993 Gas exchange of legume nodules and the regulation of nitrogease activity. Annu. Rev. Plant Physiol. Plant Mol. Biol. 44, 483–511.

Iannetta P P M, de Lorenzo C, James E K, Fernandez-Pascual M, Sprent J I, Lucas M M Witty J F, de Felipe M R and Minchin F R 1993 Oxygen diffusion in lupin nodules. J. Exp. Bot. 44, 1461–1467.

James E K, Sprent J I, Minchin F R and Brewin N J 1991 Intercellular location of glycoprotein in soybean nodules: effect of altered rhizosphere oxygen concentration. Plant Cell Environ. 14, 467–476.

Küster H, Fruhling M, Perlick A M and Pühler A 1993 The sucrose synthase gene is predominantly expressed in the root nodule tissue of *Vicia faba*. Mol. Plant-Microbe Interact. 6, 507–514.

Lauridsen P, Franssen H, Stougaard J, Bisseling T and Marcker K A 1993 Conserved regulation of the soybean early nodulin ENOD2 gene promter in determinate and indeterminate transgenic root nodules. Plant J. 3, 483–492.

Minchin F R 1996 Regulation of oxygen diffusion in legume nodules. Soil Biol. Biochem. (*In press*).

Minchin F R, Iannetta P P M, Fernandez-Pascual M, de Lorenzo C, Witty J F and Sprent J I 1992 A new procedure for the calculation of oxygen diffusion resistance in legume nodules from flow-through gas analysis data. Ann. Bot. 70, 283–289.

Morell M and Copeland L 1985 Sucrose synthase of soybean nodules. Plant Physiol. 78, 149–154.

Salanoubat M and Belliard G 1987 Molecular cloning and sequencing of sucrose synthase cDNA from potato (*Solanum tuberosum* L.): Preliminary characterization of sucrose synthase mRNA distribution. Gene 60, 47–56.

Stougaard J, Marcker K A, Otten L and Schell J 1986 Nodule-specific expression of a chimaeric soybean leghaemoglobin gene in transgenic *Lotus corniculatus*. Nature 321, 669–674.

Thummler F and Verma D P S 1987 Nodulin-100 of soybean is the subunit of sucrose synthase regulated by the availability of free heme in nodules. J. Biol. Chem. 262, 14730–14736.

Van de Wiel C, Scheres B, Franssen H, van Lierop M J, van Lammeren A, van Kammen A and Bisseling T 1990 The early nodulin transcript ENOD2 is located in the nodule parenchyma (inner cortex) of pea and soybean root nodules. EMBO J. 9, 1–7.

VandenBosch K A, Bradley D A, Knox J-P, Perotto S, Butcher G W and Brewin N J 1989 Common components of the infection thread matrix and the intercellular space identified by the immunocytochemical analysis of pea nodules and uninfected roots. EMBO J. 8, 335–342.

Webb K J, Jones S, Robbins M P and Minchin F R 1990 Characterization of transgenic root cultures of *Trifolium repens*, *Trifolium pratense* and *Lotus corniculatus* and transgenic plants of *Lotus corniculatus*. Plant Sci. 70, 243–254.

Webb K J, Skøt L and Jørgensen B 1994 Plant genes involved in nodule development and nitrogen fixation. *In* Proceedings of the 1st European Nitrogen Fixation Conference. Eds. G B Kiss and G Endre. pp 234–238. Officina Press, Szeged, Hungary.

Werr W, Frommer W B, Maas C and Starlinger P 1985 Structure of the sucrose synthase gene on chromosome 9 of *Zea mays* L. EMBO J. 4, 1373–1380.

Witty J F, Minchin F R, Skøt L and Sheehy J E 1986 Nitrogen fixation and oxygen in legume root nodules. Oxford Surv. Plant Mol. Cell Biol. 3, 274–314.

Witty J F, Skøt L and Revsbech N P 1987 Direct evidence for changes in the resistance of legume root nodules to O_2 diffusion. J. Exp. Bot. 38, 1129–1140.

Section editor: H Lambers
Guest editors: G.H. Elkan and R.G. Upchurch

Plant and Soil **186:** 107–112, 1996.

Hopanoid lipids in *Bradyrhizobium* and other plant-associated bacteria and cloning of the *Bradyrhizobium japonicum* squalene-hopene cyclase gene

E.L. Kannenberg[1,3], M. Perzl[1], P. Müller[2], T. Härtner[1] and K. Poralla[1]
[1]*Mikrobiologie/Biotechnologie, Universität Tübingen, Auf der Morgenstelle 1, D-72076 Tübingen, Germany,* [2]*Fachbereich Biologie-Botanik, Universität Marburg, D-35043 Marburg, Germany and* [3]*Complex Carbohydrate Research Center and Department of Biochemistry and Molecular Biology, The University of Georgia, 220 Riverbend Road, Athens, GA 30602–4712, USA* *

Key words: *Bradyrhizobium*, hopanoids, nitrogen fixation, plant-microbe interaction, *Rhizobium*, squalene-hopene cyclase

Abstract

Hopanoid lipids have been discovered recently in a number of nitrogen-fixing soil bacteria and in *Bradyrhizobium* bacteria which fix nitrogen in association with legume plants. We report here an investigation of the hopanoid content in an additional number of soil bacteria capable of living in close association with plants. Of the strains investigated, hopanoids were discovered in phototrophic, nitrogen-fixing bacteria and in an extended number of *Bradyrhizobium* strains. Strains in which hopanoids so far have not been found belong to the following genera: *Rhizobium, Sinorhizobium, Phyllobacterium, Agrobacterium,* and *Azoarcus.* To address the function of hopanoids in *Bradyrhizobium,* we cloned the gene coding for a key enzyme of hopanoid biosynthesis, the squalene-hopene cyclase, and expressed the gene in *E. coli.* The recombinant enzyme catalyzed in vitro the cyclization of squalene to hopanoid derivatives.

Abbreviations: SHC – squalene-hopene cyclase, *shc* – squalene-hopene cyclase gene.

Introduction

Hopanoids are a class of pentacyclic triterpenoid lipids (Figure 1) widely distributed in the domain *Bacteria* (Ourisson et al., 1987). Among nitrogen-fixing bacteria of the soil, hopanoid lipids were discovered in Gram-negative *Azotobacter* and *Beijerinckia* (Vilcheze et al., 1994) as well as in Gram-positive *Frankia* bacteria (Berry et al., 1993) which fix nitrogen in symbiotic association with plants. Recently, we discovered the presence of hopanoid lipids in Gram-negative symbiotic bacteria of the genus *Bradyrhizobium* but not in *Rhizobium* bacteria (Kannenberg et al., 1995).

Hopanoids are thought to comprise a class of bacterial membrane stabilizers which condense the lipid phase of the cell membrane (Kannenberg et al., 1980; Poralla et al., 1980). However, hopanoids may have additional functions, as a variety of structurally related molecules have been described that have not yet been explained biologically (for a recent review, see Sahm et al., 1993). For example, in bradyrhizobia besides diplopterol a series of hopanoid derivatives with a modified side chain are currently under investigation (Kannenberg et al., 1995; Figure 1). It has been suggested that a particular class of hopanoid lipids may also play a role in the oxygen protection mechanism employed by *Frankia* to protect the nitrogenase complex from oxygen destruction (Berry et al., 1993). This raises the question of whether the unique ability of bradyrhizobia to fix nitrogen ex planta could be partially ascribed to the occurrence of hopanoids in these bacteria. Preparatory to addressing this question and that of the general function of hopanoids in plant-associated bacteria, we investigated (i) the hopanoid content in additional Gram-negative bacteria capable of living in close association with plants, and (ii) the cloning and partial characterization of a gene encoding a key enzyme of hopanoid biosynthesis, the squalene-hopene cyclase.

* FAX No: + 706 542 4412.
E-mail: elmark@mond1.ccrc.uga.edu

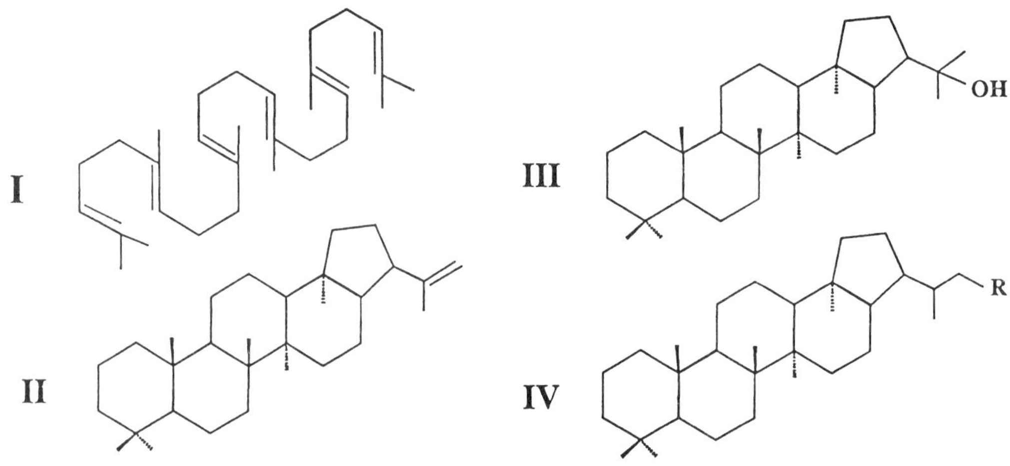

Figure 1. Hopanoid derivatives (and their biosynthetic precursor squalene) identified in *Bradyrhizobium japonicum* USDA 110. I, squalene; II, hopene; III, diplopterol; IV, elongated hopanoids (R = side chain of different polyol variants).

The cloning of this gene will allow us to investigate the function of this class of lipids via genetic studies.

Materials and methods

Bacteria were cultured in complex liquid medium (as recommended for the different strains by the Ghent Culture Collection catalogue) to estimate the hopanoid content. Lipids were extracted from bacterial cells as described by Bligh and Dyer (1959), acetylated, and analyzed by GLC (Rohmer et al., 1984). Hopanoids and squalene were identified as previously described (Kannenberg et al., 1995). The experiments were done at least twice. For cloning the squalene-hopene cyclase gene, a PCR-based screen was used to screen a cosmid clone library of strain *B. japonicum* USDA 110 for clones harboring the cyclase gene (for more details, see the legend to Figure 2). The primers for the PCR screen were based on a conserved amino acid motif of characterized cyclase genes (Poralla, 1994). The cosmid library was constructed as described in Parniske et al. (1993). For cloning and subcloning of PCR fragments, DNA was digested with restriction enzymes, ligated into the vectors pUC18 or pUC19, and transformed into *E. coli* using standard procedures.

Results and discussion

Occurrence of hopanoid lipids in plant-associated bacteria

No clear-cut patterns of hopanoid distribution have so far emerged from the bacterial strains investigated (Sahm et al., 1993). The discovery of hopanoids in nitrogen-fixing soil bacteria such as *Azotobacter, Beijerinckia,* and *Bradyrhizobium* (Kannenberg et al., 1995; Vilcheze et al., 1994) raises the question of their function in these bacteria. For example, they could play a role in protecting nitrogenase from oxygen inactivation. Hopanoids might be involved in allowing the bacteria to adapt to changes in the physiological conditions of the soil or plant micro-environments. Taxonomic relatedness may also determine their distribution. Table 1 lists the distribution of hopanoids for a number of recently investigated Gram-negative plant-associated soil bacteria compiled on the basis of the type of plant interaction, physiological considerations, and taxonomic relatedness. The picture emerging from the data is as follows:

(i) Hopanoids seem to occur in some plant-associated bacteria, including the symbiotic *Bradyrhizobium* and phototrophic, nitrogen-fixing bacteria (here called "*Photorhizobium*"). However, they have not been found in other plant-associated bacteria such as in the symbiotic *Rhizobium,* in the growth-promoting endophytes *Phyllobacterium* and *Azoarcus* (although *Azoarcus* contains squalene, the biosynthetic precursor of hopanoids and other triterpenoids), and

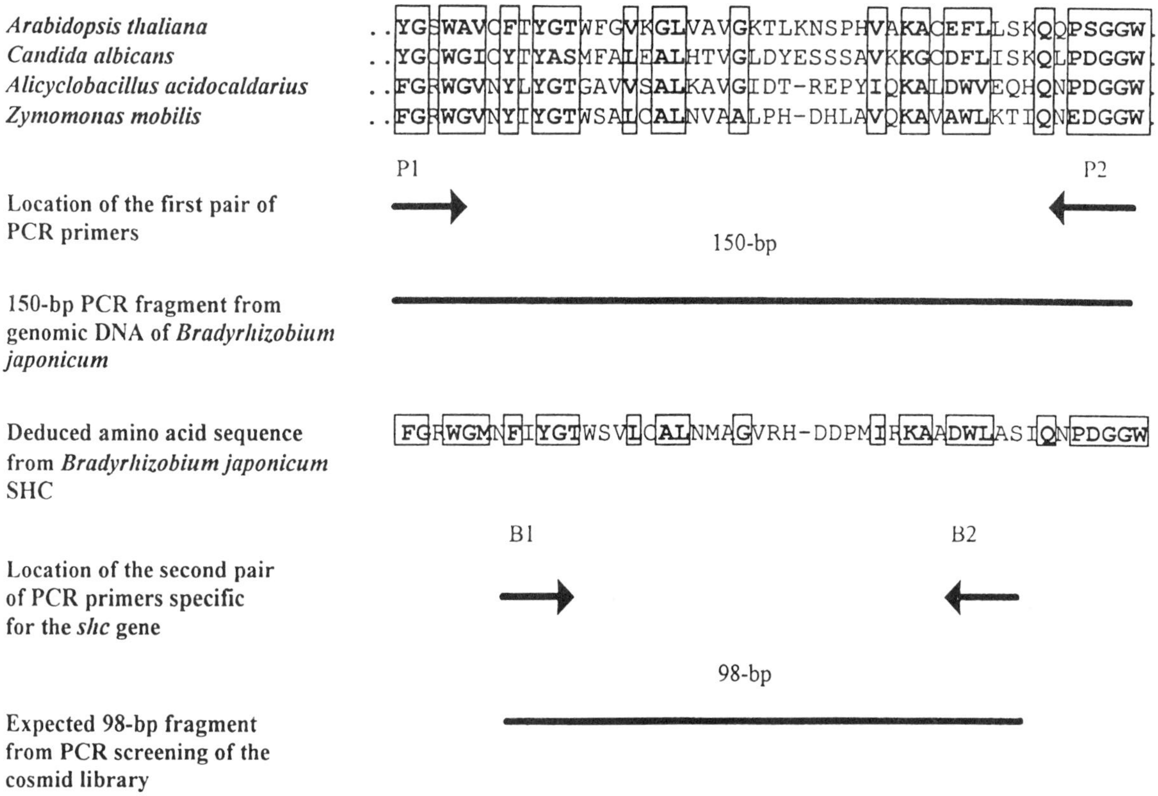

Figure 2. Alignment of a conserved region in the amino acid sequence from several cyclases and the derived PCR primers to screen for homologous DNA in a cosmid gene bank of *Bradyrhizobium japonicum* USDA 110. The two top fragments are from the oxidosqualene cyclases from *Arabidopsis thaliana* and *Candida albicans* and the two at bottom from the squalene-hopene cyclase from *Alicyclobacillus acidocaldarius* and *Zymomonas mobilis* (for a review, see Poralla et al., 1994). Similar parts of the amino acid sequences are highlighted. The screening strategy: P1/P2 were a pair of heterologous, synthetic primers with homology to the coding sequences of the conserved motifs of the various cyclases. With these primers, a 150-bp fragment was PCR-amplified from genomic DNA of *B. japonicum* USDA 110 with a deduced amino acid sequence which showed homology to variant regions of the amino acid sequences of the different cyclases. Under the assumption that this PCR fragment originated from the *Bradyrhizobium* cyclase, we constructed a pair of homologous PCR primers, B1/B2, which we used to screen gene bank clones under stringent conditions for the presence of a 98-bp PCR fragment. The presence of this 98-bp PCR fragment was taken as indicative of the presence of DNA coding for the cyclase gene of *Bradyrhizobium.* In order to avoid a tedious, time-consuming screen of individual gene bank clones, sublibraries (originally containing approximately 100 clones) were constructed from our gene bank and tested in a PCR screen for the presence of the 98-bp fragment. We investigated further only those sublibraries that were positive for the 98-bp PCR fragment by further subdividing them into even smaller clone pools (containing approximately 10 clones). As a final step, a number of individual clones of 98-bp-positive clone pools were analyzed individually by PCR amplification.

in the plant pathogen *Agrobacterium.* Although this sporadic occurrence makes it less likely that they play a generalized role in microbe-plant interactions, there are indications that the expression of triterpenoids is not yet fully understood. In bacteria in which the influence of growth conditions on lipid composition has been investigated, hopanoid content is regulated in response to the physiological conditions encountered by the bacteria (Sahm et al., 1993). An additional complication is that within the same bacteria other structurally related triterpenoid lipids can be present as well as hopanoid lipids. We recently discovered that bradyrhizobia also contain tetrahymanol (a hopanoid-related triterpenoid with a 6-carbon rather than a 5-carbon ring) and additional tetrahymanol derivatives. Overall, no firm conclusions can yet be drawn with regard to a possible role of hopanoids and related lipids in bacteria-plant associations.

(ii) Hopanoids may confer some adaptational advantage to bacteria living under tropical physiological conditions, as they seem to occur mainly in strains of tropical origin (e.g., all strains of *B. japonicum, B. elkanii* and *Bradyrhizobium* sp. tested and in *"Photorhizobium"*, putting aside the as-yet-unresolved sit-

Table 1. Occurrence of triterpenoid lipids in a range of plant-associated bacteria

Strains[a]	Triterpenoid lipids identified[b,c]			
	Squalene	Hopene	Hopane-derivatives	Source[d]
Bradyrhizobium japonicum USDA 110	+	+	+	USDA
Sinorhizobium fredii HH103 (USDA 207)	-	-	-	USDA
Rhizobium sp NGR 234 (USDA 4146)	-	-	-	USDA
Rhizobium leguminosarum B556	-	-	-	N J Brewin
Rhizobium loti NZP 2235	-	-	-	N J Brewin
Azorhizobium caulinodans LMG 6465	-	-	-[e]	LMG
"Photorhizobium" strain BTAi1[f] (USDA 4362)	+	+	+	USDA
Agrobacterium tumefaciens C58 NAL (biovar 1)	-	-	-	M I Boulton
Agrobacterium rhizogenes A4 (biovar 2)	-	-	-	M I Boulton
Phyllobacterium myrcinacearum LMG 2(t2)	-	-	-	LMG
Azoarcus communis LMG 9095	+	-	-	LMG

[a]Taxonomically, the bacteria belong in the α-subdivision of *Proteobacteria* with the exception of *Azoarcus* which is in the β-subdivision (Reinhold-Hurek and Hurek, 1993; Young et al., 1991). The *Bradyrhizobium, Sinorhizobium, Rhizobium, Azorhizobium,* and *"Photorhizobium"* strains form nitrogen-fixing nodules with a range of mainly legume host plants. *Phyllobacterium* forms leaf nodules with some plant species of the genera Rubiaceae and Myrsinaceae; the nodules have a growth-promoting effect on these plants which is poorly understood. *Azoarcus* is a growth-promoting endophyte of Kallar grass whose functional role is not well understood, and *Agrobacterium* is a tumor-forming plant pathogen found in a wide variety of plant species.

[b]For more detailed information on the occurrence of hopanoid lipids in *Rhizobium* and *Bradyrhizobium* strains, see Kannenberg et al. (1995).

[c]Bacteria were cultured and hopanoids extracted and analyzed as described in the Material and methods section. +, compound detected; -, compound not detected.

[d]USDA, United States Department of Agriculture, National *Rhizobium* Culture Collection; LMG, Ghent Culture Collection; Drs. Boulton and Brewin, both at the John Innes Institute, Norwich, UK.

[e]In initial experiments *Azorhizobium* was positive for hopanoids but negative in several subsequent experiments. This inconsistency has not been resolved and is currently under investigation.

[f]We used the phototrophic strain BTAi1 (USDA 4362) and called it *"Photorhizobium"* in these studies (see also Young et al., 1991).

uation in *Azorhizobium*; see Table 1). However, this general assumption is not supported by the finding that hopanoids are absent in the tropical strain *Rhizobium* sp. NGR 234 (Table 1), so it will be interesting to investigate other tropical rhizobia to see if the geographic distribution may play a role in hopanoid occurrence.

(iii) Hopanoid-containing strains of the α-subdivision of *Proteobacteria* seem to fall into clusters of related strains, as defined by the recently suggested phylogenetic tree for this subdivision (Martinez-Romero, 1994), such as the cluster of *Bradyrhizobium, Rhodopseudomonas,* and the phototrophic nitrogen-fixing *"Photorhizobium"* (Kannenberg et al., 1995; Kleemann et al., 1994; Table 1). In a number of other strains including *Rhizobium, Sinorhizobium,* and *Agrobacterium* biovars 1 and 2, hopanoids could not be detected (Kannenberg et al., 1995; Table 1). Since ancestors of the bacteria now grouped into the α-subdivision may have either gained or lost the ability to synthesize hopanoids, hopanoids may either confer an advantage or a disadvantage on bacteria. Much remains to be explained concerning the occurrence of hopanoids.

Cloning and preliminary characterization of DNA coding for the squalene-hopene cyclase gene of Bradyrhizobium japonicum *USDA 110*

The squalene-hopene cyclase (SHC), a key enzyme for biosynthesis, converts squalene to pentacyclic hopene and diplopterol in hopanoid-containing bacteria (for review, see Sahm et al., 1993). Several bacterial squalene-hopene cyclases (and homologous eukaryotic oxidosqualene cyclases) have been characterized (Abe et al., 1993; Reipen et al., 1995) and have been found to display a high degree of homology. Additionally, a conserved and repetitive motif has been identified within the amino acid sequence of all cyclases (Poralla et al., 1994). To see if these features occur in *Bradyrhizobium* and to lay the groundwork for the genetic analysis of hopanoid function in these bacteria, we isolated the gene (*shc*) coding for the squalene-hopene cyclase in these bacteria. We devised a powerful and fast PCR-based screening technique based on the conserved motifs (see the legend to Figure 2) which allowed the identification of two overlapping cosmid clones (BJ64 and BJ81) from a gene bank of *B. japonicum* USDA 110; these clones were further analyzed for the presence of the *shc* gene (see below).

Restriction analysis of the two overlapping clones BJ64 and BJ81 showed that they had three *Eco*RI fragments (5, 6, and 12 kb) in common (Figure 3). As explained in the legend to Figure 2, we identified a 150-bp fragment probe that is highly homologous to known cyclases. This probe hybridized with the 5-kb fragment of both cosmids in Southern hybridization experiments, suggesting that this fragment contains the gene of interest. Therefore, the 5-kb fragment was subcloned and its nucleotide sequence determined. The fragment contained an open reading frame of 1977 bp. The deduced amino acid sequence had 59% identical amino acid residues with the SHC from *Zymomonas mobilis* (Reipen et al., 1995) and 34% with that of *Alicyclobacillus acidocaldarius* (Ochs et al., 1992). It also showed the conserved and repetitive amino acid motif common in other characterized cyclases. The nucleotide sequences of the DNA region homologous to the 150-bp fragment probe in the *shc* gene and that of the 150-bp PCR fragment were identical. After cloning the DNA containing the open reading frame into *E. coli,* cell homogenates of these clones synthesized hopene and diplopterol in vitro from supplemented squalene. Our findings suggest that we have cloned the *shc* gene from *Bradyrhizobium japonicum*.

In summary, we have discovered hopanoids (and other triterpenoids) in symbiotic *Bradyrhizobium* bacteria and in the photosynthetic, nitrogen-fixing *"Photorhizobium"*. We could not detect hopanoids in a range of other plant-associated bacteria. However, no definitive picture about the occurrence of hopanoids has so far emerged. Besides the fact that a number of bacteria still have not been analyzed for hopanoid and triterpenoid content, some bacteria may also have escaped analysis due to growth conditions in which hopanoid biosynthesis was strongly reduced or repressed. As a preliminary step to address hopanoid function, we cloned the gene encoding a key enzyme in hopanoid biosynthesis, the squalene-hopene cyclase, from *B. japonicum* USDA 110. The gene showed homology to known cyclase genes from other bacteria and a conserved, repetitive motif within the amino acid sequence present in other cyclases. These traits suggest that the enzyme structure of the squalene-hopene cyclase is relatively conserved among hopanoid-containing bacteria. The cloned gene will allow us to address on a genetic level the function(s) of hopanoids in *Bradyrhizobium.*

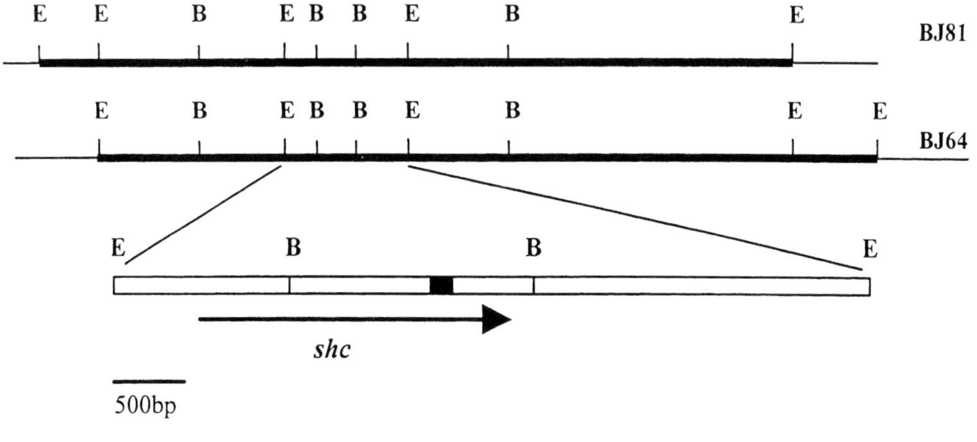

Figure 3. Restriction map of the cloned DNA in cosmids BJ64 and BJ81 containing an open reading frame (ORF) with high homology to known *shc* genes. The coding region of the *shc* gene of *B. japonicum* USDA 110 is part of a 5-kb *Eco*RI fragment of plasmid BJ64 and BJ81, respectively. The location and orientation of the ORF is indicated by the arrow. A series of restriction sites is indicated (E = *Eco*RI; B = *Bam*HI) as well as the approximate position of the 150-bp PCR fragment in the *shc* gene (black box).

Acknowledgements

We thank M Boulton and N Brewin for providing bacterial strains. We also thank R Merkle for critical comments and R Nuri for her help in preparing the manuscript. M P was supported by Graduiertenkolleg Mikrobiologie and the project by grants from the Deutsche Forschungsgemeinschaft (Sonderforschungsbereich 323 and PO 117/16-1) and the University of Tübingen.

References

Abe I, Rohmer M and Prestwich G D 1993 Enzymatic cyclization of squalene and oxidosqualene to sterols and triterpenes. Chem. Rev. 93, 2189–2206.

Berry A M, Harriott O T, Moreau R A, Osman S F, Benson D R and Jones A D 1993 Hopanoid lipids compose the *Frankia* vesicle envelope, presumptive barrier of oxygen diffusion to nitrogenase. Proc. Natl. Acad. Sci. USA 90, 6091–6094.

Bligh E G and Dyer W J 1959 A rapid method of total lipid extraction and purification. Can. J. Biochem. Physiol. 37, 911–917.

Kannenberg E, Blume A and Poralla K 1980 A hopanoid from the thermo-acidophilic *Bacillus acidocaldarius* condenses membranes. Naturwissenschaften 67, 458.

Kannenberg E L, Perzl M and Härtner T 1995 The occurrence of hopanoid lipids in *Bradyrhizobium* bacteria. FEMS Microbiol. Lett. 127, 255–262.

Kleemann G, Kellner R and Poralla K 1994 Purification and properties of the squalene-hopene cyclase from *Rhodopseudomonas palustris,* a purple non-sulfur bacterium producing hopanoids and tetrahymanol. Biochem. Biophys. Acta 1210, 317–320.

Martinez-Romero E 1994 Recent developments in *Rhizobium* taxonomy. Plant and Soil 161, 11–20.

Ochs D, Kaletta C, Entian K-D, Beck-Sickinger A and Poralla K 1992 Cloning, expression, and sequencing of squalene-hopene cyclase, a key enzyme in triterpenoid metabolism. J. Bacteriol. 174, 298–302.

Ourisson G, Rohmer M and Poralla K 1987 Prokaryotic hopanoids and other sterol surrogates. Ann. Rev. Microbiol. 41, 301–333.

Parniske M, Kosch K, Werner D and Müller P 1993 ExoB mutants of *Bradyrhizobium japonicum* with reduced competitiveness for nodulation of *Glycine max.* Mol. Plant-Microbe Interact. 6, 99–106.

Poralla K 1994 The possible role of a repetitive amino acid motif in evolution of triterpenoid cyclases. Bioorg. Med. Chem. Lett. 4, 285–290.

Poralla K, Hewelt A, Prestwich G D, Abe I, Reipen I and Sprenger G 1994 A specific amino acid repeat in squalene and oxidosqualene cyclases. Trends Biochem. Sci. 4, 157–158.

Poralla K, Kannenberg E and Blume A 1980 A glycolipid containing hopane isolated from the acidophilic, thermophilic *Bacillus acidocaldarius* has a cholesterol-like function in membranes. FEBS Lett. 113, 107–110.

Reinhold-Hurek B and Hurek T 1993 Capacities of *Azoarcus* sp., a new genus of grass-associated diazotrophs. *In* New Horizons in Nitrogen Fixation. Eds. R Palacios, J Mora and W E Newton. pp 691–694. Kluwer Academic Publishers, Dordrecht, the Netherlands.

Reipen I G, Poralla K, Sahm H and Sprenger G A 1995 *Zymomonas mobilis* squalene-hopene cyclase (*shc*): cloning, DNA sequence analysis, and expression in *Escherichia coli.* Microbiology 141, 155–161.

Rohmer M, Bouvier-Nave P and Ourisson G 1984 Distribution and hopanoid triterpenes in prokaryotes. J. Gen. Microbiol. 130, 1137–1150.

Sahm H, Rohmer M, Bringer-Meyer S, Sprenger G A and Welle R 1993 Biochemistry and physiology of hopanoids in bacteria. Adv. Microbial Physiol. 35, 247–273.

Vilcheze C, Llopiz P, Neunlist S, Poralla K and Rohmer M 1994 Prokaryotic triterpenoids: new hopanoids from the nitrogen-fixing bacteria *Azotobacter vinelandii, Beijerinckia indica* and *Beijerinckia mobilis.* Microbiology 140, 2749–2753.

Young J P W, Downer H L and Eardly B D 1991 Phylogeny of the phototrophic *Rhizobium* strain BTAi1 by polymerase chain reaction-based sequencing of a 16S rRNA gene segment. J. Bacteriol. 173, 2271–2277.

Section editor: H Lambers
Guest editors: G.H. Elkan and R.G. Upchurch

Plant and Soil **186:** 113–120, 1996.

Presence of unique repeated insertion sequences in nodulation genes of *Rhizobium 'hedysari'*

F. Meneghetti, S. Alberghini, E. Tola, A. Giacomini, F.J. Ollero, A. Squartini[1] and M.P. Nuti
Dipartimento di Biotecnologie Agrarie, Universitá di Padova, via Gradenigo 6, 35131 Padova, Italy.
[1]*Corresponding author**

Key words: Hedysarum coronarium, IS element, nodulation genes, *Rhizobium 'hedysari'*

Abstract

Cloning and sequencing of DNA from a symbiotic large plasmid in *Rhizobium 'hedysari'* strain IS 123 required for its nodulation of the mediterranean legume crop *Hedysarum coronarium* (sulla) and complementation studies of nod$^-$ mutant derivatives led to the characterization of a 30-kb region containing common and host-specific *nod* genes. This DNA region also contained at least six copies of a novel insertion sequence-like structure, some of which appeared to have suffered deletions. This 0.8 kb novel element carries two 17-bp flanking inverted repeats and an open reading frame showing homology with a transposase from *Staphylococcus aureus*. Hybridization studies revealed that several strains of *Rhizobium 'hedysari'* carry this element in various copy number. The six copies in strain IS 123 appear clustered specifically within the pSym *nod* region.

The significance of this IS element in rhizobia and its possible use as a probe for taxonomic and phylogenetic studies of Rhizobiaceae is addressed.

Introduction

Hedysarum coronarium (sulla) is a legume originating from the mediterranean basin and grown in this region as a perennial forage crop (Sarno and Stringi, 1981). This legume is appreciated for its good agronomical performance as well as its tolerance to stress conditions, such as drought, soil salinity and alkaline soil pH (Lupi et al., 1988). Sulla establishes a highly host-specific relationship with nitrogen-fixing strains of *Rhizobium 'hedysari'* (Cabrera et al., 1979; Casella et al., 1984; Ollero et al., 1989), for which the proposal as new species is in progress. Interestingly, *Rhizobium 'hedysari'* is capable of nodulating its host in soils with pH as high as 9.3. The infection process in the *Rhizobium*-sulla symbiosis has been described and includes some unique morphological features (Squartini et al., 1993).

The genetics in relation to this symbiotic interaction is one of the subjects of our research. Genes responsible for nodulation and host specificity reside on a large symbiotic plasmid (Espuny et al., 1987). In strain IS123 an in vivo recombination event during a plasmid-helped conjugation generated a smaller derivative of the symbiotic plasmid still retaining the determinants of host-specific nodulation (Ollero et al., 1993). This derivative was the starting material to clone and characterize the *nod* gene region of *R. 'hedysari'* described in this work.

A number of IS elements have been identified in rhizobia. In *Rhizobium meliloti*, ISRm1 (Ruvkun et al., 1982) displays a copy number ranging from 1 to 11 (Wheatcroft and Watson, 1988a). ISRm2 is preferentially found in the symbiotic region of *R. meliloti* and is not species-specific, being found also in several other species of *Rhizobium* (Dusha et al., 1987). ISRm3 is apparently specific for *R. meliloti* and not related to the previous ones (Wheatcroft and Watson, 1988b). A fourth element, ISRm4 has been identified (Soto et al., 1992). Strain fingerprinting by comparison of IS elements hybridization patterns have been proposed (Wheatcroft and Watson, 1988a). An element found in *Rhizobium* sp. *(Leucaena diversifolia)*, IS*Rld*TAL1145-1, has been tested to identify cross-inoculation group relationships (Rice et al., 1994). Kosier et al. (1993) have reported the use of rhizobial IS elements for genotyping of field iso-

* FAX No: +39498070517. E-mail: biotmpn@ipdunidx.unipd.it

lates. These elements found in rhizobia often do not share DNA homologies among themselves nor with published sequences, but their ORFs contain conserved aminoacid residues matching those of transposase enzymes (Wheatcroft and Watson, 1991).

Materials and methods

Bacterial strains, plasmids, and culture conditions

Strains and plasmids are listed in Table 1. Rhizobia were grown in TY (Beringer, 1974) or BIII (Dazzo, 1984) medium at 28 °C. *E. coli* was grown in LB at 37 °C. Antibiotics were used at the following concentrations: ampicillin, 50 $\mu g\ mL^{-1}$, kanamycin, 30 $\mu g\ mL^{-1}$.

DNA manipulations and sequencing

Standard cloning and restriction mapping techniques described by Sambrook et al. (1989) were adopted. DNA hybridizations made use of digoxygenin labelled probes obtained by a DIG nonradioactive DNA labelling and hybridization kit from Boehringer Mannheim. Both labelling and hybridization were carried out according to the manufacturer. Double-stranded DNA fragments were cloned into pUC 19 and sequenced by the di-deoxy nucleotide chain termination method (Sanger et al., 1977) using fluorescent di-deoxy nudeotides and an Applied Biosystems 373A DNA sequencer. Sequences were analyzed using the IG Suite Molecular Biology Software System, release 5.4 (Intelligenetics Inc).

Results and discussion

The nodulation genes in R. 'hedysari' *IS123 are clustered within a 35 kb region of its sym plasmid*

Plasmid pR2, originated from in vivo recombination of the symbiotic plasmid of strain IS 123 and the helper plasmid pJB3JI during a conjugal event, still contains the necessary information for *Hedysarum* nodulation (Ollero et al., 1993). This 85-90 kb plasmid was partially digested with *Bgl*II and the fragments were cloned into the vector pRL497. The resulting clones were screened by dot blot hybridization with a 2.6 kb *Bam*HI fragment probe containing *nodDABC* from *R. leguminosarum* bv. *viciae* strain 1001 (Squartini et al., 1988), or with a *nodFE* probe from *R. leguminosarum* bv. *trifolii* strain ANU843 which consisted of a 1.8 *Cla*I fragment from plasmid pAYS14. Two clones containing adjacent *Bgl*II fragments of 16 and 14 kb were identified (plasmids pBPZ14 and pBPZ18, respectively). Their mapped inserts are represented as a single region in Figure 1. The sequencing strategy of the region included the sonication of the purified *Bgl*II fragments to generate a series of random small fragments in the range of 500-700 bp. Areas encompassing common and host-specific nod genes are indicated by boxes. A detailed map of each ORF present in this region is under construction (Tola et al.).

The pSym nod *region of* R. 'hedysari' *IS123 contains repeated copies of insertion elements*

A DNA sequence of 0.8 kb (Figure 2) is repeated in several positions, at least six of which have been mapped in the *nod* region. The element is present in both orientations and some copies are incomplete. The extremities of this element have a conserved pair of 17 bp inverted repeats containing a single base mismatch. Homology search using BLAST (Altschul et al., 1990) analysis in GenBank database did not reveal homologies of this element with known DNA sequences. At the aminoacid level, however, this element contains an ORF encompassing almost its entire length, with 55% homology to transposases TRA1 and TRA2 of IS431MEC from *S. aureus* (Figure 3). These findings suggest that this sequence is a novel insertion element, which has been tentatively named ISRh1.

The sequenced region (Figure 1) also includes a stretch of 2.1 kb which shows significant homology with an IS element in *Burkholderia cepacea.*

The element is unique to Rhizobium 'hedysari'

In order to test the distribution of ISRh1 in rhizobia, a *Bam*HI-*Eco*RI internal fragment was separated by agarose gel electrophoresis and purified by electroelution. This fragment was digoxygenin-labelled and used to probe the genomes of other rhizobia. Hybridization to *Bam*HI-digested total DNA from the type-strains of different *Rhizobium* species plus 9 isolates of *Rhizobium 'hedysari'* from various sites clearly showed that the ISRh1 insertion element Figure 4 is found only in *R. 'hedysari'* strains. The number of hybridizing bands ranged from a minimum of one for strain RH19, isolated in Sicily, to a maximum of 20.

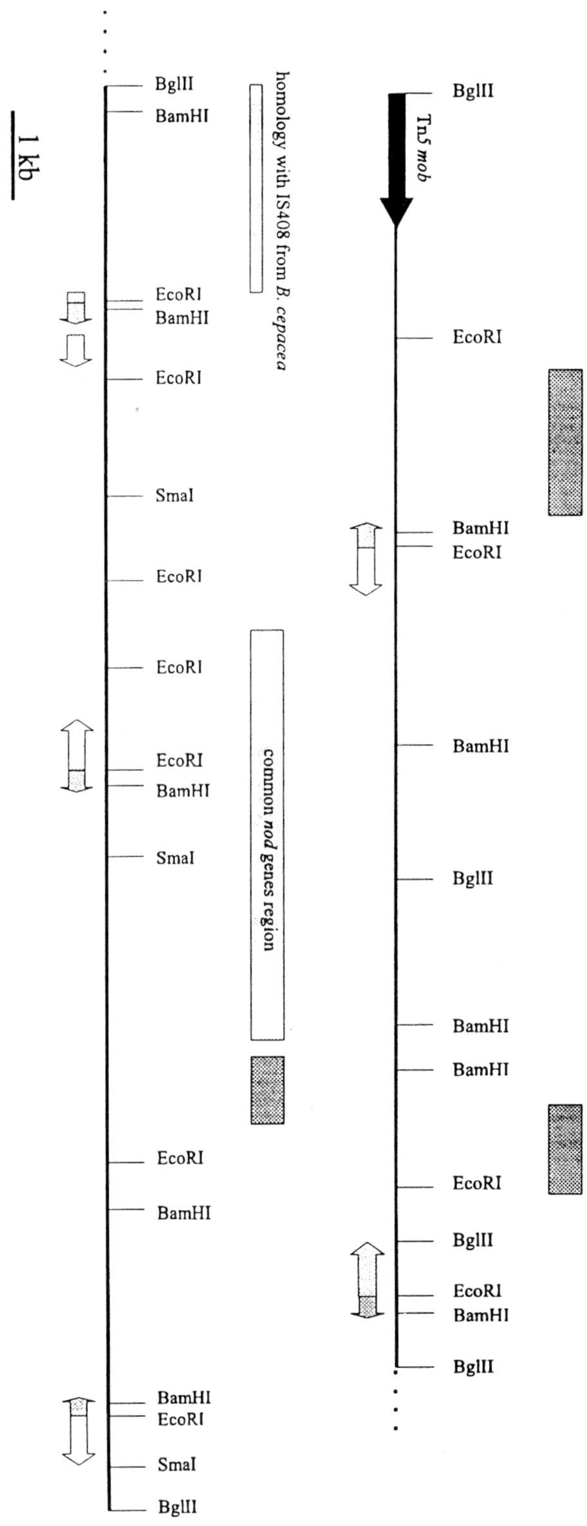

Figure 1. Region of about 35 kb from *R. "hedysari"* IS123 sym plasmid containing common and host specific *nod* genes. Light shaded box indicates the common *nod* genes region, dark shaded boxes indicate positions of genes related to host-specificity. Arrowed boxes indicate the newly idenified IS element ISRhl or portions of it. The double tone in the arrows is used to indicate the orientation.

Table 1. Bacterial strains and plasmids

Strain or plasmid	Relevant traits	Source or reference
Rhizobium 'hedysari' IS 123	Wild type, isolated from southern Spain	F J Ollero/F Temprano
Rhizobium 'hedysari' CC1335	Wild type, isolated from southern Spain	J Brockwell
Rhizobium 'hedysari' HCNT1	Wild type, isolated from Volterra, Italy	S Casella
*Rhizobium 'hedysari'*RHF	Wild type, isolated from Pisa, Italy	S Casella
Rhizobium 'hedysari' RH19	Wild type, isolated from Sicily, Italy	S Casella
Rhizobium 'hedysari' RH 100	Wild type, isolated from the Balearic Islands	A Toffanin
Rhizobium 'hedysari' RH44	Wild type, isolated from Southern Spain	A Toffanin
Rhizobium 'hedysari' A6	Wild type, isolated from Algeria	A Benguedouar
Rhizobium 'hedysari' A10	Wild type, isolated from Algeria	A Benguedouar
Rhizobium 'hedysari' RJ243	pSym-cured derivative of IS123	Ollero et al. (1989)
R. meliloti ATCC9930^{T}	Type Strain	P Van Berkum
R. leguminosarum bv. *trifolii* ATCC 14480^{T}	Type Strain	P Van Berkum
R. leguminosurum bv. *viciae* ATCC 10004^{T}	Type Strain	P Van Berkum
R. leguminosarum bv. *phaseoli* RCR 3644^{T}	Type Strain	P Van Berkum
R. etli CFN42^{T}	Type Strain	P Van Berkum
R. tropici IIB CIAT899^{T}	Type Strain	P Van Berkum
R. tropici IIA CFN299^{T}	Type Strain	P Van Berkum
R. galegae HAMBI 503		K Lindström
pR2	R-prime bearing *R. 'hedysari' nod* genes	Ollero et al. (1993)
pRL497	IncQ, broad host range vector	Elhai and Wolk (!988)
pBPZ18	cloned *R. 'hedysari' nod* region	This work
pBPZ14	cloned *R. 'hedysari' nod* region	This work
pAYS14	14 kb *Hin*dIII fragment from *R.l.* b.v. *trifolii* ANU843 cloned in pRL497. Source of *nodFE* probe	A Squartini

Insertion element ISRh1 is typically located on plasmids, and clusters preferentially among nod *genes*

The comparison of wild type IS123 and its pSym-cured derivative (lanes 1 and i of Figure 4) shows that ten out of the eleven hybridizing fragments are located on the 160Md sym plasmid. Furthermore, plasmid gel hybridization indicated that the single hybridizing band in the cured strain (lane i) comes from a copy of the element located on a smaller plasmid (data not shown). Therefore, all the copies of ISRh1 in strain IS123 are plasmid-borne. Sequencing of the *nod* gene region from this strain (Figure. 1) shows that at least six copies of ISRh1 are clustered in the *nod* gene region.

Is this element capable of transposition?

Although no transposition test was performed, the presence of several dispersed copies within the genomes of different *Rhizobium 'hedysari'* strains, the integrity of the inverted repeats, and the presence of a transposase-like ORF, are all indicative of illegitimate recombination of ISRh1. At least it could be postulated that the element had undergone transposition in

```
        10         20         30         40         50         60         70
 CGGCGATGTC ACGTTGTTTG ATCAAGGTCG GAACAAGGCG CTTATGTGAG ACGTCAGGCG GTTGCGCCGG

        80         90        100        110        120        130        140
 TCACGGCTTT CCACTGCGCC ATTGAGCGGA TCCGATGGAT GTGGATGGCG AGGGCTGAGT TTTTCTGGTG

       150        160        170        180        190        200        210
 CGGGGGGACG AAGAGATTTC GGAGTGCCGA AAAGATCGAT ATGAACCGTT GCAAGCCGCC GACGGATCGA

       220        230        240        250        260        270        280
 AATCTCTGCA TCATCCGCTC CCGTTTTCGA AGCGGCACGT GAGAATTCTC GGCTCGATTG TTCAGGCCCT

       290        300        310        320        330        340        350
 TGTGCGATCG ATGTTCGACG GCGGGCATCA CCTCCCGTCT TGCAGCACCA TATGAGCGCA ATTTGTCGGT

       360        370        380        390        400        410        420
 GACGATCCGC TTCGGCGTCA GGCCTTGCTT CTTCAGCAGC CTGACCAGCA ATCGCCTGGC GGCTTGGGTA

       430        440        450        460        470        480        490
 TCGCGGCGGG CTTGAACGAT CTCGTCGAGA ACGTAACCGT CTTGGTCAAC GGCACGCCAG AGCCAATGTT

       500        510        520        530        540        550        560
 TGCGGCCGCC GATGGAAATC ACCACCTCGT CCAGATGCCA GATATCCTTT CGCGAAGGCC TCTTTCTGCA

       570        580        590        600        610        620        630
 CAACTGCCTG GCATAAGCCG CTCCGAATTT GGGACCCCAT CTCCGGATCG TCTCATGGGA GACGACGATA

       640        650        660        670        680        690        700
 CCGCGCTCCA GCAGCATCTC CTCGACCATC CTCAGGCTCA AAGGGAACCG AAAATACAGC CAGACCGCAC

       710        720        730        740        750        760        770
 GCGCGATAAT CTTGGGTGGG AAAGCGGTGG TTCTTGGTAG GTTACGGGCG GATTGTTCAT CCCAACCCAT

       780        790        800        810
 TATCCGACAA CCGTTAAGCC GCTGACAACG TGACATCACC G
```

Figure 2. Nucleotide sequence of the putative insertion element ISRh1. The 17 bp inverted repeats are indicated by arrowed boxes. The ORF showing homology to a transposase is encoded on the complementary strand starting at position 806 and ending at position 57.

the past. The sequencing done in the *nod* gene region allowed the analysis of the DNA junctions surrounding the different copies. Although the 17 bp inverted repeats are precisely conserved, no directly repeated target duplications were found bordering the element, even around its intact copies. This suggested that the copies located in this region may be vestiges of a once-active insertion sequence. Alternatively, even if the element remained potentially active, these sites experienced insertion long enough ago to allow mutations to erode short tandem duplications at the target site.

Why does ISRh1 transpose preferentially on pSym?

Location of ISRh1 on an extrachromosomal replicon guarantees vertical (progeny) gene transfer and greatly enhances the possibility of horizontal (conjugal) exchange. Furthermore, its residence on a symbiotic plasmid reinforces its maintenance, due to the critical importance of pSym genes for the species' ecology. In this respect, finding most copies inserted within the *nod* genes seems to support this hypothesis. Furthermore, being located on a very host-specific DNA domain such as these genes, a mobile element could gain access to a unique ecological niche, in this case represented by the root nodule of *H. coronarium*. Of relevance to this is the fact that, when the host specific *nod* genes are introduced into a heterologous *Rhizobium* species, an instability results in deletion of one set of host-specific genes, with the host plant determining such outcome (Ollero et al., 1993). This mechanism ensures the preservation of the specific plant/niche system, typical of the temperate zone rhizobia. Residence on a host-specific plasmid, that is incompatible

```
TRA2_STAAU    M--------------------NYFRYKQFNKDVITVAVGYYLRYTLSYRDISEILRERGV
TRA1_STAAU    M--------------------NYFRYKQFNKDVITVAVGYYLRYALSYRDISEILRGRGV
ORF8_ISRH1    MSRCQRLNGCRIMGWDEQSARNLPRTTAFPPKIIARAVWLYFRFPLSLRMVEEMLLERGI
              *                    *  *   *   .*. **  *.*. ** * . *.*  **.

TRA2_STAAU    NVHHSTVYRWVQEYAPILYQIWKKKHKKAYYKWRIDETYIKIKGKWSYLYRAIDAEGHTL
TRA1_STAAU    NVHHSTVYRWVQEYAPILYQIWKKKHKKAYYKWRIDETYIKIKGKWSYLYRAIDAEGHTL
ORF8_ISRH1    VVSHETIRRWGPKFGAAYARQLCRKRPSRKDIWHLDEVVISIGGRKHWLWRAVDQDGYVL
               * * *. **   ...   .   .*.      *..**  * * *.   * **.* .*  *

TRA2_STAAU    DIWLRKQRDNHSAYAFIKRLIKQFG-KPQKVITDQAPSTKVAMAKVIKAFKLKPDCHCTS
TRA1_STAAU    DIWLRKQRDNHSAYAFIKRLIKQFG-KPQKVITDQAPSTKVAMAKVIKAFKLKPDCHCTS
ORF8_ISRH1    DEIVQARRDTQAARRLLVRLLKKQGLTPKRIVTDKLRSYGAARREVMPAVEHRSH-----
              *  .. .** ..*  .. **.*. *  *....**.  *   *   *. *   ...

TRA2_STAAU    KYLNNLIEQDHRHIKVRKT---RYQSINTAKNTLKGIECIYALY---KKNRRSLQIYGFS
TRA1_STAAU    KYLNNLIEQDHRHIKVRKT---RYQSINTAKNTLKGIECIYALY---KKNRRSLQIYGFS
ORF8_ISRH1    KGLNNRAENSHVPLRKRERMMQRFRSVGGLQRFISIFSALRNLFVPPHQKNSALAIH---
              * ***  *. *  .. *     *..*.   .  .  .  .  *.    ..  .* *

TRA2_STAAU    PCHEISIML--------AS
TRA1_STAAU    PCHEISIML--------AS
ORF8_ISRH1    -IHRIRSMAQWKAVTGATA
                * *  *          ..
```

Figure 3. CLUSTALV multiple sequence alignment of the longest open reading frame (ORF8) found in the IS element with two transposases from IS elements found in *Staphylococcus aureus*. Aminoacid identity is indicated by asterisks; dots represent conservative substitution.

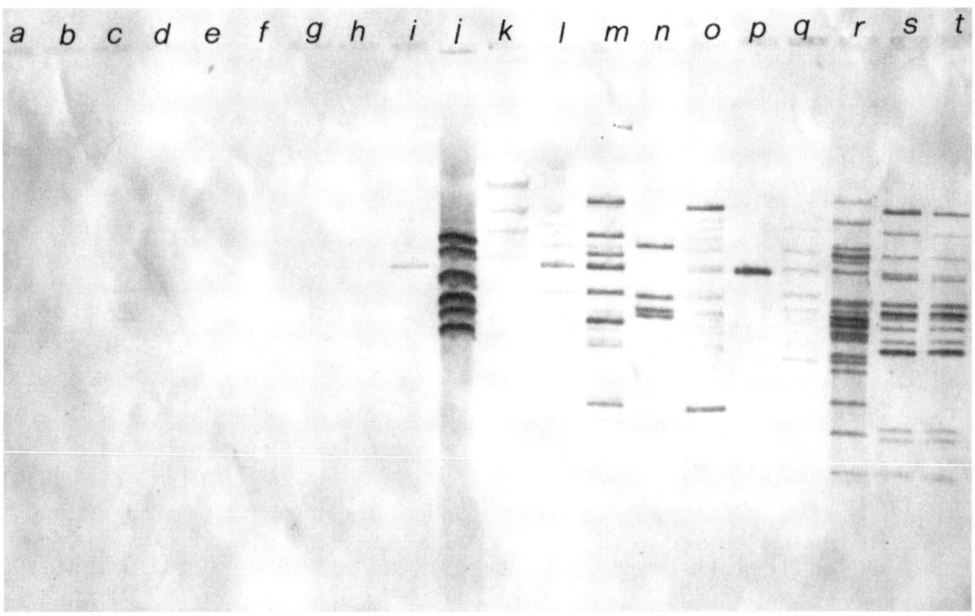

Figure 4. Hybridization of digoxygenin-labelled *Bam*HI-*Eco*RI fragment from ISRH1 to *Bam*HI-digested total DNA from different species of *Rhizobium*: **a.** *R. meliloti* ATCC9930^T; **b.** *R. leguminosarum* bv. *trifolii* ATCC14480^T; **c.** *R. leguminosarum* bv. *viciae* ATCC10004^T; **d.** *R. leguminosarum* bv. *phaseoli* RCR3644^T; **e.** *R. etli* CFN42^T; **f.** *R. tropici* IIB CIAT899^T; **g.** *R. tropici* IIA CFN299^T; **h.** *R. galegae* HAMBI503; **i.** *R. 'hedysari'* RJ243 (= strain IS123 cured of its sym plasmid); **j.** *Bam*HI-digested purified plasmid pR2, a smaller derivative of sym plasmid from strain IS123 still carrying the region of the *nod* genes; **k.** lambda-*Hind*III molecular weight dig-labelled standard; **l.** *R. 'hedysari'* IS123; **m.** *R. 'hedysari'* CC133: (835); **n.** *R. 'hedysari'* HCNT1; **o.** *R. 'hedysari'* RHF; **p.** *R. 'hedysari'* RH19; **q.** *R. 'hedysari'* RH100; **r.** *R. 'hedysari'* RH44; **s.** *R. 'heysari'* A6; **t.** *R. 'hedysari'* A10.

witlh closely related species, could therefore be an explanation for the presence of ISRhl exclusively in *R. 'hedysari'*.

Possible practical applications

The availability of a species-specific, conserved, variable-copy number DNA insertion element suggests several practical uses. First, it could provide species-specific probing. All strains of *R. 'hedysari'* harbour, to our knowledge, at least one copy of ISRh 1, while no other organism has been found to contain this element. Therefore, hybridization with the labelled element or PCR amplifications, using the flanking inverted repeats as primers, could quickly discriminate this *Rhizobium* species. Given the surprising constance of such element in strains from different geographical origins, it is temptative to speculate that other rhizobia could contain their own yet-to-discover specific insertion elements, which would facilitate their detection. Second it could provide strain fingerprinting. Within the *R. 'hedysari'* strains, the hybridization profiles (Figure 4) clearly allow the comparison and distinction of each strain, also pointing out the similarities or identities, as in the case of lanes s) and t) which contain DNA from two strains from Algeria. These profiles were known to match exactly also from Pulsed-Field electrophoresis and PCR fingerprinting. Indeed a RAPD amplification using a primer from the highly repetitive insertion element could offer (uniquely) a species-specific and fingerprinting application at the same time. Third, it could serve taxonomical and evolutionary studies. Postulating a conservative transposition mechanism, the interesting cue that there are strains with one single copy of ISRh1, e.g. RH19 from Sicily, and others harbouring increasing numbers of this element, whose positions could match or be totally different, opens the possibility of tracking the center of origin of the bacterial species in relation to that of the host plant. In this case, *Hedysarum coronarium* which is believed to originate in the center of the mediterranean basin.

Acknowledgements

Frank B Dazzo is gratefully acknowledged for critically reviewing the manuscript. This work was supported by M A F, P F "Biotecnologie Avanzate Applicate alle Piante", and in part by the BRIDGE program, contract BIOT-CT91-0283.

References

Altschul S F, Gish W, Myers E W and Lipman D J 1990 Basic local alignment search tool. J. Mol. Biol. 215, 403–410.

Cabrera E, Villa A and Ruiz-Argüeso T 1979 Diversidad y caraterizaciòn de la flora nativa de rizobios que nodulan en zulla (*Hedysarum coronarium*). *In* VII Congreso Nacional de Microbiologia, Cadiz (Abstr. Vol.). p 294.

Casella S, Gault R R, Reynolds K C, Dyson J R and Brockwell J 1984 Nodulation studies on legumes exotic to Australia: *Hedysarum coronarium*. FEMS Microbiol. Lett. 22, 37–45.

Dazzo F 1984 Leguminous root nodules. *In* Experimental Microbial Ecology. Eds. R Burns and J Slater. pp 431–446. Blackwell Sci. Publ., Oxford, UK.

Dusha I, Kovalenko S, Banfalvi Z and Kondorosi A 1987 *Rhizobium meliloti* insertion element ISRm2 and its use for identification of the *fixX* gene. J. Bacteriol. 169, 1403–1409.

Elhai J and Wolk C P 1988 A versatile class of positive selection vectors based on the nonviability of palindrome-containing plasmids that allows cloning into long polylinkers. Gene 68, 119–138.

Espuny M R, Ollero F J, Bellogin R A, Ruiz-Sainz J E and Perez-Silva J 1987 Transfer of the *Rhizobium leguminosarum* biovar. *trifolii* symbiotic plasmid pRtr5a to a strain of *Rhizobium* sp. that nodulates on *Hedysarum coronarium*. J. Appl. Bacteriol. 63, 13–20.

Higgins DG and Sharp PM 1988 Clustal: a package for performing multiple seqence alignment on a microcomputer. Gene 73, 237–244.

Kosier B, Puhler A, and Simon R 1993 Monitoring the diversity of *Rhizobium meliloti* field and microcosm isolates with a novel rapid genotyping method using insertion elements. Mol. Ecol. 2, 35–46.

Lupi F, Casella S, Tofanin A and Squartini A 1988 Introduction of *Rhizobium "hedysari"* in alkaline clay-loam soil by different inoculation techniques. Arid Soil Res. Rehabil. 2, 19–28.

Ollero F J, Espuny M R and Bellogin R A 1989 Mobilization of symbiotic plasmid from a strain of *Rhizobium* sp. *(Hedysarum coronarium)*. System. Appl. Microbiol. 11, 217–222.

Ollero F J, Valverde M A, Espuny M R and Bellogin R A 1993 *In vivo* formation of R-prime plasmids harbouring *nod* genes of *Rhizobium "hedysari"*. FEMS Microbiol. Lett. 110, 269–174.

Rice J D, Somasegaran P, MacGlashnan K and Bohlool B B 1994 Isolation of insertion sequence IS*RLd*TAL1145-1 from a *Rhizobium* sp. *(Leucaena diversifolia)* and distribution of homologous sequences identifying cross-inoculation group relationships. Appl. Environ. Microbiol. 60, 4394–4403.

Ruvkan G B, Long S R, Meade H M, van den Bos R C and Ausubel F M 1982 ISRm1: a *Rhizobium meliloti* insertion sequence that transposes preferentially into nitrogen fixation genes. J. Mol. Appl. Genet. 1, 405–418.

Sambrook J, Fritsch E F and Maniatis T 1989 Molecular Cloning. A Laboratory Manual, Second Edition. Cold Spring Harbor Laboratory, Cold Spring Harbor, NY, USA.

Sanger F, Nicklen S and Coulson A R 1977 DNA sequencing with chain terminating inhibitors. Proc. Natl. Acad. Sci. USA 74, 5463–5467.

Sarno R and Stringi L 1981 Sulla (*Hedysarum coronarium* L.) in coltivazioni Erbacee. Eds. R Baldoni and L Giardini pp 897–907. Patron, Bologna, Italy.

Soto M, Zorzano A, Olivares J and Toro N 1992 Sequence of IS*Rm4* from *Rhizobium meliloti* strain GR4. Gene 120, 125–126.

Squartini A, Van Veen R J M, Regensburg-Tuink T, Hooykaas P J J and Nuti M P 1988 Identification and characterization of the *nod*D gene in *Rhizobium leguminosarum* strain 1001. Mol. Plant-Microbe Interact. 1, 145–149.

Squartini A, Dazzo F B, Casella S and Nuti M P 1993 The root nodule symbiosis between *Rhizobium "hedysari"* and its drought-tolerant host *Hedysarum coronarium*. Symbiosis 15, 227–238.

Wheatcroft R and Watson R J 1988a A positive strain identification method for *Rhizobium meliloti*. Appl. Environ. Microbiol. 54, 574-576.

Wheatcroft R and Watson R J 1988b Distribution of insertion sequence IS*Rm 1* in *Rhizobium meliloti* and other gram-negative bacteria. J. Gen. Microbiol. 134, 113–121.

Wheatcroft R and Laberge S 1991 Identification and nucleotide sequence of *Rhizobium meliloti* insertion squence IS*Rm 3*: similarity between the putative transposase encoded by IS*Rm 3* and those encoded by *Staphylococcus aureus* IS256 and *Thiobacillus ferrooxidans* IS*T 2*. J. Bacteriol. 173, 2530–2538.

Section editor: H. Lambers
Guest editors: G.H. Elkan and R.G. Upchurch

Plant and Soil **186:** 121–125, 1996.

Enhanced competitiveness of a *Bradyrhizobium japonicum* mutant strain improved for nodulation and nitrogen fixation

L.D. Kuykendall[1], F.M. Hashem[1] and W.J. Hunter[2]
[1] *USDA-ARS, Soybean and Alfalfa Research Laboratory, 10300 Baltimore Avenue, Beltsville, MD 20705-2350, USA and* [2] *USDA-ARS, Soil-Plant Nutrient Research Unit, 1701 Center Avenue, Fort Collins, CO 80526-2083, USA**

Key words: competitiveness, inoculant, nitrogen fixation, nodulation, soybean, yield

Abstract

Bradyrhizobium japonicum strain TA-11NOD$^+$, with altered indole biosynthesis, exhibited enhanced nodulation and nitrogen fixation on soybean in previous greenhouse studies. In this study, field experiments were conducted at Upper Marlboro, Maryland, in the summers of 1988 and 1993. In 1988, the site used was essentially free of soybean-nodulating bacteria and seed yield in plots inoculated with either I-110ARS or TA-11NOD$^+$ was significantly higher by 12 or 20%, respectively, than that of the uninoculated controls. The 1993 site had an indigenous soil population (about 10^4 cells g^{-1}) of symbiotically ineffective soybean-nodulating bacteria. Nevertheless, six-week-old 'Morgan' soybean plants inoculated with strain TA-11NOD$^+$ had 44% more nodules and exhibited 50% more nitrogen fixation by acetylene reduction when compared with plants that received the parental strain I-110ARS. Nodule occupancy, as determined using genetic markers for rifampicin and streptomycin resistance, was significantly higher for strain TA-11NOD$^+$ than for strain I-110ARS. Overall, for the two years and the two soybean genotypes, the yield obtained with TA-11NOD$^+$ was 6% higher than that obtained with I-110ARS. Competition experiments were conducted in the greenhouse and strain TA-11NOD$^+$ was significantly more competitive than strain I-110ARS in competition with strains USDA 6 or USDA 438.

Introduction

The literature on the *Rhizobium*-plant symbiosis has recently been reviewed (Van Rhijn and Vanderleyden, 1995). *Bradyrhizobium japonicum* strain TA-11NOD$^+$ was obtained as a prototrophic revertant of a non-nodulating tryptophan auxotroph, TA-11 (Kuykendall and Hunter, 1991). This strain has an enhanced ability to nodulate soybean (Huater and Kuykendall, 1990). Under greenhouse conditions, soybean cv. 'Tracy-M' inoculated with strain TA-11NOD$^+$ had 41% more nodule mass and 50% more nitrogen content than plants inoculated with the wild-type *B. japonicum* progenitor strain, I-110ARS (Hunter and Kuykendall, 1990; Kuykendall and Hunter, 1991). Altered indole biosynthesis is probably involved in enhanced nodulation since the parent strain of the prototrophic revertant was specifically deficient in indole glycerol phosphate synthase (IGPS). Also, the activity of this enzyme was restored in the revertant, and strain TA-11NOD$^+$ can be distinguished from its progenitor strain, I-110ARS, by its greater resistance to inhibitory amounts of excess tryptophan (Kuykendall and Hunter, 1991).

Our working hypothesis is that the alterations in symbiotic properties of strains TA-11 and TA-11NOD$^+$ are due specifically to changes in the gene encoding indole glycerol phosphate synthase, *trpC*. To study this phenomenon further, the *trpCD* region of *B. japonicum* has been identified using the *trpCD* region of *Rhizobium meliloti* as a probe (Kuykendall and Hunter, 1992). Site-directed marker exchange mutagenesis was performed and new Δ *trp* deletion mutants were obtained (Kuykendall and Hunter, 1995). These new mutants lacked the ability to nodulate soybean and were missing activities for the *trpC*- and *trpD*-encoded enzymes, IGPS and phosphoribosyl anthranilate transferase, respectively. They had a higher specific activity for the *trpEG*-encoded enzyme, anthranilate synthase (Kuykendall and Hunter, 1995).

* FAX No: +13015045728

Inoculant strains are often poor competitors with indigenous strains which usually occupy the vast majority (80–95%) of nodules on soybean (Johnson et al., 1965; Kuykendall and Weber, 1978). Several approaches to solving this problem have been reviewed (Dowling and Broughton, 1986; Streeter, 1994; Triplett and Sadowsky, 1992), but, to date, no successful solution has been found, although some competitively superior *Rhizobium fredii* mutants were reported (Zdor and Pueppke, 1991).

The objectives of the present study were to determine whether the enhanced nodulation phenomenon exhibited by strain TA-11NOD$^+$ in greenhouse studies would also occur in the field and to determine the relative competitiveness of this strain compared with its progenitor strain when introduced into a field soil containing soybean-nodulating bacteria.

Materials and methods

Bacterial strains and growth conditions

Strain TA-11NOD$^+$ is designated NRRL B-18466 in the ARS Culture Collection in Peoria, IL 61604. The genetic markers of the progenitor strain, I-110ARS, were previously described (Kuykendall and Weber, 1978). Strain 6 of the USDA Beltsville Culture Collection is the same as ATCC 10324, the type strain of the species. USDA 438 is a highly competitive South Dakota isolate originally described as strain SD6-1c (Keyser and Cregan, 1987). Cultures were grown in A1E medium (Kuykendall, 1987) and incubated on a rotary shaker at 100 rpm for 4 days at 30 °C.

Field testing

Field experiments were conducted at the University of Maryland Tobacco Research Farm, Upper Marlboro, Maryland. The soil was an Adelphia fine sandy loam classified as Aquic Hapludult. The site used in 1988 was essentially free of indigenous soybean-nodulating bacteria as determined by tests in the greenhouse. The site used in 1993 had soil that was determined by most-probable-number estimation, using dilutions of soil for nodulation tests, to contain about 10^4 soybean-nodulating bacterial cells g^{-1} (Weaver and Frederick, 1971). Replicates were in a randomized complete block design. *Glycine max* (L.) Merr. 'Morgan', or A5043 (in 1988 only), seeds were planted in six 5-meter-long rows (0.75 m row space; 30 seeds m^{-1}). After sowing, each row received about one liter of a 5×10^{-2} dilution of culture containing either 6.0 or 5.5×10^7 viable cells mL^{-1} of strain I-110ARS or strain TA-11NOD$^+$, respectively. Control plots were left uninoculated. Also, the periphery of the field was planted with a two-meter-wide border of uninoculated soybean. In 1988, the field test was essentially for yield and seed quality. In the 1993 test, the soil was tested and found to have 0.25 g kg^{-1} total N and 0.05 g kg^{-1} P. Various traits of symbiotic fitness were examined. At six weeks after planting, five plants randomly selected from the middle of an outer row of each plot were tested for acetylene reduction, nodule number and mass, and shoot dry weight. At maturity, all of the plants from each row were harvested and seed yield was determined. Seed protein and oil content was determined in the USDA Laboratory at Peoria, IL.

Greenhouse tests

Soybean cultivar 'Kent' seeds were surface-sterilized with 50% ethanol for 30 seconds, swirled, and rinsed three times with deionized water. Ten eight-inch plastic pots were washed thoroughly and filled with a mixture of unsterilized soil and perlite (8:1). The soil was a sandy loam with 0.39 g kg^{-1} total N and with 0.17 g kg^{-1} available P. The pots were watered with tap water to allow the soil mixture to hydrate overnight. Six seeds per pot were sown, hylems down, about one half inch deep and left exposed for inoculation. Using late-log to early stationary phase, 4- to 5-day-old cultures containing about 2.5 $\times$ 10^9 cells mL^{-1}, 12 mL of either I-110ARS or TA-11NOD$^+$ was mixed with 108 mL of USDA 6. Thirty milliliters of these mixtures were added to each of four pots, giving four replications of each treatment. The remaining two pots were each inoculated with 30 mL of USDA 6 only and served as controls. Since strain I-110, or its derivatives, and strain USDA 6 are approximately equal in their competitive ability when mixed in equal numbers (data not shown), the ratio of I-110ARS or TA-11NOD$^+$:USDA 6 was adjusted to 1:9 in order to challenge the test strains. Experiments involving highly competitive strain USDA 438 received mixtures in equal numbers with either I-110ARS or TA-11NOD$^+$. Seeds were covered with the surrounding soil and plants were grown for 40 days with supplemental lighting extending the light period to 16 h. During the growth period, the pots were watered as needed every three to four days with tap water.

Table 1. 1993 field tests on 'Morgan' soybean of various parameters of symbiotic fitness of *Bradyrhizobium japonicum* mutant and wild-type strains in soil containing an indigenous population of soybean-nodulating bacteria

Parameter[a]	Uninoculated	*B. japonicum* I-110ARS	TA-11NOD+ mutant
Nodule number plant^{-1}	11.0A	25.0B	36.0C
Nodule mass (mg plant^{-1})	14.0A	44.0B	65.0C
Nitrogenase activity (μmol hour^{-1} plant^{-1})	0.2A	1.0B	1.6C
Shoot dry weight (g plant^{-1})	2.3A	3.5B	4.8C
Nodule occupancy (%Rif^r Str^r)	0.0	20.0A	33.0B

[a]The mean of four replications at 42d after planting. Data were analyzed as randomized complete block design using SAS Institute's PROC.GLM. The strains were the treatment factor and replicates were the blocks. When necessary, the iterated re-weighted least square techniques was used to correct for variance heterogeneity. Parameter means followed by a different letter were significantly different ($p<0.01$).

Table 2. Nodule occupancy by strain TA-11NOD+ or its progenitor strain I-110ARS in competition with either strain USDA 6 or USDA 438

Strain	Competing strain[a]	Ratio	Number of nodules tested	%Nodule occupancy[b]
USDA 6	None	9:0	200	0.0%
I-110ARS	USDA 6	1:9	399	7.8% A
TA-11NOD+	USDA 6	1:9	400	39.5% B
USDA 438	None	1:0	120	0.0%
I-110ARS	USDA 438	1:1	133	3.8%A
TA-11NOD+	USDA 438	1:1	156	9.6% B

[a]Treatments involving strain USDA 6 were harvested at 40 days after planting (dap) while those involving USDA 438 were harvested at 47 dap.
[b]Means followed by a different letter were significant at $p = 0.0005$ for strains competing with USDA 6 and at $p < 0.01$ for USDA 438 mixtures.

Nitrogenase activity analysis

At six weeks after planting, five plants from each field plot were carefully uprooted. The plant tops were cut and each root system was gently separated from the soil and placed in a reaction jar fitted with a rubber septum. The air in each jar was amended with 10% acetylene and the jars were incubated at room temperature (25 °C) for 20 minutes. A sample of gas from each jar was analyzed for ethylene production by gas chromatography (Kummer and Kuykendall, 1989).

Nodule occupancy determinations

Procedures for screening genetic markers of bacteria within nodules were previously described (Kuykendall and Weber, 1978). Selective A1E medium (Kuykendall, 1987) contained 500 μg rifampicin SV (Sigma Chemical Co., St. Louis, MO) mL^{-1} and 1.0 mg streptomycin sulfate (Sigma) mL^{-1} and was used for unambiguous identification of the inoculum strain. To determine strain competitiveness among indigenous soybean-nodulating bacteria in the field, at least 20 randomly selected nodules from each replicate plot were surface-sterilized in 3% H_2O_2 for 1 hour, rinsed three times in sterile distilled water, then aseptically and individually crushed with a sterile glass rod. Some of the contents of each nodule was patched, using a sterile loop, onto both selective and nonselective A1E medium plates. The plates were incubated at 30 °C for seven days and scored by checking for growth of the patches on both media. Positives for growth on the selective medium were counted and taken as a fraction

Table 3. Field evaluation of *B. japonicum* strains at Upper Marlboro, Maryland, in 1988 and 1993

Cultiver	Inoculum strain	Seed protein (%)	Seed oil (%)	Seed yield (kg ha^{-1})
A5043[a]	None	35.7	22.2	2544 A
	I-110ARS	40.8	20.5	2860 B
	TA-11NOD$^+$	42.3	19.9	3053 B
Morgan[a]	None	40.1	22.4	2544 A
	I-110ARS	43.5	20.5	2888 B
	TA-11NOD$^+$	44.0	20.8	3025 B
Morgan[b]	None	41.2	19.2	1450 A
	I-110ARS	41.9	19.4	2732 B
	TA-11NOD$^+$	44.0	18.6	2942 B
Main Effect[c]	None	39.0	21.3	2179 A
	I-110ARS	42.1	20.1	2827 B
	TA-11NOD$^+$	43.4	19.8	3007 C

[a]In 1988, there were five replications. Within a cultivar, means followed by the same letter were not significantly different ($p < 0.05$).
[b]In 1993, there were four replications. Within a cultivar, means followed by the same letter were not significantly different ($p < 0.05$).
[c]The mean of the means for the two cultivars and the two years.

of the total number of nodules analyzed in either case and 'percent nodule occupancy' was calculated.

Results and discussion

Traditionally, it has long been recognized that nodule number is generally not correlated with the effectiveness of a strain since ineffective strains often form numerous small nodules. Our discovery of the ability of strain TA-11NOD$^+$ to form more nodules and fix more nitrogen per plant was different. Nodule mass increased concurrent with nodule number, and specific activity in fixing nitrogen was constant (Hunter and Kuykendall, 1990). We wanted to see if similar results could be obtained in the field. After six weeks of growth, even in soil already populated with soybean-nodulating *Bradyrhizobium*, plants inoculated with strain TA-11NOD$^+$ were significantly better nodulated than those inoculated with strain I-110ARS (Table 1). Forty-four percent more nodules were produced by the mutant strain compared to its progenitor. This result agreed well with our earlier findings in greenhouse studies under single-strain conditions (Hunter and Kuykendall, 1990). Acetylene reduction assays (Table 1) indicated that about 50% more nitrogen was being fixed, at the time the assays were performed, by plants inoculated with strain TA-11NOD$^+$ compared with those inoculated with its progenitor strain, I-110ARS. The difference between TA-11NOD$^+$ and I-110ARS in nitrogen fixation ability substantiates similar results obtained earlier in the greenhouse (Hunter and Kuykendall, 1990). A new finding was obtained from nodule occupancy determinations (Table 1). Since TA-11NOD$^+$ has the same Rif^r Str^r markers as its progenitor strain, I-110ARS (Kuykendall and Weber, 1978), it was possible to determine nodule occupancy using these genetic markers. Strain TA-11NOD$^+$ occupied a significantly greater proportion of nodules than did strain I-110ARS when these strains were introduced into a field already populated with soybean-nodulating bacteria. The mutant strain was thus shown for the first time to be more competitive than its progenitor.

In order to rule out the possibility that higher nodule occupancy was simply due to the formation of more nodules, greenhouse studies were conducted. These studies involved either strain USDA 6 or USDA 438 as the competing reference strain. Data on the relative competitive ability of strains TA-11NOD$^+$ and

I-110ARS is presented in Table 2. Interestingly, when mixed with strain USDA 6, strain TA-11NOD$^+$ produced five-fold greater percent nodule occupancy than did strain I-110ARS. In another experiment using equal mixtures with the highly competitive strain USDA 438, strain TA-11NOD$^+$ was present in a 2.5-fold higher proportion of the nodules. It is important to note that, although TA-11NOD$^+$ in direct competition with USDA 438 only occupied 10% of the nodules, the highly competitive field isolate, USDA 438, was present in numbers that were high, many fold greater than would be encountered in a soybean production field. In any case, regardless of whether either strain was very competitive with strain USDA 438, these data can be interpreted as demonstrating unambiguously that strain TA-11NOD$^+$ is more competitive than its progenitor strain I-110ARS.

Perhaps due to its enhanced competitiveness, strain TA-11NOD$^+$ was demonstrably superior to the progenitor strain I-110ARS in terms of producing significantly higher soybean seed yield (Table 3). Strain TA-11NOD$^+$ produced 6% more seed yield than strain I-110ARS on the two varieties in the two years. This finding is similar to the report (Bosworth et al., 1994) that genetically modified *Rhizobium meliloti* increased alfalfa yields. We plan to do more experiments with different varieties of soybean since Pinochet et al. (1993) reported cultivar-specific differences in interstrain competitiveness. These studies will help form the basis for future development of improved strains that will reliably increase yields of both alfalfa and soybean, while protecting the environment and decreasing our reliance on fossil fuel energy which is consumed in industrial production of nitrogen fertilizers.

Acknowledgements

We are grateful to Michael Behler and Robin Montenieri for their expert technical assistance. The help of Drs R C Leffel and D F Weber in conducting the 1988 field experiment is gratefully acknowledged. We also gratefully acknowledge the help of Professor Bill Kenworthy at the University of Maryland in conducting the 1993 field test. The authors also appreciate Dr Tom Wacek who shared the results of field experiments conducted by Urbana Laboratories of St Joseph, Missouri. The USDA has, recently licensed Urbana Laboratories to sell strain TA-11NOD$^+$ for use as a new soybean inoculant. We thank Mary Camp for performing the statistical analyses.

References

Bosworth A H, Williams M K, Albrecht K A, Kwiatkowski R, Beynon J, Hankiuson T R, Ronson C W, Cannon F, Wacek T J and Triplett E W 1994 Alfalfa yield response to inoulation with recombinant strains of *Rhizobium meliloti* with an extra copy of *dctABD* and/or modified *nifA* expression. Appl. Environ. Microbiol. 60, 3815–3832.

Dowling D N and Broughton W J 1986 Competition for nodulation of legumes. Annu. Rev. Microbiol. 40, 131–157.

Hunter W J and Kuykendall L D 1990 Enhanced nodulation and nitrogen fixation by a revertant of a nodulation-defective *Bradyrhizobium japonicum* tryptophan auxotroph. Appl. Environ. Microbiol. 56, 2399–2403.

Johnson H W, Means U M and Weber C R 1965 Competition for nodule sites between strains of *Rhizobium japonicum* applied as inoculum and strains in the soil. Agron. J. 57, 178–185.

Keyser H H and Cregan P B 1987 Nodulation and competition for nodulation of selected soybean genotypes among *Bradyrhizobium japonicum* serogroup 123 isolates. Appl. Environ. Microbiol. 53, 2631–2635.

Kummer R M and Kuykendall L D 1989 Symbiotic properties of amino acid auxotrophs of *Bradyrhizobium japonicum*. Soil Biol. Biochem. 21, 779–782.

Kuykendall L D 1987 Isolation and identification of genetically marked strains of nitrogen-fixing microsymbionts of soybeans. *In* Symbiotic Nitrogen Fixation Technology. Ed. G H Elkan. pp 205–220. Marcel Dekker, New York, USA,

Kuykendall L D and Hunter W J 1991 U.S. patent 5,021, 076.

Kuykendall L D and Hunter W J 1992 Altered tryptophan biosynthesis in *Bradyrhizobium japonicum* gives enhanced nodulation and nitrogen fixation. *In* Plant Biotechnology and Development. Ed. P M Gresshoff. pp 71–79. CRC Press, Ann Arbor, MI, USA.

Kuykendall L D and Hunter W J 1995 Symbiotic ineffectiveness of *trpCD* deletion mutants of *Bradyrhizobium japonicum*. Soil Biol. Biochem. 27, 1035–1039.

Kuykendall L D and Weber D F 1978 Genetically marked *Rhizobium* identifiable as inoculum strain in nodules of soybean plants grown in fields populated with *Rhizobium japonicum*. Appl. Environ. Microbiol. 36, 915–919.

Pinochet X, Arnaud F and Cleyet-Marel J C 1993 Competition for nodule occupancy of introduced *Bradyrhizobium japonicum* strain SMGS1 in French soils already containing *Bradyrhizobium japonicum* strain G49. Can. J. Microbiol. 39, 1022–1028.

Streeter J G 1994 Failure of inoculant rhizobia to overcome the dominance of indigenous strains for nodule formation. Can. J. Microbiol. 40, 513–522.

Triplett E W and Sadowsky M J 1992 Genetics of competition for nodulation of legumes. Annu. Rev. Microbiol. 46, 399–428.

Van Rhijn P and Vanderleyden J 1995 The *Rhizobium*-plant symbiosis. Microbiol. Rev. 59, 124–142.

Weaver R W and Frederick L R 1971 A new technique for most-probable-number of rhizobia. Plant and Soil 36, 219–222.

Zdor R E and Pueppke S G 1991 Nodulation competitiveness of Tn-5 induced mutants of *Rhizobium fredii* USDA 208 that are altered in motility and extracellular polysaccharide production. Can. J. Microbiol. 37, 52–58.

Section editor: H Lambers
Guest editors: G.H. Elkan and R.G. Upchurch

Plant and Soil **186:** 127–134, 1996.

Phage susceptibility and plasmid profile analysis of *Sinorhizobium fredii*

F.M. Hashem, L.D. Kuykendall, S.E. Udell[1] and P.M. Thomas[2]
Soybean and Alfalfa Research Laboratory, USDA/ARS, 10300 Baltimore Avenue, Building 011, HH-19, Beltsville, MD 20705-2350, USA. Present addresses:* [1]*Department of Microbiology, North Carolina State University, Raleigh, NC 27695, USA and* [2]*Department of Biology, San Diego State University, San Diego, CA 92182, USA*

Key words: host range, megaplasmid, phage typing, rhizobiophage, *Rhizobium fredii*

Abstract

This is the first report identifying bacteriophages and documenting megaplasmids of *Sinorhizobium fredii*. Plasmid DNA content and bacteriophage typing of eighteen strains of *S. fredii* were determined. *S. fredii* strains fell into ten plasmid profile groups containing 1 to 6 plasmids, some evidently larger than 1000 MDa. Twenty-three *S. fredii* lytic phages were isolated from soil, and they lysed six different *S. fredii* strains. The host range and plaque morphology of these phages were studied. Susceptibility to *S. fredii* phages was examined for *S. meliloti*; *Rhizobium leguminosarum* bvs. *viceae*, *trifolii* and *phaseoli*; *R. loti*; *Bradyrhizobium japonicum*; *B. elkanii* and *Bradyrhizobium* sp. (*Arachis*). Several phages that originally lysed *S. fredii* strain USDA 206 also lysed strains of all three *S. fredii* serogroups described originally by Sadowsky et al. Phages that infected *S. fredii* strains USDA 191 and USDA 257 were highly specific and lysed only serogroup 193 strains. *S. meliloti* strains L5-30 and USDA 1005 were lysed by three of the phages that lysed *S. fredii* strain USDA 217. No other *Rhizobium* or *Bradyrhizobium* strain tested was susceptible to lysis by any of the *S. fredii* phages. The present investigation indicates that phage susceptibility in conjunction with plasmid profile analysis may provide a rapid method for identification and characterization of strains of *S. fredii*.

Introduction

Studies of bacterial viruses and of plasmids have been useful for advancing bacterial genetics, taxonomy and ecology. Specifically, bacteriophage susceptibility (Kankila and Lindstrom, 1994; Werquin et al., 1988) and plasmid DNA content have proven to be useful tools for differentiating *Rhizobium* strains and studying their ecology (Brockman and Bezdicek, 1989; Buendia-Claveria et al., 1989; Hartmann and Amarger, 1991; Hashem and Kuykendall, 1994; Kuykendall et al., 1996; Moenne-Loccoz et al., 1994; Shishido and Pepper, 1990; Zahran, 1992). Their ability to promote genetic exchange, or transduction, between host strains has also been useful (Kinkle and Schmidt, 1991; Pretorius-Guth et al., 1990). Recently, both *R. fredii* and *R. meliloti* were reclassified into a new genus, and their new names are *Sinorhizobium fredii* (Chen et al., 1988) and *Sinorhizobium meliloti* (De Lajudie et al., 1994), respectively.

Plasmid profiles analysis and the identification of symbiotic plasmids (Sym) were recently performed on some agronomically important *Rhizobium* species that nodulate alfalfa, berseem clover, or *Leucaena* (Hashem and Kuykendall, 1994). Individual *Rhizobium* strains within a species varied widely in plasmid DNA content, including the presence and sizes of Sym plasmids. A similar diversity within serogroups of *R. leguminosarum* bv. *viceae* from the Palouse region of eastern Washington was observed using plasmid profiles, intrinsic antibiotic resistance and topography (Brockman and Bezdicek, 1989). Moenne-Loccoz and Weaver (1995) and Thurman et al. (1985) found a specific relationship between plasmid number and growth, survival, acid tolerance and symbiotic efficiency of some *R. leguminosarum* bv. *trifolii* strains in the rhizosphere of clover. Plasmids also influence motility (Baladani and Weaver, 1992) and the ability of *Rhizobium* strains to utilize particular organic compounds (Baladani et al., 1992). One or more plasmids that carry genes controlling symbiotic functions are

* FAX No: + 13015045728. E-mail: fhashem@asrr.arsusda.gov

found in certain strains of *R. leguminosarum* bv. *trifolii* (Hashem and Kuykendall, 1994; Hynes and McGregor, 1990), *R. meliloti* (Hynes et al., 1986; Hashem and Kuykendall, 1994), and *Rhizobium* species that nodulate *Leucaena* (Hashem and Kuykendall, 1994; Kuykendall et al., 1996). In addition to symbiotic plasmids, *Rhizobium* strains may carry between one and ten cryptic plasmids (Hashem and Kuykendall, 1994; Thurman et al., 1985) ranging in size from < 30 MDa to > 1000 MDa (Baladani and Weaver, 1992; Hashem and Kuykendall, 1994). In early studies that characterized some of the smaller plasmids of *S. fredii* strains (Heron and Pueppke, 1984; Masterson et al., 1982), the plasmids ranged in number from one to four and were up to 300 MDa in size. Because symbiosis genes are generally plasmid-borne in *Rhizobium*, and some interesting megaplasmids of up to and perhaps greater than 1000 MDa have been identified for some related *S. meliloti* strains (Hashem and Kuykendall, 1994; Hynes et al., 1986), this study was carried out to determine plasmid profiles and to identify megaplasmid content of eighteen *S. fredii* strains.

Rhizobiophages vary considerably in their morphology, plaque characteristics, and host range (Barnet, 1972; Dhar and Ramkrishna, 1987; Hashem et al., 1986; Kankila and Lindstrom, 1994; Werquin et al., 1988). Bacteriophages have been isolated for most *Rhizobium* and *Bradyrhizobium* species including *B. japonicum* (Hashem et al., 1986; Stacey et al., 1984), *R. leguminosarum* bv. *trifolii* (Barnet, 1972; Kankila and Lindstrom, 1994), and *S. meliloti* (Werquin et al., 1988). Most rhizobiophages studied are virulent (Hashem et al., 1986), although lysogeny has been found in *Rhizobium* (Kowalski, 1966) and *Bradyrhizobium* (Abebe et al., 1992).

Bacteriophages of slow-growing *Bradyrhizobium* strains that nodulate soybean have been investigated in several studies including phage typing of *B. japonicum* and *B. elkanii* strains, their roles in the survival and symbiotic effectiveness of their hosts, in the competition between strains nodulating soybean, and their effects on soybean growth and yield (Hashem and Angle, 1988a,b, 1990; Hashem et al., 1986). Abebe et al. (1992) isolated one lytic and two temperate phages specific for *B. japonicum* USDA 123 and studied their effects on soybean nodulation. These phages differed in host range and virulence, and strains carrying prophages were unable to nodulate soybean plants. Schmidt et al. (1986) isolated 12 phages from soil and used them to characterize a group of 79 isolates of *B. japonicum* serocluster 123. Stacey et al. (1984) isolated, from Tennessee soil, a bacteriophage named TN1 that lysed *B. japonicum* strain USDA 110, but could not lyse a particular non-nodulating mutant of that same strain.

Rhizobiophages infective for *S. fredii* have not been reported, and the plasmid DNA content of *S. fredii* has not been extensively studied. Therefore, our main objectives were as follows: (1) to analyze plasmid DNA profiles and megaplasmid content of strains of *S. fredii*, (2) to isolate and characterize phages specific for *S. fredii*, and (3) to determine the susceptibility of diverse *Rhizobium*, *Sinorhizobium* and *Bradyrhizobium* strains to the phages.

Materials and methods

Bacterial strains and growth conditions

Forty *Rhizobium*, *Sinorhizobium* and *Bradyrhizobium* strains were obtained from the United States Department of Agriculture, USDA-ARS National *Rhizobium* Culture Collection, Beltsville, Maryland. *S. fredii* strains were used as standard hosts for rhizobiophage isolation and phage typing, and for plasmid DNA isolation and plasmid profile analysis. *Rhizobium*, *Sinorhizobium* and *Bradyrhizobium* strains were maintained on yeast extract mannitol agar (YEMA) slants (Vincent, 1970) or tryptone yeast agar (TY) slants (Beringer, 1974). Bacterial strains used were: (a) *S. fredii* USDA strains 191, 192, 193, 194, 201, 205, 206, 208, 214, 217, 257, HH 303, and NifTAL strains 1832, 1899, 1901, 1902, 1903; (b) *S. meliloti* USDA strains 1002, 1005, and L5-30; (c) *R. leguminosarum* bv. *viceae* USDA strains 2370, 2335, bv. *trifolii* USDA strain 2048, and bv. *phaseoli* USDA strains 2667, 2669, and strain DB1; (d) *R. loti* USDA strains 3228 and 3339; (e) *Bradyrhizobium* sp. (*Arachis*) strain USDA 3275; (f) *B. japonicum* USDA strains 6, 110, 122, 123, 124, 243 and 258, and (g) *B. elkanii* USDA strains 31, 46, 76, and 94.

Plasmid DNA content and plasmid profile

Plasmid DNAs were visualized on horizontal agarose gels using an in-well lysis procedure (Hashem and Kuykendall, 1994). Gels were prepared in TBE buffer (89 m*M* Tris-borate, 2 m*M* EDTA, pH 8.0) with 0.75% (w:v) agarose and 1% (w:v) SDS. An aliquot of 0.2 mL early log-phase culture was transferred to a sterile Eppendorf tube on ice, then five volumes of 0.3%

(w:v) N-lauroylsarcosine in TBE buffer was added to the culture and mixed gently by tapping. The tube was kept on ice for at least 10 min. The cells were pelleted by centrifugation in a benchtop microcentrifuge at 17,000 rpm at 4°C for 5 min and the supernatant was removed as completely as possible without disturbing the pellet. Pellets were suspended in 25 μL of lysis solution containing lysozyme (0.2 mg mL^{-1}), RNAase (10 mg mL^{-1}), and 10% (w:v) sucrose in TBE, then immediately loaded into the wells in the gel. Five volts were applied for 30–45 min, or until turbidity disappeared, then the gel was run at 85 volts for 8 h. Gels were stained in deionized water containing 1 μL of a 1% (w:v) ethidium bromide stock solution per 100 mL for 1 h at room temperature, and destained in deionized water for 1 h before photographing on a UV transilluminator. Molecular sizes for plasmids were determined as described by Shishido and Pepper (1990), using strain USDA 205 as a reference (Hashem and Kuykendall, 1994).

Bacteriophage isolation

Phages infecting *S. fredii* strains were isolated from the rhizosphere soil of soybean plants grown in the Shanghai region of China by the enrichment technique of Barnet (1972) or by the direct double-layer agar technique of Hashem et al. (1986). The initial isolation was conducted by suspending a soil sample in a phosphate buffer. The suspension was shaken in the presence of 0.5% chloroform for 30 min. The soil suspension was then centrifuged at 500 × *g* for 15 min. The supernatant was filtered through a sterile 0.22-μm Millipore filter. The resulting cell-free soil extract was tested for specific phages against eighteen *S. fredii* strains. Whenever a phage could not be detected with a potential host strain, an enrichment technique (Barnet, 1972) was used. In this case, 10 g soil was transferred into a 250-mL flask containing a 100-mL *Sinorhizobium* culture in the early log phase. The mixture was incubated at 30°C for five days, then 0.5% chloroform was added to the culture. The mixture was centrifuged to pellet the soil and bacterial debris, and the suspension was filtered as described before. One mL of each *Sinorhizobium* strain and 0.5 mL of a diluted phage suspension were mixed with 5 mL of molten TY soft agar (0.7% agar). The mixture was overlaid on solid TY agar plates. Plates were incubated at 30°C for three days. Twenty-three phages infectious for six of the *S. fredii* strains were isolated and purified through five successive single-plaque isolations (Hashem et al., 1986). No phages infective on any of the other *S. fredii* strains were isolated. Rhizobiophage stocks were obtained by infecting an exponentially growing broth culture of the original host with a suspension of a phage. By using this procedure, titres of about 1 × 10^{11} plaque-forming units (PFU) mL^{-1} were obtained for each phage. The phages were stored in both TY broth containing 0.5% chloroform at 4°C and in TYB containing 0.5% chloroform and 50% glycerol at -70°C. Phage storage with chloroform maintained viability over a period of 28 months (Hashem et al., 1986).

Phage typing, phage host specificity and plaque morphology

Forty *Rhizobium*, *Sinorhizobium* and *Bradyrhizobium* strains, mentioned before, were examined for phage susceptibility. Samples (0.5 mL) of exponentially growing cultures of each strain and 0.5 mL of diluted phage stock were suspended in 5.0 mL molten TYA or YEMA (0.7% agar) and overlaid onto TYA or YEMA plates. Plates were incubated at 30°C and were examined for plaque formation after two days for *Sinorhizobium* and *Rhizobium* and after six days for *Bradyrhizobium.* Plaque size and morphology for each phage were determined daily for seven days. Three different stocks of each phage, prepared at different times, were assayed. The average sizes of 90 plaques for each phage were recorded daily until the plaque size remained unchanged for several days.

Results

Plasmid DNA content and plasmid profiles

Plasmids were analyzed from each of eighteen *S. fredii* strains and varied in number from 1 to 6 and in size from < 50 to > 1000 MDa (Table 1). *S. fredii* strains had ten different plasmid profiles, based on number and size of plasmids, containing 1, 2, 3, 4, or 6 plasmids (Table 1). Strains differing in either the number or size of plasmids represented different plasmid profiles. For example, strain USDA 205 and TAL 1899 harbored four plasmids each, but differed significantly in the size of two plasmids, and were placed into two different plasmid profile groups. All strains except HH 303 and USDA 257 harbored at least one cryptic megaplasmid as large as 1000 MDa. Strain HH 303 had two large plasmids of 700 and 800 MDa and strain USDA 257 evidently had only one large plasmid of 308 MDa. The

Table 1. Plasmid DNA content of *Sinorhizobium fredii* strains

Strain	Number of Plasmids	Size (MDa) a	b	c	d	e	f	Plasmid profile group
USDA 191	4	1000	230	70	< 50			7
USDA 192	3	1000	250	80				6
USDA 193	2	1000	250					3
USDA 194	2	1000	140					4
USDA 201	3	1000	210	120				6
USDA 205	4	1000	248	127	83			9
USDA 206	4	1000	230	65	56			7
USDA 208	2	1000	240					3
USDA 214	2	1000	235					3
USDA 217	3	1000	245	95				6
USDA 257	1	308						2
TAL 1832	2	1000	270					3
TAL 1899	4	1000	300	83	< 50			7
TAL 1900	6	> 1000	270	83	50	< 50	<< 50	10
TAL 1901	1	1000						1
TAL 1902	2	> 1000	250					3
TAL 1903	1	> 1000						1
HH 303	2	800	700					5

plasmid profiles of strains USDA 193 and TAL 1902 were evidently identical.

Rhizobiophage isolation and characterization

Phages were isolated for six of the *S. fredii* strains tested. Phages were lytic for USDA strains 191, 205, 206, 208, 217 and 257 (Table 2). Plaques ranged in size from 0.5 to 3.0 mm in diameter and were either clear, turbid with a clear center, or turbid. Margins were either sharp, spattered, fuzzy or fuzzy with a halo. Plaques exhibited some diversity in morphology and size, with phages 205B and 257B having the largest and smallest plaque sizes, respectively. Whereas plaques formed by phage 205B were clear with sharp margins, plaques formed by phage 257B were turbid with indistinct margins.

Host range and phage typing

The host ranges of the twenty-three phage isolates were examined on the forty *Rhizobium*, *Sinorhizobium* and *Bradyrhizobium* strains mentioned before. A strain was judged as highly susceptible if the phage produced a number of plaques similar to that obtained on the original host; less susceptible if the phage produced plaques at several log numbers lower than those obtained on the original host; or resistant if no lysis was observed (Table 3). Some phages were highly specific and lysed only their original hosts, while others were found to be less specific, infecting more than one *S. fredii* strain. Some of the phages, however, had an extended host range within strains of *S. fredii* (Table 3). Phages isolated on USDA 206 also lysed other strains within all serogroups of *S. fredii* (Table 4). Phage 206D had a wide host range, infecting and lysing *S. fredii* USDA strains 193, 201, 206, 214 and 217. In contrast, phages isolated on USDA strains 191 and 257 were highly specific, infecting and lysing only USDA strains 191 and 257, respectively. *Sinorhizobium fredii* USDA strains 193, 194, 201 and 214, for which no phages were originally isolated, were susceptible to phages isolated on other *S. fredii* strains. The phages, with the exception of three USDA 217 phages, did not lyse any of the strains tested as hosts other than *S. fredii* strains. Phage isolates 217B, 217C and 217D also infected *S. meliloti* strains L5-30 and USDA 1005. *S. meliloti* strain USDA 1002 was not lysed by any of the phages. Phages fell into 11 phage clusters based on their plaque morphology, plaque size and host specificity (Table 3).

Table 2. Description of plaque morphologies on various strains of *S. fredii*

Host strain and phage	Plaque size (mm)	Clarity	Margins
191 A	1.0	Clear	Sharp
191 B	1.0	Clear	Fuzzy
191 C	0.5	Slightly turbid	Fuzzy with halo
191 D	3.0	Clear	Sharp
205 A	2.0	Clear	Sharp
205 B	3.0	Clear	Sharp
206 A	1.2	Clear	Fuzzy with halo
206 B	2.0	Clear	Sharp
206 C	2.0	Clear	Fuzzy with halo
206 D	1.5	Clear	Fuzzy with halo
208 A	1.0	Clear	Sharp
208 B	0.8	Clear	Sharp
208 C	2.0	Clear	Fuzzy
208 D	2.0	Clear	Sharp
217 A	1.8	Clear	Sharp
217 B	2.0	Clear	Spattered
217 C	0.5	Clear center	Turbid with halo
217 D	0.5	Clear center	Turbid
217 E	2.5	Clear	Spattered
257 A	0.5	Clear	Sharp
257 B	0.5	Turbid	Fuzzy
257 C	1.8	Clear	Sharp
257 D	1.5	Clear	Sharp

Discussion

This study identifies megaplasmids greater than 1000 MDa in the fast-growing *S. fredii* strains that nodulate soybean. The *S. fredii* strains examined had unique plasmid profiles, except strains USDA 193 and TAL 1902 which had indistinguishable profiles. The presence of 1000-MDa megaplasmids, as well as the large number of plasmids that many *S. fredii* strains harbored, is similar to data reported for some *S. meliloti* strains (Hashem and Kuykendall, 1994; Hynes et al., 1986). Although some plasmids of *S. fredii* have been studied before (Barbour and Elkan, 1989; Buendia-Claveria et al., 1989; Heron and Pueppke, 1984; Masterson et al., 1982; Sadowsky and Bohlool, 1983), these studies focused on one or few *S. fredii* strains. These investigations reported narrow plasmid profile groups consisting of up to four plasmids; the largest plasmid detected was 300 MDa in size. The present study used a modified in-well lysis procedure (Hashem and Kuykendall, 1994) which identified a wide range of plasmid profile groups for eighteen *S. fredii* strains containing up to six plasmids, and each strain, except strains USDA 257 and HH 303, carried a megaplasmid approximately 1000 MDa in size. Smaller megaplasmids of about 600 MDa have been identified in *R. tropici* and were specific for subspecies A and B. These plasmids have been used to differentiate between the two subtypes of *R. tropici* (Geniaux et al., 1995). The present investigation, and that of Heron and Pueppke (1984), shows that plasmid profile analysis can be used to distinguish *S. fredii* strains. This finding is in agreement with Thomas et al. (1994), who used plasmid profile analysis and the presence of megaplasmids as large as 900 MDa to distinguish among groups of mesquite rhizobia and bradyrhizobia. Other studies have also reported a wide range of plasmid profiles for *R. leguminosarum* bv. *trifolii* (Moenne-Loccoz et al.,

Table 3. Cross-infection of *Sinorhizobium fredii* phages with other *S. fredii* strains

Phage	Phage cluster	*Sinorhizobium fredii* strains										
		191	192	193	194	201	205	206	208	214	217	257
191 A	1	+[a]	0	0	0	0	0	0	0	0	0	0
191 B	1	+	0	0	0	0	0	0	0	0	0	0
191 C	1	+	0	0	0	0	0	0	0	0	0	0
191 D	1	+	0	0	0	0	0	0	0	0	0	0
205 A	2	0	0	0	0	0	+	0	+	0	0	+
205 B	2	0	0	0	0	0	+	0	+	0	0	+
206 A	4	0	0	+	0	+	0	+	0	0	+	0
206 B	4	0	0	+	0	+	0	+	0	0	+	0
206 C	4	0	0	+	0	+	0	+	0	0	+	0
206 D	5	0	0	+	0	+	0	+	0	±	+	0
208 A	2	0	0	0	0	0	+	0	+	0	0	+
208 B	2	0	0	0	0	0	+	0	+	0	0	+
208 C	2	0	0	0	0	0	+	0	+	0	0	+
208 D	3	0	0	0	0	0	±	0	+	0	0	+
217 A	6	0	0	0	0	0	0	0	0	0	+	0
217 B	8	0	0	0	+	+	0	0	0	0	+	0
217 C	10	0	0	0	±	±	0	0	0	0	+	0
217 D	9	0	0	0	±	+	0	0	0	0	+	0
217 E	7	0	0	0	+	0	0	0	0	0	+	0
257 A	11	0	0	0	0	0	0	0	0	0	0	+
257 B	11	0	0	0	0	0	0	0	0	0	0	+
257 C	11	0	0	0	0	0	0	0	0	0	0	+
257 D	11	0	0	0	0	0	0	0	0	0	0	+

[a]+: indicates that number of PFUs mL^{-1} is similar to that obtained on original host; ±: indicates that number of PFUs mL^{-1} is several log numbers lower than those obtained on original host; 0: no visible infection.

1994), *S. meliloti* (Hashem and Kuykendall, 1994) and *R. leguminosarum* strains that nodulate *Vicia faba* (Van Berkum et al., 1995). Furthermore, this technique has been used to discriminate between *R. galegae* and *S. meliloti* (Salenska-Trajkowa et al., 1990). Plasmid profiles, thus, can be more discriminating for *Rhizobium* and *Sinorhizobium* strains than insertion sequence (IS) fingerprints (Hartmann and Amagar, 1991) and RFLP analysis of total DNA (Kuykendall et al., 1996).

This investigation is also the first to isolate, characterize and document bacteriophages infective for *S. fredii* strains, and also the first to examine bacterial susceptibility to these phages. Several monovalent and polyvalent lytic phages for *S. fredii* were identified. The bacterial viruses exhibited some diversity in plaque size, plaque morphology and host range. The

Table 4. Phage cross-infection with *Sinorhizobium fredii* serogroups

Phage	*S. fredii* Serogroup (Sadowsky et al., 1987)		
	192	193	194
191 A, B, C, D	0[a]	+	0
205 A, B	+	+	0
206 A, B, C, D	+	+	+
208 A, B, C, D	+	+	0
217 A	+	0	0
217 B, C, D, E	+	0	+
257 A, B, C, D	0	+	0

+: indicates infection and lysis; 0: indicates no visible infection.

plaque sizes and morphologies for these phages were similar to those of phages described for other species of *Rhizobium*, *Sinorhizobium* and *Bradyrhizobium* (Dhar and Ramkrihana, 1987; Hashem et al., 1986; Werquin et al., 1988). Some phages were highly specific to a certain *S. fredii* serogroup, whereas others lysed strains within all three serogroups. Interestingly, three phages isolated on strain USDA 217 had an extended host range beyond *S. fredii*, including *S. meliloti* strains L5-30 and USDA 1005. The diversity of host range found among *Sinorhizobium* phages has been reported for bacteriophages infecting *R. leguminosarum* bv. *trifolii* (Kankila and Lindstrom, 1994), whereas *B. japonicum* phages were found to be highly specific to their hosts (Hashem et al., 1986). Our results also indicate a clear phage cluster pattern for *S. fredii* phages since eleven phage host-range clusters were identified.

In conclusion, the present study indicates that both plasmid profile analysis and phage typing were more discriminatory of strains of *S. fredii* than were serological assays (Krishnan and Pueppke, 1994; Sadowsky et al., 1987), host range for plant species, or restriction fragment length polymorphism (RFLP) analyses (Krishnan and Pueppke, 1994). Therefore, phage typing, in conjunction with plasmid profile analysis, can be used to identify and characterize *S. fredii* strains.

Acknowledgements

We thank Dr Peter van Berkum for helpful discussion, and Mr Mike Behler for proofreading the manuscript.

References

Abebe H M, Sadowsky M J, Kinkle B K and Schmidt E L 1992 Lysogeny in *Bradyrhizobium japonicum* and its effect on soybean nodulation. Appl. Environ. Microbiol. 58, 3360–3366.

Baladani J I and Weaver R W 1992 Survival of clover rhizobia and their plasmid-cured derivatives in soil under heat and drought stress. Soil Biol. Biochem. 24, 737–742.

Baladani J I, Weaver, R W, Hynes M F and Erdly B D 1992 Utilization of carbon substrates, electrophoretic enzyme patterns, and symbiotic performance of plasmid-cured clover rhizobia. Appl. Environ. Microbiol. 58, 2308–2314.

Barbour W M and Elkan, G H 1989 Relationship of the presence and copy number of plasmids to symbiotic effectiveness in *Rhizobium fredii*. Appl. Environ. Microbiol. 55, 813–818.

Barnet Y M 1972 Bacteriophages of *Rhizobium trifolii*. I. Morphology and host range. J. Gen. Virol. 15, 1–15.

Beringer J E 1974 R factor transfer in *Rhizobium leguminosarum*. J. Gen. Microbiol. 84, 188–198.

Brockman F J and Bezdicek D F 1989 Diversity within serogroups of *Rhizobium leguminosarum* biovar *viceae* in the Palouse region of eastern Washington as indicated by plasmid profiles, intrinsic antibiotic resistance, and topography. Appl. Environ. Microbiol. 55, 109–115.

Buendia-Claveria A M, Chamber M and Ruiz-Sainz J E 1989 A comparative study of the physiological characteristics, plasmid content and symbiotic properties of different *Rhizobium fredii* strains. Syst. Appl. Microbiol. 12, 203–209.

Chen W X, Yan G H and Li J L 1988 Numerical taxonomic study of fast-growing soybean rhizobia and a proposal that *Rhizobium fredii* be assigned to *Sinorhizobium* gen. nov. Int. J. Syst. Bacteriol. 28, 392–397.

Dhar B and Ramkrishna K 1987 Morphology and general characteristics of phages of chickpea rhizobia. Arch. Microbiol. 147, 121–125.

De Lajudie P, Willems A, Pot B, Dewettinck D, Maestrojuan G, Neyra M, Collins M D, Dreyfus B, Kersters K and Gillis M 1994 Polyphasic taxonomy of rhizobia: Emendation of the genus *Sinorhizobium* and description of *Sinorhizobium meliloti* comb. nov., *Sinorhizobium saheli* sp. nov., and *Sinorhizobium teranga* sp. nov. Int. J. Syst. Bacteriol. 44, 715–733.

Geniaux E, Flores M, Palcios R, and Martinez E 1995 Presence of megaplasmids in *Rhizobium tropici* and further evidence of differences between the two *R. tropici* subtypes. Int. J. Syst. Bacteriol. 45, 392–394.

Hartmann A and Amarger N 1991 Genotypic diversity of an indigenous *Rhizobium meliloti* field population assessed by plasmid profiles, DNA fingerprinting, and insertion sequence typing. Can. J. Microbiol. 37, 600–608.

Hashem F M and Angle J S 1988a Rhizobiophage effects on *Bradyrhizobium japonicum* nodulation and soybean growth. Soil Biol. Biochem. 20, 69–73.

Hashem F M and Angle J S 1988b Rhizobiophage enhancement of nodule occupancy by inoculated soybean rhizobia. *In* The Role of Biological Nitrogen Fixation for Development of Rural Africa. Eds. A M Abdel-Hafez, M E El-Haddad and M N Magdoub. pp 209–218. Cairo Mircin Publishers, Cairo, Egypt.

Hashem F M and Angle J S 1990 Rhizobiophage effects on nodulation, nitrogen fixation, and yield of field-grown soybeans (*Glycine max* L. Merr.). Biol. Fertil. Soils 9, 330–334.

Hashem F M, Angle J S and Ristiano P A 1986 Isolation and characterization of rhizobiophage specific for *Rhizobium japonicum* USDA117. Can. J. Microbiol. 32, 326–329.

Hashem F M and Kuykendall L D 1994 Plasmid DNA content of several agronomically important *Rhizobium* species that nodulate alfalfa, berseem clover, and *Leucaena*. *In* Symbiotic Nitrogen Fixation. Eds. P H Graham, M J Sadowsky and C P Vance. pp 181–188. Kluwer Academic Publishers, Dordrecht, The Netherlands.

Heron D S and Pueppke S G 1984 Mode of infection, nodulation specificity, and indigenous plasmids of 11 fast-growing *Rhizobium japonicum* strains. J. Bacteriol. 160, 1061–1066.

Hynes M F and McGregor N F 1990 Two plasmids other than the nodulation plasmids are necessary for formation of nitrogen-fixing nodules by *Rhizobium leguminosarum*. Mol. Microbiol. 4, 567–574.

Hynes M F, Simon R, Muller P, Niehaus K, Labs M and Puhler A 1986 The two megaplasmids of *Rhizobium meliloti* are involved in the effective nodulation of alfalfa. Mol. Gen. Genet. 202, 356–362.

Kankila J and Lindstrom K 1994 Host range, morphology and DNA restriction patterns of bacteriophage isolates infecting *Rhizobium leguminosarum* bv. *trifolii*. Soil Biol. Biochem. 26, 429–437.

Kinkle B K and Schmidt E L 1991 Transfer of the pea symbiotic plasmid pJB5JI in nonsterile soil. Appl. Environ. Microbiol. 57, 3264–3269.

Kowalski M 1966 Lysogeny in *Rhizobium meliloti*. Acta Microbiol. Pol. 15, 119–128.

Krishnan H B and Pueppke S G 1994 Host range, RFLP, and antigenic relationships between *Rhizobium fredii* and *Rhizobium* sp. NGR234. Plant and Soil 161, 21–29.

Kuykendall L D, Swelim D M, Hashem F M, Abdel-Wahab S M and Hegazi N I 1996 Symbiotic competence, genetic diversity, and plasmid profiles of Egyptian isolates of a *Rhizobium* sp. from *Leucaena leucocephala* (Lam.) Dewit. Lett. Appl. Microbiol. 22, 347–352.

Masterson R V, Russel P R, and Atherly A G 1982 Nitrogen fixation (*nif*) genes and large plasmids of *Rhizobium japonicum*. J. Bacteriol. 152, 928–931.

Moenne-Loccoz Y, Sen D, Krause E S and Weaver R W 1994 Plasmid profiles of rhizobia used in inoculants and isolated from clover. Agron. J. 86, 117–121.

Moenne-Loccoz Y and Weaver R W 1995 Plasmids influence growth of rhizobia in the rhizosphere of clover. Soil Biol. Biochem. 27, 1001–1004.

Pretorius-Guth I M, Puhler A and Simon R 1990 Conjugal transfer of megaplasmid between *Rhizobium meliloti* strains in alfalfa nodules. Appl. Environ. Microbiol. 56, 2354–2359.

Sadowsky M J and Bohlool B B 1983 Possible involvement of a megaplasmid in nodulation of soybean by fast-growing rhizobia from China. Appl. Environ. Microbiol. 46, 906–911.

Sadowsky M J, Bohlool B B and Keyser H H 1987 Serological relatedness of *Rhizobium fredii* to other rhizobia and bradyrhizobia. Appl. Environ. Microbiol. 53, 1785–1789.

Schmidt E L, Zidwick M J, and Abebe H M 1986 *Bradyrhizobium japonicum* serocluster 123 and diversity among member isolates. Appl. Environ. Microbiol. 51, 1212–1215.

Salenska-Trajkowa S, Radewa G and Markov K 1990 Comparison between *Rhizobium galegae* and *Rhizobium meliloti* plasmid content. Lett. Appl. Microbiol. 10, 123–126.

Shishido M and Pepper I L 1990 Identification of dominant indigenous *Rhizobium meliloti* by plasmid profiles and intrinsic antibiotic resistance. Soil Biol. Biochem. 22, 11–16.

Stacey G, Poctratatsky L and Puvanesarajah V 1984 Bacteriophage that can distinguish between wild type *Rhizobium japonicum* and a non-nodulating mutant. Appl. Environ. Microbiol. 48, 68–72.

Thomas P M, Golly K F, Zyskind J W and Virginia R A 1994 Variation of colonal, mesquite-associated rhizobial and bradyrhizobial populations from surface and deep soils by symbiotic gene region restriction fragment length polymorphism and plasmid profile analysis. Appl. Environ. Microbiol. 60, 1146–1153.

Thurman N P, Lewis D M and Jones D G 1985 The relationship of plasmid number to growth, acid tolerance and symbiotic efficiency in isolates of *Rhizobium trifolii*. J. Appl. Bacteriol. 58, 1–6.

Van Berkum P, Beyene D, Vera F T and Keyser H H 1995 Variability among *Rhizobium* strains originating from nodules of *Vicia faba*. Appl. Environ. Microbiol. 61, 2649–2653.

Vincent J M 1970 A Manual for the Practical Study of Root-nodule Bacteria. Blackwell Scientific Publications, Oxford, UK.

Werquin M, Ackermann H W and Levesque R C 1988 A study of 33 bacteriophages of *Rhizobium meliloti*. Appl. Environ. Microbiol. 54, 188–196.

Werquin M, Ackermann H W and Levesque R C 1989 Characteristics and comparative study of five *Rhizobium meliloti* bacteriophages. Curr. Microbiol. 18, 307–311.

Zahran H H 1992 Characterization of root-nodule bacteria indigenous in the salt-affected soils of Egypt by lipopolysaccharides, protein and plasmid profiles. J. Basic Microbiol. 32, 279–287.

Section editor: H Lambers
Guest editors: G.H. Elkan and R.G. Upchurch

Plant and Soil **186:** 135–142, 1996.

A diazotrophic bacterial endophyte isolated from stems of *Zea mays* L. and *Zea luxurians* Iltis and Doebley

James A. Palus[1,3], James Borneman[1,3], Paul W. Ludden[2,3] and Eric W. Triplett[1,3,4]
[1]*Department of Agronomy* [2]*Department of Biochemistry and* [3]*the Center for the Study of Nitrogen Fixation, University of Wisconsin-Madison, Madison, 1575 Linden Drive, WI 53706–1597, USA**. [4]*Corresponding author*

Key words: diazotrophs, endophytes, Klebsiella, sweet corn *Zea*

Abstract

The apoplastic fluids of field-grown *Zea mays* and *Zea luxurians* plants were isolated from surface sterilized stem tissue by centrifugation and spread on agar plates containing a nitrogen-free, defined medium. The predominant bacterium isolated from these plates was characterized further. The ability of this bacterium to fix nitrogen was confirmed by its ability to grow on a semi-solid, nitrogen-free medium and reduce $^{15}N_2$ to $^{15}NH_3$ and acetylene to ethylene. Portions of the *nifH* and 16S rRNA genes from this organism were amplified by PCR and sequenced. The *nifH* gene, which codes for dinitrogenase reductase, from this organism is closely related to *nifH* from *Klebsiella pneumoniae.* Similarly, the 16S rRNA gene sequences and carbon utilization tests grouped it closely with *K. pneumoniae.* Based an these data, the isolates from *Z. mays* and *Z. luxurians* are tentatively classified as *Klebsiella* spp. (*Zea*). The ability of this bacterium to contribute to the nitrogen economy of the corn plant is unknown.

Introduction

Biologists at EMBRAPA in Rio de Janiero have found large numbers of the bacterium *Acetobacter diazotrophicus* in the stem tissue of certain cultivars of sugarcane (Boddey et al., 1991; Reis et al., 1990). *A. diazotrophicus* is believed to fix nitrogen in the stem tissue of sugarcane using the large amount of sucrose present as a carbon source. Certain cultivars of sugarcane, not fertilized with nitrogen but infected with *Acetobacter*, do not require the addition of any fertilizer nitrogen to achieve optimum yields (Boddey et al., 1991; Urquiaga et al., 1992). However, no one has made the appropriate *nif* mutants to demonstrate that *A. diazotrophicus* is a source of fixed nitrogen for sugarcane. *Acetobacter* has also been shown to be an endophyte of sugarcane cultivars in Australia, Cuba, and Mexico (Dong et al., 1994; Fuentes-Ramirez et al., 1993; Li and MacRae, 1992).

Acetobacter enters sugarcane through mycorrhizae (Boddey et al., 1991; Paula et al., 1991; Reis et al., 1990). Once established in the stem, the endophyte is spread by vegetative propagation, the primary method of sugarcane establishment.

Our primary objective is to determine whether a similar diazotrophic association can be demonstrated or developed for sweet corn. In support of this notion, Fisher et al. (1992) have found that healthy maize plants in England possess endophytic bacteria and fungi. The most common bacterial endophytes in maize were identified using the Biolog system as strains of *Enterobacter agglomerans, Klebsiella terrigena, Pseudomonas corrugata, P. fluorescens, P. marginalis, P.* sp., 4, *P.* sp. 5, and *Vibrio* sp. 1. Strains of *Vibrio, Pseudomonas, E. agglomerans* and *K. terrigena* have been shown to be diazotrophs (for review see Young, 1992). Thus, it seems likely that diazotrophs can inhabit maize. However, the ability of the Biolog system to correctly identify strains of *Enterobacter* and *Klebsiella* is in doubt (Klingler et al., 1992).

McInroy and Kloepper (1991, 1995) have also isolated and provided preliminary identification for bacterial endophytes of maize. These workers found 29 genera of bacteria in field-grown stems and roots of maize. Among these genera were several that include diazotrophic members such as *Bacillus, Enterobacter,*

* FAX No: + (608) 262–5217.
E-mail: ewtriple@facstaff.wisc.edu

Erwinia, Klebsiella, and *Xanthobacter* (Young, 1992). The bacteria in this work were identified by gas chromatographic analysis of fatty acid methyl esters.

McInroy and Kloepper (1991, 1995) have also made four other interesting observations. First, the number of bacterial genera found in maize stems decreased with the age of the plant. At plant maturity, *Pseudomonas* and *Bacillus* strains were dominant. Second, maize tissue was found to contain about 10^4 bacteria per gram fresh weight of tissue. Third, surface-sterilized seeds possessed 10^5 to 10^6 bacteria per gram fresh weight. Fourth, plants grown under sterile conditions from surface-sterilized seeds also contained stem bacteria but at levels 100-fold lower than plants cultured in non-sterile soil. This suggests that the endophytic bacteria can be transmitted either through seed or soil. The observation of seed transmission is potentially important for annual species such as maize and suggests a simple method of inoculation of plants via seed coating or infiltration.

No one to date has shown whether any bacterial endophytes of maize are capable of nitrogen fixation either in vitro or in planta. The objective of the work described here was to determine whether a diazotroph could be isolated from field grown maize from Wisconsin.

Materials and methods

Source of plants and harvesting of tissue

Mature sweet corn, *Zea mays* L. cv. Excellency, and teosinte, *Z. luxurians* Iltis and Doebley, plants were harvested from the University of Wisconsin West Madison Farm in 1994. The sweet corn and teosinte were harvested at sites located at 43°03.68N, 89°32.54W and 43°03.76N, 89°32.54W, respectively. Mature plants of each species was cut near the base, placed in water, and cut under water to limit air from entering the stem.

Media used for bacterial isolation and culture

Isolation of diazotrophic bacteria was done using a modified LGI medium prepared as described by Cavalcante and Döbereiner (1988) except that 10% sucrose replaced crystalized cane sugar and the yeast extract was omitted. One an one-half percent agar was added for solid media and 0.05% agar was added for semisolid media (sLGI). Tryptic Soy Agar (TSA, Difco) was used to increase cell numbers rapidly. For determination of nitrogenase activity, the bacteria were cultured in AcD medium developed by R H Burris (unpublished), which includes (per liter): 3 g $K_2HPO_4 \cdot 3H_2O$, 0.2 g $MgSO_4 \cdot 7H_2O$, 0.04 g $CaSO_4 \cdot 2H_2O$, 0.2 g NaCl, 5 mg $(NH_4)_2SO_4$, 3 mg $FeCl_3$, 1 mg $Na_2MoO_4 \cdot 2H_2O$, 0.1 g yeast extract, and 20 g sucrose adjusted to pH 6.5 with 85% H_3PO_4.

Surface sterilization of stem tissue and culture of bacteria

A section of stem, roughly six cm in length, was cut near the base of the mature plants, from which the leaf was removed and the section was immersed in 70% ethanol and flamed as described by Dong et al. (1994). Using a sterile scalpel, two cm of stem cross section was removed from the sterilized stem. This step, in addition to the removal of the surface leaf tissue were done in order to avoid inadvertent contamination of bacterial surface inhabitants in our apoplastic fluid samples. Apoplastic fluid from each maize stem was removed by centrifuging stem sections at 3000 *g* for 20 min as described by Dong et al. (1994). Apoplastic fluid, 0.3 mL, was recovered from each, diluted 1:1 with 30% glycerol and stored at -75 °C. The apoplastic fluid from each maize stem was serially diluted, plated on LGI, and incubated at 28 °C for up to five days. Nineteen separate colonies were kept and cleaned on LGI from the mature maize, cultured thickly on TSA and stored in 15% glycerol at -75 °C.

All 19 isolates from the mature maize were inoculated with a needle into 5 mL sLGI and incubated at 28 °C. Cultures were reinoculated into fresh sLGI after one week and this transfer was repeated twice. Seven of the 19 isolates grew visibly during the final incubation. These seven isolates were identical with respect to colony morphology, cell shape and size, and antibiotic sensitivity.

PCR amplification and sequencing of partial 16S rRNA and nifH genes

Genomic DNA was extracted from the isolates by the method of Sambrook et al. (1989) and the 16S rRNA gene was amplified from four isolates, three taken from *Z. mays* and one taken from *Z. luxurians,* and *K. pneumoniae* URl with PCR in an air thermocycler. The oligonucleotide primer sequences (Angert et al., 1993) used were TR7 (5′-GCTGCAGAGTTTGATCCTGGCTCAG-3′) and

TR8 (5′-CGGGTTACCTTGTTACGACTT-3′) with reactants as described by Wittwer (1992) using the following conditions: denaturation at 92 °C for one minute; 35 cycles of denaturation at 92 °C for 10s, annealing at 50 °C for 30s, and elongation at 72 °C for 60s; and a final elongation at 72 °C for three minutes. PCR products were purified using the Wizard DNA purification kit (Promega).

PCR products were sequenced using ABI's (Perkin-Elmer) dideoxy terminator cycle sequencing kit. Sequence analysis was done using an automated ABI sequencer at the University of Florida Interdisciplinary Center for Biotechnclogy Research or at the University of Wisconsin-Madison Department of Horticulture. Primers used for the sequencing reactions were TR7, TR8, and six additional primers: 16SF2 (5′-CTGAGACACGGCCCAGACTC-3′), 16SF3 (5′-GCGTAGAGATCTGGAGGAATA-3′), SR2R (5′-TCCCGCAACGAGCGCAACCC-3′), 16SR2 (5′-GGGTTGCGCTCGTTGCGGGA-3′), 16SR3 (5′-TTCGCCACCGGTATTCCTCCA-3′), and SF2R (5′-GAGTCTGGGCCGTGTCTCAG-3′).

A partial *nifH* sequence of one isolate was amplified with PGR using universal primers described by Ueda et al. (1995) and later with primers based on *K. pneumonia* sequence nifh.f (5′-ACCATGCGTCAATGCGCATT-3′) and nifh.r (5′-TCGGTCTGACGTGAGTTACA-3′). Reaction conditions were 94 °C for one minute; 35 cycles at 94 °C for 10s, 55 °C for 30s, and 75 °C for 60s; and a final elongation of 75 °C for three minutes. Processing of the PCR products was identical to that for the 16S rDNA and primers used for sequencing were the universal forward and the Nifh.f primers. All sequences were entered in the GenBank database and given accession numbers.

Phylogenetic analysis of the nifH and 16S rRNA genes

All sequence analyses were performed using programs of the Genetics Computer Group Inc. (Madison, WI) unless otherwise stated. Sequences of 16S rDNA with high homology to the new sequences described here were identified using the BLAST program. The *nifH* sequences used were compiled by Ueda et al. (1995). These sequences were then aligned using the PILE-UP program. Using MEGA (Kumar et al., 1993) the aligned sequences were constructed into phylogenetic trees with a pairwise distance matrix, using the Jukes-Cantor algorithm and the Neighbor-Joining method.

Acetylene and $^{15}N_2$ reduction

One bacterium that grew in sLGI was inoculated into 50 ml of AcD and cultured without shaking for 36 hours. Five milliliters of culture were added to each of six 38 mL serum bottles. Three milliliters acetylene was added and samples (1.0 mL) were removed after incubation for zero, four, and eight hours. The amount of ethylene in a 0.5 mL sample was measured by using a Shimadzu gas chromotograph. Reduction of $^{15}N_2$ to $^{15}NH_3$ was as described by Burris (1972) using 5 mL cultures of cells cultured in AcD medium in a 30 mL serum vial. Total cell protein was determined by a modified Lowry procedure (1977) using bovine serum albumin as a standard.

Physiological characteristics of isolates

Many physiological characteristics of the maize endophytes were determined using the API 20E kit (bioMÉRIEUX VITEK, Hazelwood, MO).

Quantification of the number of diaztrophic isolates in maize tissue

The apoplastic fluids from *Z. mays* and *Z. luxurians* were diluted appropriately to observe single colonies and the dilutions spread on the modified LGI medium. The number of colonies which grew on this medium were counted. About thirty of these colonies were then inoculated into semi-solid modified LGI medium to determine the number of these that could grow in the N-free medium. Based on the proportion of colonies that grew in the semisolid medium compared to the number that grew on solid medium and based on the fresh weight of the maize samples used for the isolation of the apoplastic fluids, the number of diazotrophs that grow on the semi-solid modified LGI medium per g fresh weight of maize stem tissue was determined.

Results

Strain characterization

Seven of the 19 isolates kept from the apoplastic fluids of *Zea mays* and *Z. luxurians* grew in semi-solid LGI. These seven isolates had identical phenotypes with respect to the following characteristics: a) small orange colony morphology on LGI supplemented with 1 g L^{-1} NH_4Cl; b) large white colony morphology on TSA; c) rod shape when simple stained with gram violet and viewed at 1000×; d) acidification of Kligler's Iron Agar; e) resistance to 25 μg mL^{-1} penicillin, carbenicillin, rifamycin, ampicillin, erythromycin, cyclohexamide, trimethoprin, and fusaric acid; f) sensitivity to 10 μg mL^{-1} gentamycin, spectinomycin, kanamycin, naladixic acid, tetracycline, streptomycin, and chloramphenicol; and h) growth on TSA with 5% NaCl. Two *Z. mays* isolates were analyzed in detail and are referred to as strains zmmo and zmvsy. One *Z. luxurians* isolate was also examined and is referred to as strain zlmy.

Physiological characteristics of the maize isolates

Seventeen physiological characteristics of zmvsy and zlmy were determined and were found to be identical in all cases to each other and to *Klebsiella pneumoniae.* The maize isolates were capable of utilizing all carbon sources tested including citrate, glucose, mannitol, inositol, sorbitol, rhamnose, sucrose, and arabinose. They also tested positive for ONPG hydrolysis, lysine decarboxylation, and urease production; tests for arginine dihydrolase and production of H_2S, indole, and acetoin were negative. They did not liquify gelatin.

Quantification of the number of diazotrophic isolates in maize tissue

The number of colonies capable of growing on semisolid N-free medium was found to be 2. 2 × 10^3 and 1. 4 × 10^1 per gram of fresh weight in *Z. mays* and *Z. luxurians,* respectively. The colony morphology of the cells which grew in this medium were very similar to each other suggesting that they are the same as, or very similar to, strains zmmo, zmvsy, and zlmy. The number of these diazotrophic bacteria found in *Zea mays,* 2.2 × 10^3, is at the lower limit of the amount found in sugarcane (Dobereiner et al., 1988). The number found in *Z. luxurians* is considerably lower.

Sequence analysis of 16S rRNA genes

As the isolates obtained from the N-free medium were identical by the above criteria, three of the strains, zmmo, and zmvsy from *Zea mays* and zlmy from *Zea luxurians,* were characterized in more detail. These three isolates had 16S rRNA gene sequences which differed in only one or three bases among ca. 1500 bp sequenced. The nearly complete sequences of the 16S rRNA gene from these isolates were also compared with those of other 16S rDNA sequences in the database (Figure 1). The sequences reported in this paper have been deposited in the GenBank data base (16S accession numbers: zmvsy, U31076; zmmo, U31075; zlmy, U32868; *K. pneumoniae,* U33121)

Sequence analysis of nifH

The partial *nifH* sequence from zmvsy was amplified and sequenced. In comparison with other *nifH* sequences in the database, the zmvsy *nifH* sequence was most similar to that of *Klebsiella pneumoniae* with 84.5% identity over 251 bp, (Figure 2). The *nifH* accession number for zmvsy is U32183).

Nitrogenase activity

The nitrogenase activity of zmvsy was determined by acetylene reduction and $^{15}N_2$ reduction. The acetylene reduction rate of zmvsy under the conditions described above was 1.1 μmol ethylene produced per hr per mg protein. This strain assimilated 4. 04% of its total nitrogen in 220 minutes as determined by $^{15}N_2$ reduction.

Discussion

This work demonstrates that the stems of *Zea* spp. are capable of maintaining a population of diazotraphic bacteria. The diazotrophic nature of these isolates was shown by the reduction of $^{15}N_2$ and acetylene and by their ability to grow well in a semi-solid nitrogen-free medium.

Under the conditions used here for the isolation of diazotrophs, only strains with nearly identical characteristics were identified. Given that corn and sugarcane are closely related, we expected to find diazotrophs that were similar to those found in sugarcane, such as *A. diazotrophicus.* For that reason, we chose a medium for diazotroph isolation that was used by Dong et al. (1994) in the isolation of *Acetobacter* from sugarcane. No strains similar to *Acetobacter* were found in maize stems in this work. Instead, a new diazotroph was discovered with characteristics very similar to *Klebsiella.*

Several lines of evidence show that zmmo, zmvsy, and zlmy are in the γ-subdivision of the proteobacteria. The *nifH* and 16S rDNA sequence data both show that these isolates from maize are enterics which are close relatives of *Klebsiella, Serratia, Citrobacter* and *Erwinia.* The isolates also acidified Kligler's iron agar and grew on a medium containing a high salt concen-

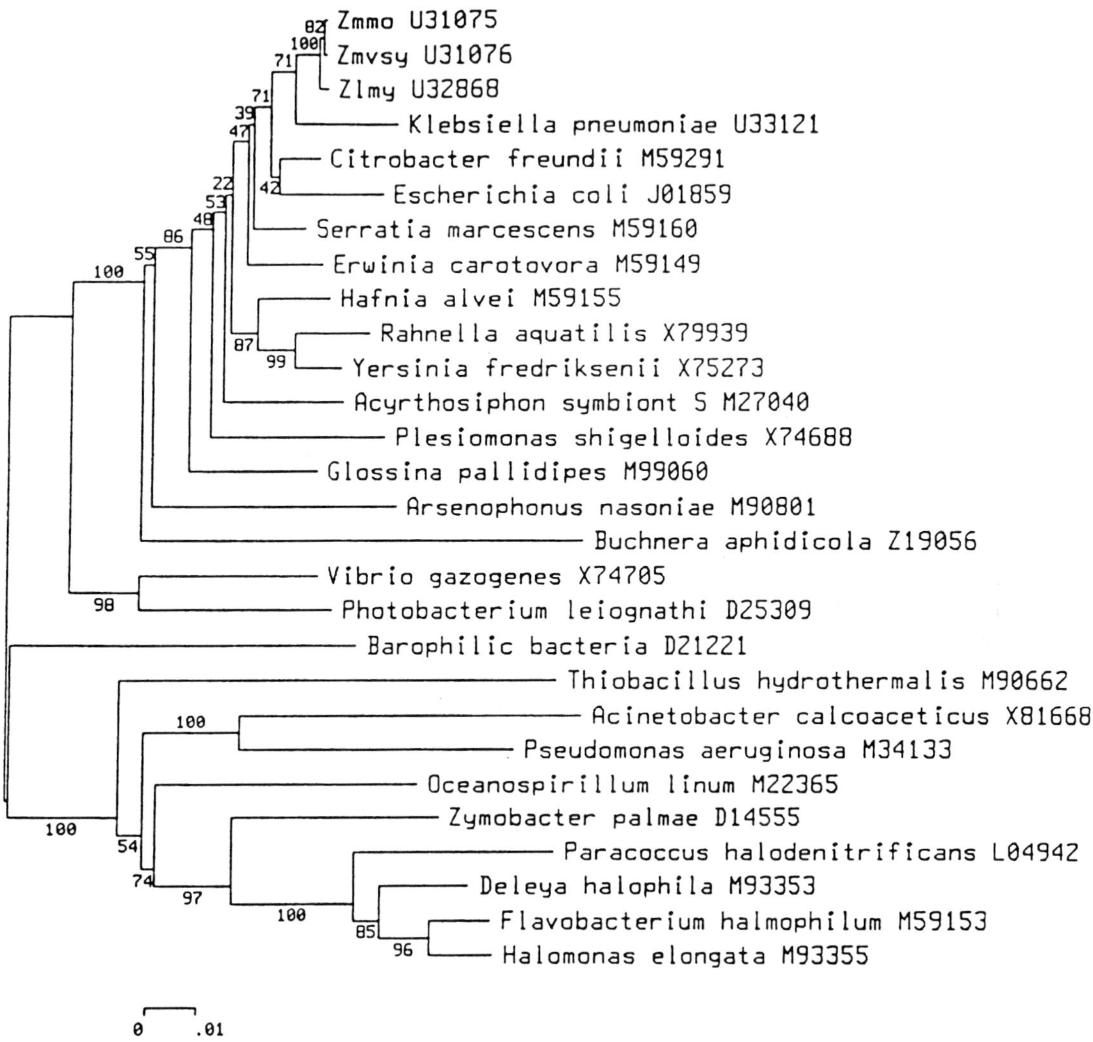

Figure 1. Phylogenetic tree of 16S rDNA sequences of the *Klebsiella* spp. (*Zea*) isolates and selected bacteria from the GenBank database. Confidence levels (%) above each node were generated from 500 bootstrap trees. The distance scale represents the number of changes per sequence position with a median rate of change.

tration, both qualities supporting their enteric nature. Although more work is necessary to propose a taxon for these strains, their close relationship to *K. pneumoniae,* as illustrated by 16S rDNA sequence comparisons and by physiological characteristics, supports the notion of giving these strains the preliminary designation *Klebsiella* spp. (*Zea*).

The development of nitrogen-fixing corn has often been touted as one of the ultimate goals of nitrogen fixation research. Several approaches have been described in the past for the development of nitrogen-fixing maize. These include: a) the transfer of nodulation genes from a legume to maize for the purpose of inducing nitrogen-fixing nodules on maize (Dixon et al., 1993); b) the expression of the bacterial *nif* regulon in maize organelles (Dixon et al., 1993); and c) the development of maize lines with the ability to accept fixed nitrogen from diazotrophs in the rhizophere (Ela et al., 1982). All of these proposals have enormous technical problems to overcome such that the development of nitrogen-fixing maize in the near term has been considered unlikely.

However, by adapting the sugarcane paradigm for a similar strategy for maize, nitrogen-fixing maize may be possible without large technical impediments. This paper represents the first step in this strategy which is the identification of endophytic diazotrophs in maize.

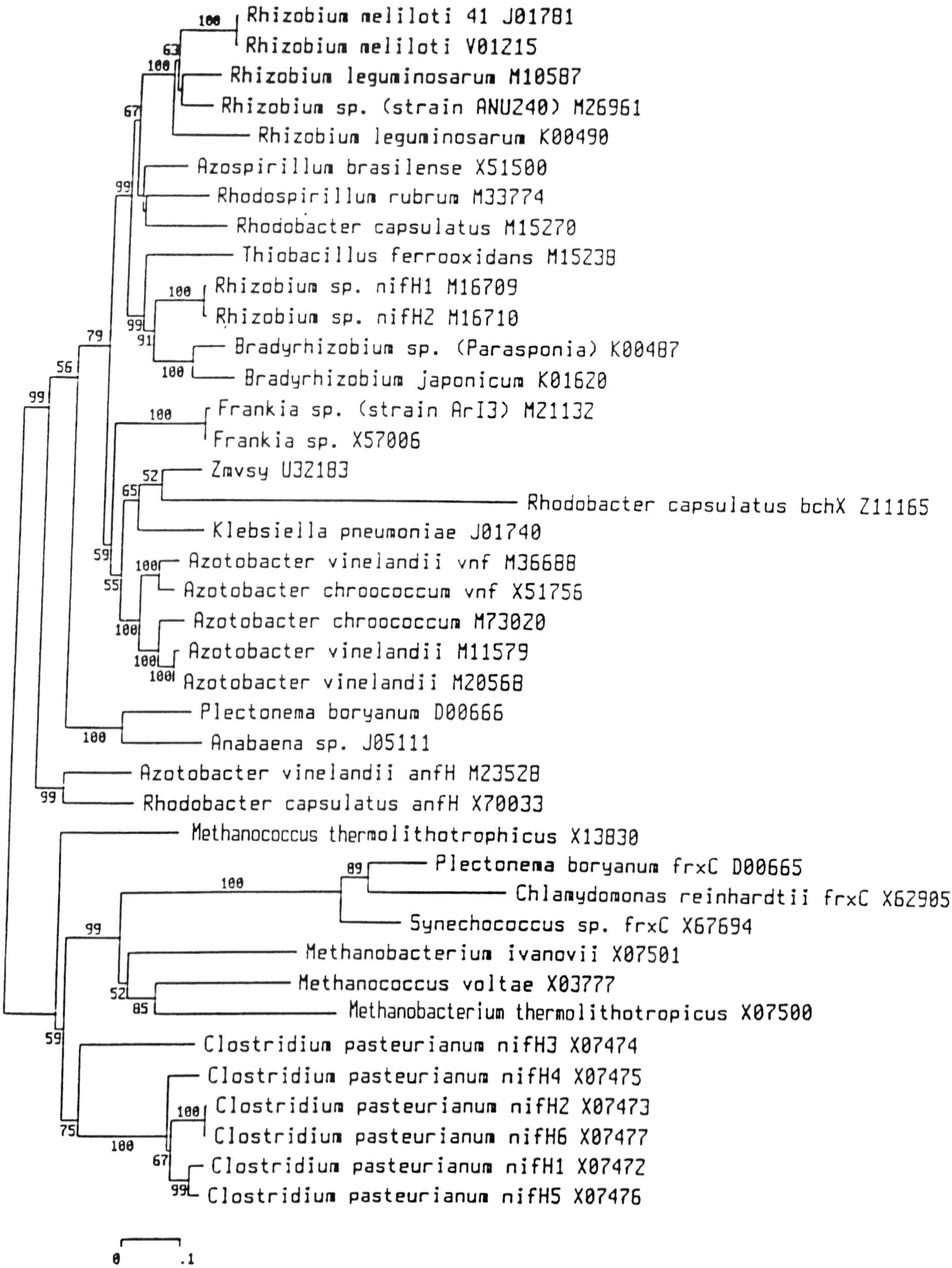

Figure 2. Phlogenetic tree of selected *nifH* sequences (Ueda et al., 1995) and *Klebsiella* spp. (*Zea*) strain zmvs. Confidence levels (%) above each node were generated from 500 bootstrap trees. The distance scale represents the number of changes per sequence position with a median rate of change.

The mechanism by which these bacteria enter the plant and the ability of these strains to provide any benefit to the nitrogen economy of maize is unknown. The plants chosen for bacterial isolation were cultured under conditions of high nitrogen fertilization so it is reasonable to assume that the nitrogenase system of these endophytes was repressed. The apoplast is likely to contain that nitrogen source, most likely nitrate, absorbed by the plant from the soil.

Herbaspirillum seropedicae is a diazotroph which has also been described as a maize endophyte. To date, it has been found in roots only (Baldani et al., 1986, 1992; Döbereiner et al., 1993). *Herbaspirillum seropedicae* is pathogenic on *Sorghum* and *Pennisetum* (Döbereiner et al., 1993; Pimentel et al., 1991). Thus, its exploitation as a potential inoculant for corn is doubtful. Also, in contrast to the *Klebsiella* spp. (*Zea*) strains described here, *H. seropedicae* cannot use sucrose as a carbon source (Baldani et al., 1986). Because sucrose is in abundance in corn (Bertolini et al., 1993), the prediction could be made that a more successful diazotroph in corn would be one, such as *Klebsiella* spp. (*Zea*), that thrives on sucrose as a carbon source.

Acknowledgements

We thank Prof R H Burris for conducting the $^{15}N_2$ reduction assay and Holly Simon for donating *K pneumoniae* DNA. This work was supported by funds from USDA NRI grant no 94–37305–0767 and by Hatch project no 5201 from the University of Wisconsin-Madison.

References

Angert E R, Clements K D and Pace N R 1993 The largest bacterium. Nature 362, 239–241.

Baldani J I, Baldani V L D, Seldin L and Döbereiner J 1986 Characterization of *Herbaspirillum seropedicae* gen. nov., sp. nov., a root-associated nitrogen-fixing bacterium. Int. J. Syst. Bacteriol. 36, 86–93.

Baldani V L D, Baldani J I, Olivares F and Döbereiner J 1992 Identification and ecology of *Herbaspirillum seropedicae* and the closely related *Pseudomonas rubrisubalbicans*. Symbiosis 13, 65–73

Bertolini M, Lorenzoni C, Marocco A and Maggiore T 1993 Soluble solids content in the stakl of maize (*Zea mays* L.) lines and hybrids. Maydica 8, 321–324.

Boddey R M, Urquiaga S, Ries V and Dobereiner J 1991 Biological nitrogen fixation associated with sugar cane. Plant and Soil 137, 111–117.

Burris R H 1972 Nitrogen fixation - assay methods and techniques. Meth. Enzymol. 24B, 415–431.

Cavalcante V A and Döbereiner J 1988 A new acid-tolerant nitrogen-fixing bacterium associated with sugarcane. Plant and Soil 108, 23–31.

Dixon R, Vanderleyden J and Romero D 1993 Extension of nitrogen fixation to other crops. *In* New New Horizons in Nitrogen Fixation. Eds. R Palacios, J Mora and W E Newton. pp 765–768. Kluwer Academic Publishers, Dordrecht, the Netherlands.

Döbereiner J, Reis V and Lararini, A C 1988 New N_2 fixing bacteria in association with cereals and sugarcane. *In* Nitrogen Fixation: Hundred Years After. Eds. H Bothe, F J de Bruijn and W E Newton. pp 717–722. Gustav Fischer, Stuttgart, Germany.

Döbereiner J, Reis V M, Paula M A and Olivares F 1993 Endophytic diazotrophs in sugar cane, cereals and tuber plants. *In* New Horizons in Nitrogen Fixation. Eds. R Palacios, J Mora and W E Newton. pp 671–676. Kluwer Academic Publishers, Dordrecht, the Netherlands.

Dong Z, Canny M J, McCully M E, Roboredo M R Cabadilla C F, Ortega E and Rodés R 1994 nitrogen-fixing endophyte of sugarcane stems. A new role for the apoplast. Plant Physiol. 105, 1139–1147.

Ela S W, Anderson M A and Brill W J 1982 Screening and selection of maize to enhance associative bacterial nitrogen fixation. Plant Physiol. 70, 1564–1567.

Fisher P J, Petrini O and Scott H M L 1992 The distribution of some fungal and bacterial endophytes in maize (*Zea mays* L.). New Phytol. 122, 299–305.

Fuentes-Ramirez L E, Jimenez-Salgado T and Abarca-Ocampo I R 1993 *Acetobacter diazotrophicus,* an indoleacetic acid producing bacterium isolated from sugarcane cultivars of Mexico. Plant and Soil 154, 145–150.

Klingler J M, Stowe R P, Obenhuber D C, Groves T O, Mishra S K and Pierson D L 1992 Evaluation of the Biolog automated microbial identification system. Appl. Environ. Microbiol. 58, 2089–2092.

Kumar S, Tamura K and Nei M 1993 MEGA: molecular evolutionary genetics analysis, version 1.0. Pennsylvania State University, University Park, USA.

Li R P and MacRae I C 1991 Specific association of diazotrophic acetobacters with sugarcane. Soil Biol. Biochem. 23, 999–1002.

Li R P and MacRae I C 1992 Specific identification and numeration of *Acetobacter diazotrophicus* in sugarcane. Soil Biol. Biochem. 24, 413–419.

McInroy J A and Kloepper J W 1991 Analysis of population densities and identification of endophytic bacteria of maize and cotton in the field. *In* Plant Growth-Promoting Rhizobacteria-Progress and Prospects. Eds. C Keel, B Koller and G Défago. Bull. SROP 14, 328–331.

McInroy J A and Kloepper J W 1995 Population dynamics of endophytic bacteria in field-grown sweet corn and cotton. Can. J. Microbiol. 41, 895–901.

Paula M A, Reis V M and Döbereiner J 1991 Interactions of *Glomus clarum* with *Acetobacter diazotrophicus* in infection of sweet potato (*Ipomoea batatas*), sugarcane (*Saccharum* spp.) and sweet sorghum (*Sorghum vulgare*). Biol. Fert. Soils 11, 111–115.

Peterson G L 1977 A simplification of the protein assay method of Lowry et al. which is more generally applicable. Anal. Biochem. 83, 346–356.

Pimentel J P, Olivares F, Pitard R M, Urquiaga S, Akiba F and Döbereiner J 1991 Dinitrogen fixation and infection of grass leaves by *Pseudomonas rubrisubalbicans* and *Herbaspirillum seropedicae.* Plant and Soil 137, 61–65.

Reis V M, de Paula M A, Urquiaga S and Dobereiner J 1990 Infection of sugar cane by *Acetobacter diazotrophicus* is aided by VA mycorrhyzae. *In* Nitrogen Fixation: Achievements and Objectives. Eds. P M Gresshoff, L E Roth, G Stacey and W E Newton. p 688. Chapman and Hall, New York, USA.

Sambrook J, Fritsch E F and Maniatis T 1989 Molecular Cloning: a laboratory manual. Cold Spring Harbor Laboratory, Cold Spring Harbor, New York, USA.

Ueda T, Suga Y, Yahiro N and Matsuguchi T 1995 Remarkable N_2-fixing bacterial diversity detected in rice roots by molecular evolutionary analysis of *nifH* gene sequences. J. Bacteriol. 177, 1414–1417.

Urquiaga S, Cruz K H S and Buddey R M 1992 Contribution of nitrogen fixation to sugar cane. N-15 and nitrogen balance estimates. Soil Sci. Soc. Am. J. 56, 105–114.

Wittwer Carl T 1992 Buffers and reaction components for rapid cycling. Rapid Cyclist l, 1–6.

Young J P W 1992 Phylogenetic classification of nitrogen-fixing organisms. *In* Biological Nitrogen Fixation. Eds. G Stacey, R H Burris and H J Evans. pp 43–86. Chapman and Hall, New York, USA.

Section editor: H Lambers
Guest editors: G.H. Elkan and R.G. Upchurch

Plant and Soil **186**: 143–149, 1996.

Identification of enzymes involved in indole-3-acetic acid degradation

Marianne R. Olesen and Bjarne U. Jochimsen
Department of Molecular Biology, University of Aarhus, DK 8000 Aarhus, Denmark *

Key words: amidohydrolase, *Bradyrhizobium japonicum*, IAA, isatin, pea, reductase, *Rhizobium*

Abstract

Strains of *Bradyrhizobium japonicum* with the ability to catabolize indole-3-acetic acid (IAA) and strains of *B. japonicum, Rhizobium loti,* and *Rhizobium galegae*, unable to catabolize IAA, were analyzed for enzymes involved in the pathway for IAA degradation. Two enzymes having isatin as substrate were detected. An isatin amidohydrolase catalyzing the hydrolysis of isatin into isatinic acid was found in some *B. japonicum* strains and in two *Rhizobium* species, *R loti* and *R. galegae*. The enzyme was inducible (4-5-fold) by its substrate, isatin, and the partially purified enzyme from *R. loti* showed an apparent K_M of 11 μM for isatin. A NADPH-dependent isatin reductase was measured in extracts from a strain of *B. japonicum* lacking the isatin amidohydrolase. The structure of the reaction product, dioxindole was verified by NMR spectroscopy. Isatin reductase activity was also detected in extracts of dry pea seeds, and present in at least two isoforms. A low K_M of 10 μM for isatin was found with a partially purified preparation of the pea enzyme. The presence of such an enzyme activity in pea indicates dioxindole and isatin as possible intermediates in IAA degradation in pea.

Introduction

Although the chemical structure of the naturally occurring plant hormone Indole-3-Acetic Acid (IAA) has been known since 1934, our knowledge about the metabolism of this compound in plants is still limited. Using *Escherichia coli* and *Salmonella typhimurium* as model systems to study IAA degradation, Moyed and Tuli (1968) identified 3-hydroxymethyloxindole, 3-methyleneoxindole and 3-methyloxindole as metabolites of IAA. 3-hydroxymethyloxindole was formed from IAA in intact pea seedlings (Tuli and Moyed, 1967) and an enzymatic activity converting 3-methyleneoxindole into 3-methyloxindole was partially purified and characterized from an extract of pea seeds (Moyed and Williamson, 1967). In this study, we took advantage of the finding that strains of the root nodule-inducing *Bradyrhizobium japonicum* were able to degrade IAA, 5-hydroxy-IAA, 4- and 5- chloro-IAA (Egebo et al., 1991; Jensen et al., 1995). The chloro-derivatives were, however, only metabolized to a limited extent, making the identification of intermediates in the pathway possible. 4-Cl-IAA metabolism resulted in accumulation of 4-chlorodioxindole, and 5-chloro-IAA metabolism stopped at 5-chloroanthranilic acid. Addition of isatin (indole-2,3-dione) to suspension cultures of *B. japonicum* 110 results in formation of isatinic acid and what is proposed to be dioxindole (Jensen et al., 1995). Isatinic acid is readily converted into anthranilic acid. Anthranilic acid could either be used in the biosynthesis of tryptophan, or catabolized further by pyrrole ring fission. Having established this new pathway with strains of *B. japonicum* (Figure 1), we initiated a search for the individual enzymes involved in order to obtain enzymatic evidence in support of the proposed pathway. Isatin was originally found to be an intermediate in the breakdown of IAA in *B. japonicum* (Jensen et al., 1995). In this study, we report on the development of assays for two enzymes involved in the metabolism of isatin; namely isatin amidohydrolase and isatin reductase and on the partial purification and characterization of isatin amidohydrolase from *R. loti* and isatin reductase from pea seeds.

* FAX No: +4586196500. E-mail:

Table 1. Effect of IAA and isatin addition on isatin amidohydrolase levels

Strain	Specific activity of isatin amidohydrolase[a]		
	None[b]	IAA	Isatin
B. japonicum USDA122	150	430	610
B. japonicum 61A24	60	500	340
B. japonicum 110	80	490	410
B. japonicum 61A101	< 1	< 1	< 1
B. lupini 3045	< 1	< 1	< 1
R. loti NZP 2037	120	360	640
R. galegae HAMBI 540	100	320	690

[a]Given as nmol isatinic acid formed min^{-1} mg $protein^{-1}$. Values are the average of two experiments. Due to the instability of the enzyme in extracts values might deviate as much as 50% from the mean.
[b]Bacteria grown in YB (yeast broth) medium without addition and with 0.2 m*M* IAA or 0.2 m*M* isatin respectively.

Materials and methods

Bacterial strains and media

Strains and species (listed in Table 1) were obtained from the following sources, USDA 122: U S Department of Agriculture, Beltsville, MD, USA; 6 1A24 and 61A101: D Werner, Marburg, Germany; 110: H Hennecke, Zürich, Switzerland; 3045: A B Legocki, Poznan, Poland; NZP2037: D B Scott, Palmerston North, New Zealand; HAMBI540: A K Lindström, Helsinki, Finland. Bacteria were grown and processed for HPLC analysis as described by Jensen et al. (1995).

Isolation of dioxindole or NMR analysis

Three hours after the addition of 0.2 m*M* isatin to suspension cultures of *B. japonicum* 61A101, the bacteria were removed by centrifugation. The supernatant was passed through a Sep Pac C_{18} cartridge (Waters). The cartridge was washed with water, and dioxindole was eluted with 100% methanol. The methanol was evaporated under a stream of N_2 gas to ensure that the dioxindole was not oxidized. The sample was dissolved in chloroform ($CDCl_3$) and analyzed in a Varian Gemini 200 NMR spectrometer. Proton resonance was performed at 200 MHz.

Preparation of extracts for in vitro assays

Bacterial cultures were centrifuged 10 min at 10,000 × *g*, washed twice with 50 m*M* Tris-HC1 (pH 7.4), and resuspended in 1 mL of the same buffer supplemented with 7 m*M* 2-mercaptoethanol. Cells were broken by two bursts of 30 sec sonication. The sonicated extract was clarified by centrifugation for 10 min at 10,000 × *g*.

Extracts of dry pea seeds (*Pisum sativum* cv. Bodil) were prepared essentially as described by Moyed and Williamson (1967).

Enzyme assays

All enzyme assays were performed in a Shimadzu UV2101PC spectrophotometer with a temperature controlled cuvette compartment set at 32 °C. For isatin amidohydrolase, a 1 mL reaction mixture contained 90 m*M* Tris-HCl (pH 9.0), 0.2 m*M* isatin and extract. Increase in absorbance at 368 nm represents isatinic acid production. An increase in absorbance of 4.5 (1 cm light path) is equivalent to a change in concentration of 1 m*M*. For isatin reductase, a 1 mL assay mixture contained 45 m*M* Tris-HCl (pH 7.4), 0.2 m*M* NADPH, 0.2 m*M* isatin and extract. Change in absorbance at 340 nm due to oxidation of NADPH was measured. A molar extinction coefficient of 6.22×10^3 dm^3 mol^{-1} cm^{-1} for NADPH was used in calculating activity. Protein concentration was determined as described by Bradford (1976) using bovine serum albumin as standard.

Results

Reduction of isatin and identification of dioxindole

Following addition of isatin to a suspension culture of *B. japonicum* 61A101, culture supernatants were analyzed by HPLC. The integrated peak areas from the

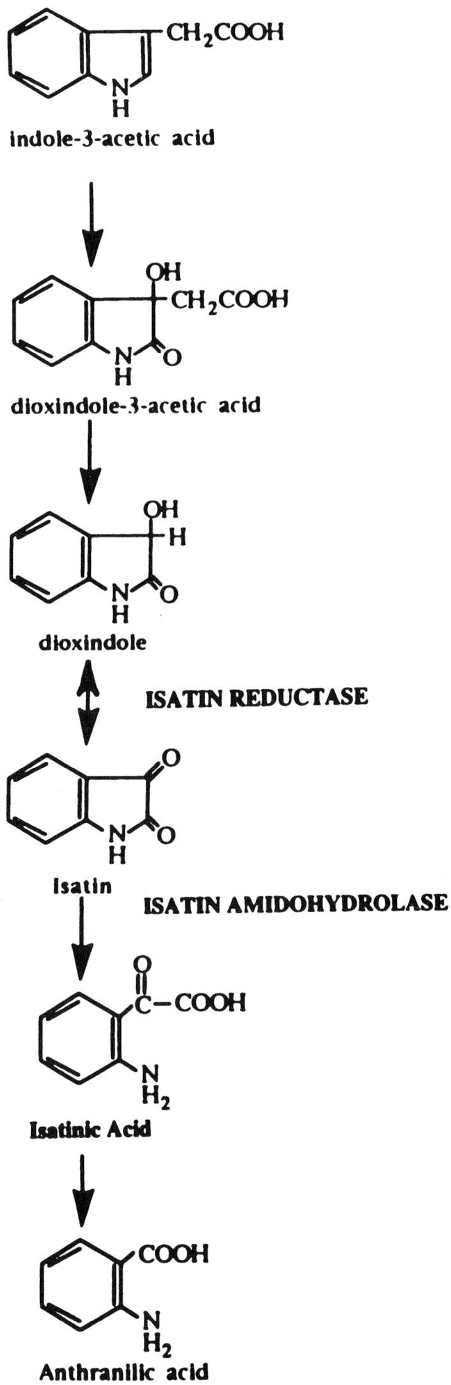

Figure 1. IAA degradation pathway as identified in *Bradyrhizobium japonicum* by Jensen et al. (1995). Enzymes identified in this study are marked with tentative names.

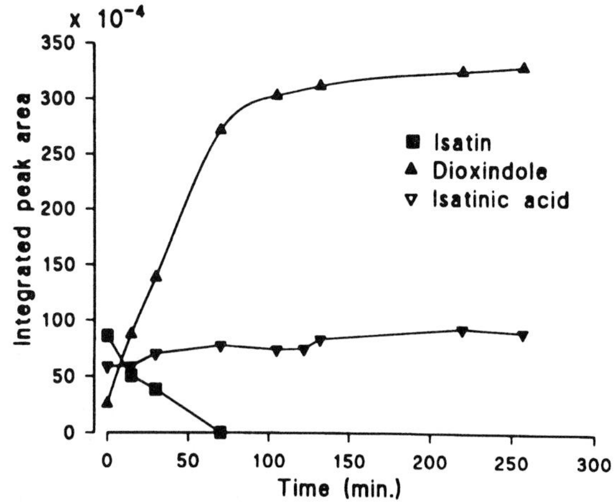

Figure 2. Metabolism of isatin by a suspension culture of *B. japonicum* 61A101. After addition of isatin samples of the supernatant were removed and analyzed by HPLC chromatography. The presence of a constant amount of isatinic acid indicated a contamination of the added isatin.

HPLC elution were plotted versus time (Figure 2). One major compound (RT, 3.9 min) more hydrophilic than isatin and more hydrophobic than isatinic acid accumulated. This compound was isolated and analyzed as described in Materials and Methods. The ^{1}H NMR spectrum (Figure 3) revealed a compound consisting of an aromatic ring structure with two substituents in ortho position (insert a) and a C-H group (a doublet at 5.07 ppm, insert b) coupled to a O-H group (a dublet at 2.74 ppm, insert c), with a coupling constant between the two hydrogens of 4.7 Hz. Together with the UV spectrum (not shown), the NMR spectrum verified that reduction of isatin takes place at a single carbon, thereby giving rise to dioxindole.

Isatin amidohydrolase

Since the absorption spectra of isatin and isatinic acid are different, a spectrophotometric assay based on the absorbance difference at 368 nm was established. Different strains and species of *Bradyrhizobia* and *Rhizobia* were grown in yeast broth medium with and without a supplement of IAA or isatin. The specific activity of isatin amidohydrolase measured in sonicated extracts are given in Table 1. The enzyme was present in *B. japonicum* strains USDA 122, 61A24, 110 and absent in 61A101 and *B. lupini* 3045. In the *Rhizobium* strains tested it was present in NZP 2037 and HAMBI 540. The enzyme activity is increased 3 to 8 -fold in media supplemented with IAA and 4 to 5 -fold

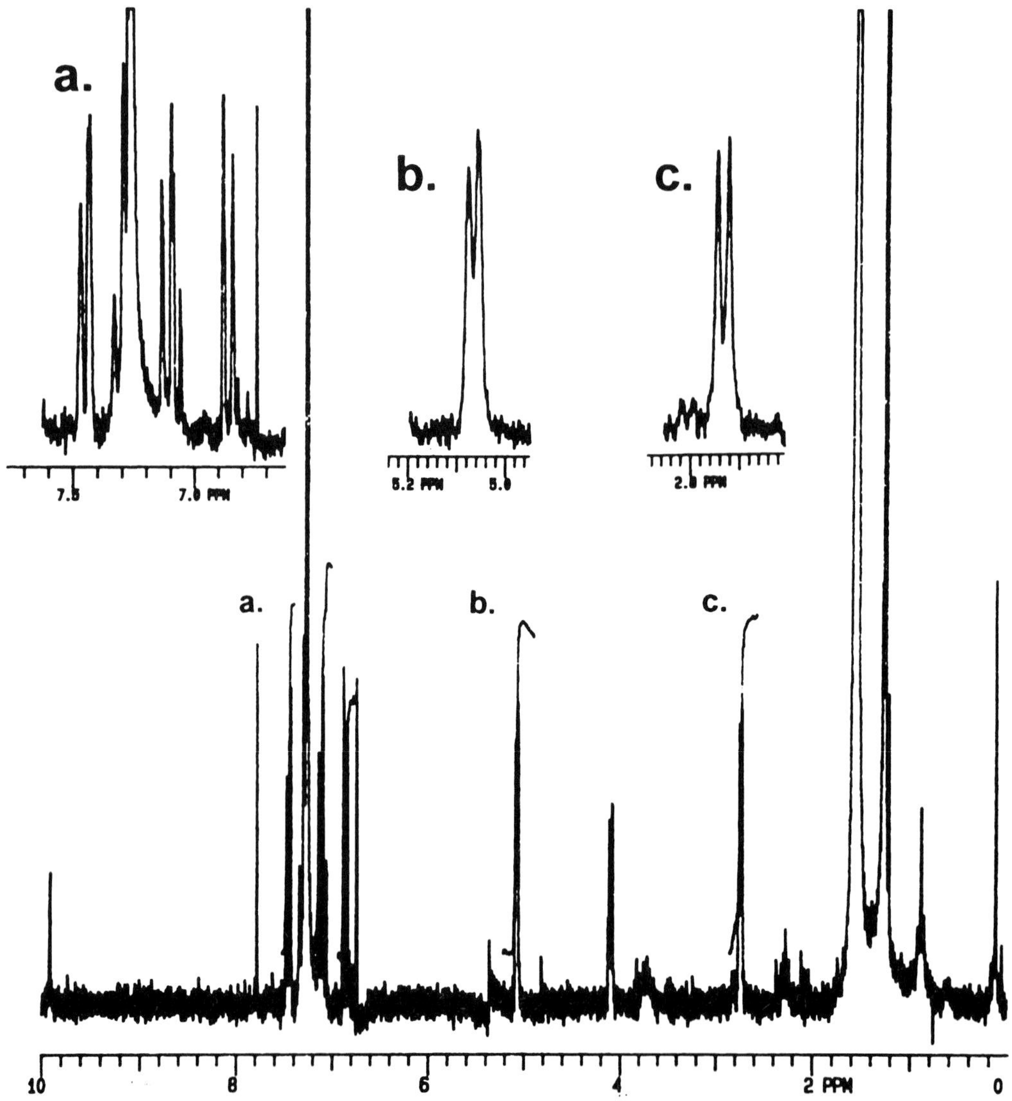

Figure 3. ^{1}H NMR spectrum of the compound isolated from isatin metabolism in Figure 3. The dominant peak at 7.25 ppm together with 4 minor peaks (2 spinning sidebands and 2 C^{13}-satellites) represent a $CHCl_3$ contamination in the solvent, $CDCl_3$. Panel a represents the 4 hydrogens of the aromatic ring and the above mentioned contamination of $CHCl_3$. Panel b represents a single hydrogen associated with carbon and panel c represents a hydrogen associated with oxygen. Peaks at 4 ppm, 1.3 ppm and 1.5 ppm represent impurities.

when supplemented with isatin. Strain 61A101, lacking isatin amidohydrolase, was used to accumulate the dioxindole used for structural verification (see Section Reduction of isatin and identification of dioxindole).

Partial purification and characterization of isatin amidohydrolase from R. loti

Isatin amidohydrolase from *R. loti* NZP 2037 turned out to be much more stable during purification than the enzyme from *B. japonicum* 110. Based on this fact the enzme from *R. loti* were used in a preliminary characterization. Crude extracts of *R. loti* NZP 2037 were loaded onto a Phenyl-Sepharose column. The enzyme eluted after addition of 40% ethyleneglycol (Figure 4). Maximum isatin amidohydrolase activity was found at pH 9 and above and a K_M value of 11 μM for isatin was established with this partially purified enzyme preparation (data not shown).

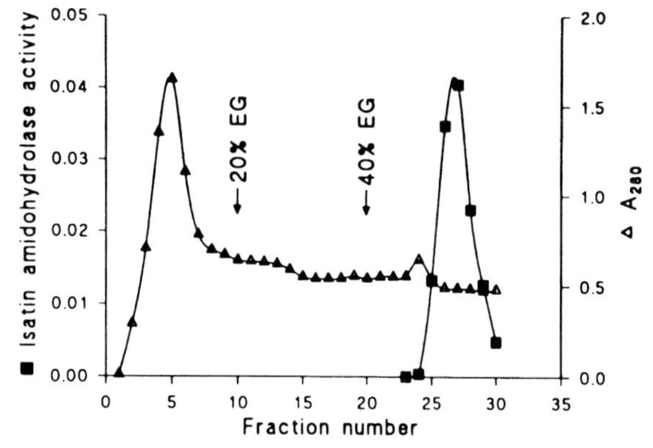

Figure 4. Elution profile of isatin amidohydrolase from a Phenyl Sepharose column. A crude extract of *R. loti* NZP2037 was prepared in 20 m*M* potassium phosphate buffer (pH 7.1) and pumped directly onto the column. Isatin amidohydrolase activity eluted after the stepwise addition of 40% ethyleneglycol (EG).

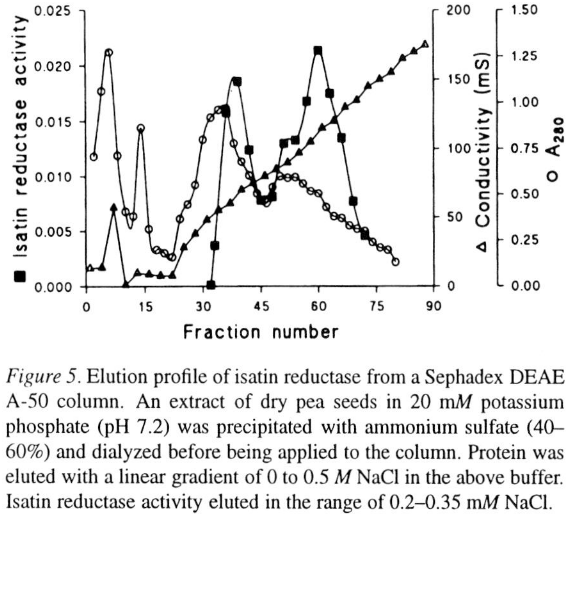

Figure 5. Elution profile of isatin reductase from a Sephadex DEAE A-50 column. An extract of dry pea seeds in 20 m*M* potassium phosphate (pH 7.2) was precipitated with ammonium sulfate (40–60%) and dialyzed before being applied to the column. Protein was eluted with a linear gradient of 0 to 0.5 *M* NaCl in the above buffer. Isatin reductase activity eluted in the range of 0.2–0.35 m*M* NaCl.

Isatin reductase

The assay used to measure isatin reductase activity in extracts was developed for a fungal NADPH-dependent carbonyl reductase (Shimizu et al., 1988). In extracts of *B. japonicum* 61A101 an isatin reductase was found with a specific activity of 35 nmoles min^{-1} mg $protein^{-1}$ and a K_M of 115 μM for isatin. Analysis of the in vitro reaction product gave a compound with the same HPLC retention time as dioxindole.

Partial purification and characterization of isatin reductase from pea seeds

In order to test for the presence of an isatin reductase activity in plants, we analyzed extracts from different tissues of pea and found the highest isatin reductase activity in dry pea seeds. Pea seeds were used as starting material in a purification scheme that included ammonium sulfate precipitation, hydrophobic chromatography on Phenyl Sepharose, Sephadex G-100 gel filtration and DEAE Sephadex anion exchange chromatography (purification data not shown). The elution profile of the ammonium sulfate precipitate applied to a DEAE column is shown in Figure 5 and the pattern indicates the presence of two, possibly three isoforms. A K_M value of 10 μM for isatin was established with the ammonium sulfate precipitate. The most purified extract was used to isolate enough of the in vitro product for analysis. The NMR spectrum and retention time on HPLC verified that the product of isatin reduction catalyzed by the plant enzyme was dioxindole (not shown).

Discussion

Addition of isatin to suspension cultures of *B. japonicum* 110 resulted in the formation of dioxindole and isatinic acid (Jensen et al., 1995), while addition of isatin to suspension cultures of *B. japonicum* 61A101 resulted in formation of dioxindole, verified by NMR spectrometry. Isatin was originally found to be an intermediate in the breakdown of IAA in *B. japonicum* 110 (Jensen et al., 1995) and to verify this pathway we searched for the enzymes involved. The presence (Table 1) of an inducible isatin amidohydrolase activity, which catalyzes the hydrolysis of isatin to isatinic acid, was established in *Bradyrhizobium* strains USDA 122, 110 and 61A24, all with the ability to catabolize IAA. The lack of isatin amidohydrolase activity in strain 61A101 and *B. lupini* 3045 was in accordance with the inability of these strains to catabolize IAA (unpublished). Two strains of *Rhizobium* species, *R. loti* NZP 2037 and *R. galegae* HAMBI 540, both of which are unable to catabolize IAA, however do posses isatin amidohydrolase activity at levels comparable to the positive *B. japonicum* strains. The low K_M of 11 μM for isatin obtained with the enzyme partially purified from *R. loti* indicates isatin as a natural substrate. From the above preliminary analysis of only a few species and strains, it seems that the presence of isatin amidohydrolase is not essential for the establishment of effective nitrogen fixing root nodules, since both *B. japonicum* 61A101 and *B. lupini* 3045 form nitrogen fixing nodules on soybean and lupin, respectively. Isatin amidohydrolase activity was only detectable in nodule extracts from plants infected with strains possessing isatin amidohydrolase, while no activity was detectable in uninfected plant parts.

Using the assay developed by Shimizu et al. (1988), we were able to measure a NADPH dependent isatin reductase activity in extracts of *B. japonicum* 61A101. The apparent K_M for isatin was 115 μM for the bacterial enzyme, versus 50 μM and 14 μM for the enzymes from two species of fungi, *Mucor ambiguus* (Shimizu et al., 1988) and *Candida parapsilosis* (Hata et al., 1989). This indicates the presence of enzymes with a similar function in both bacteria and fungi.

With the knowledge of IAA being a plant hormone, combined with the fact that the above eucaryotes are able to metabolize isatin, it was obvious to look for an isatin reductase activity in plants. Extracts of pea seeds were found to have a NADPH-dependent isatin reductase activity, and a low K_M value of 10 μM for isatin indicates that isatin could be a substrate for the enzyme under natural conditions. A NADPH-dependent 3-methyleneoxindole reductase which reduces 3-methyleneoxindole to 3-methyloxindole has been partially purified from pea seeds (Moyed and Williamson, 1967). This reductase had a relatively high K_M value of 0.5 mM for 3-methyleneoxindole. Since the purification procedure and assay used for the NADPH-dependent 3-methyleneoxindole reductase purified by Moyed and Williamson (1967) and the isatin reductase partially purified from extracts of pea seeds in the present study are almost identical we find it likely that the two enzymes might be identical, and we also find it plausible that isatin might be the natural substrate for the enzyme rather than 3-methyleneoxindole, due to the higher affinity for isatin.

The two compounds, 3-methyleneoxindole and 3-methyloxindole were identified as degradation products, of IAA in extracts of pea seedlings (Tuli and Moyed, 1967). We find it likely that isatin and dioxindole are more natural degradation products of IAA, than 3-methyleneoxindole and 3-methyloxindole. Maybe due to the structural resemblance with isatin and dioxindole, the structures might originally have been misinterpreted.

Earlier studies report that isatin act like an auxin. Addition of isatin to etiolated and green pea stem sections caused elongation (Galston and Chen, 1965) and addition of isatin as an auxin to thin layer explants of tomato pedicel resulted in the production of calli with shoots (Applewhite et al., 1994). Both papers suggest that the auxin effect may be related to a slow conversion of isatin to an active auxin and not to isatin itself. ^{14}C-labeled isatin added to pea stem sections was converted to anthranilic acid and isatinic acid (Kutacek and Galston, 1968). This result, together with our results for the degradation of IAA in *B. japonicum* and for the conversion of isatin to dioxindole with extract of pea seeds, could explain the auxin effect of isatin in two ways. Isatin could either be converted to anthranilic acid, that serves as a precursor for IAA formation or isatin could be converted to dioxindole, as seen with the pea seed extract, and end up as IAA by reversal of the reactions in the pathway found with strains of *B. japonicum* (Figure 1).

The first step in the *B, japonicum* IAA degradation pathway (Figure 1) is the oxidation of IAA to form dioxindoleacetic acid. In three plants species; rice, corn and broad bean, dioxindole-3-acetic acid has been identified as a degradation product after IAA addition (reviewed by Reinecke and Bandurski, 1987). Since the enzymology of the oxidation of IAA to dioxindole-3-acetic acid in plants has not been fully worked out, we look forward to establish the in vitro conditions for the bacterial IAA oxidation and apply these to the plant species above.

Acknowledgements

We thank Henrik Bildsøe for operating the NMR spectrometer and explaining the spectra and Annette Rommelhoff for technical assistance. This work was supported by a grants from the Danish Agricultural and Veterinary Research Council and the NOVO Foundation.

References

Applewhite P B, K Sawhney R and Galston A W 1994 Isatin as an auxin source; favoring floral and vegetative shoot regeneration from calli produced by thin layer explants of tomato pedicel. Plant Growth Regul. 15, 17–21.

Bradford M M 1976 A rapid and sensitive detection method for the quantitation of microgram quantities of protein utilising the principle of protein-dye binding. Anal. Biochem. 72, 248–254.

Egebo L A, Nielsen S V S and Jochimsen B U 1991 Oxygen-dependent catabolism of indole 3-acetic aid in *Bradyrhizobium japonicum*. J. Bacteriol. 173. 4897–4901.

Galston A W and Chen H R 1965 Auxin activity of isatin and oxindole-3-acetic acid. Plant Physiol. 40, 699–705.

Hata H, Shimuzi S, Hattori S and Yamada H 1989 Ketopantoyl-lactone reductase from *Canidida parapsilosis* purification and characterization as a conjugated polyketone reductase. Biochem. Biophys. Acta. 990, 175–181.

Jensen J B, Egsgaard H, van Onckelen H and Jochimsen B U 1995 Catabolism of indole-3-acetic acid and 4-and 5-chloroindole-3-acetic acid in *Bradyrhizobium japonicum*. J. Bacteriol. 177, 5762–5766.

Kutacek M aad Galston A W 1968 The metabolism of ^{14}C-labeled isatin and anthranilate in *Pisum* stem sections. Plant Physiol. 43, 1793–1798.

Moyed H S and Williamson V 1967 3-methyleneoxindole reductase of peas. Plant Physiol. 42, 510–514.

Moyed H S and Tuli V 1968 The oxindole pathway of 3-indoleacetic acid metabolism and the action of auxins. *In* Biochemistry and Physiology of Plant Growth Substances. Eds. F Wightman and G Setterfield. pp 289–300. Runge Press Ltd, Ottawa, Canada.

Reinecke D M and Bandurski R S 1987 Auxin biosynthesis and metabolism. *In* Plant Hormones and their Role in Plant Growth and Development. Ed. P J Davis. pp 24–42. Martinus Nijhoff Publishers, Dordrecht, The Netherlands.

Shimizu S, Hattori S, Hata H and Yamada H 1988 A novel fungal enzyme, NADPH-dependent carbonyl reductase, showing high specificity to conjugated polyketones. Eur. J. Biochem. 174, 37–44.

Tuli V and Moyed H S 1967 Inhibitory oxidation products of indole-3-acetic acid: 3-hydroxy methyloxidole and 3-methyleneoxindole as plant metabolites. Plant Physiol. 42, 425–430.

Section editor: H Lambers
Guest editors: G.H. Elkan and R.G. Upchurch

Plant and Soil **186:** 151–159, 1996.

Toward a new concept of the evolution of symbiotic nitrogen fixation in the Leguminosae

James A. Bryan, Graeme P. Berlyn and John C. Gordon
*205 Prospect Street, School of Forestry and Environmental Studies, Yale University, New Haven, CT 06511, USA**

Key words: Caesalpinioideae, nitrogenase activity, nodulation, root cortex, taxonomy

Abstract

It is generally believed that only the nodulating species of the Leguminosae fix atmospheric nitrogen; however, anatomical, ecological and taxonomic considerations indicate that non-nodulating legume species may also fix nitrogen. To test whether nitrogen-fixing symbioses in the Leguminosae might extend to the non-nodulating species, a survey of the Leguminosae was conducted: living plants of non-nodulating species were assayed using acetylene reduction. Ethylene evolution, indicating apparent nitrogenase activity, was detected in non-nodulating species representing the major taxonomic groups of Caesalpinioideae as well as in non-nodulating species of the Papilionoideae and Mimosoideae. Non-nodules nitrogen fixation appears to have provided evolutionary precursors for the nodular symbiosis in the Leguminosae.

Introduction

Even before the discovery that bacteria in the root nodules of legumes provide atmospheric nitrogen for plant growth (Hellriegel and Wilfarth, 1888), it had been observed that some species of the Leguminosae do not form nodules (Lachmann, 1858; reviewed by Leonard, 1925 and Fred et al., 1932). It has long been assumed that the non-nodulating species do not fix atmospheric nitrogen. Consequently, on the basis of their observed failure to produce root nodules in either natural settings or in the presence of introduced rhizobia, approximately 71% of the species of the Caesalpinioideae, 10% of the Mimosoideae and 3% of the Papilionoideae are thought to lack the capacity to fix nitrogen symbiotically (Allen and Allen, 1981; Faria et al., 1989).

Both the Mimosoideae and the Papilionoideae are considered to have evolved, separately, from species resembling primitive Caesalpinioideae (Polhill and Raven, 1981). The apparent relationships between non-nodulating and nodulating taxa are illustrated in Figure 1, in which the reported nodulation data of Allen and Allen (1981) and Faria et al. (1989) have been overlaid on the putative evolutionary relationships of the Leguminosae, as presented by Polhill et al. (1981). Taxonomic groups with only non-nodulating species are shown in white; those in which nodulation is the rule are shown in dark gray; and taxonomic groups with rare or mixed reports of nodulation or with a mix of both nodulating and nodulating species are shown with stripes. In all three subfamilies of the Leguminosae, some or most advanced species nodulate, while other species, generally the less specialized, lack nodulation capacity.

The assumption that non-nodulating species of the Leguminosae do not fix atmospheric nitrogen has profoundly affected our concepts of the evolution of the nitrogen-fixing symbioses between legumes and rhizobial bacteria. Based on present understanding of the taxonomy of the Leguminosae (Figure 1), the symbiosis would have had to evolve, separately, in at least three taxonomic groups of the Leguminosae: in the *Chamaecrista* or their ancestors; in the ancestors of the Mimosoideae such as the *Dimorphandra* group of the Caesalpinioideae; in ancestors of the Papilionoideae such as the *Sclerolobium* group of the Caesalpinioideae (Doyle, 1994; Sprent, 1994; Young and Johnston, 1989); and probably also in several other taxa such as the Detarieae-Amherstieae and the Bauhiniinae, where nodulation has been reported to occur occasionally (Allen and Allen, 1981; Athar and Mahmood, 1980; Bonnier, 1957; Bonnier, 1958; Ding et al., 1986;

* FAX No: +1203432 5942.
E-mail: BRYJAMA@YALEVM.YCC.YALE.EDU

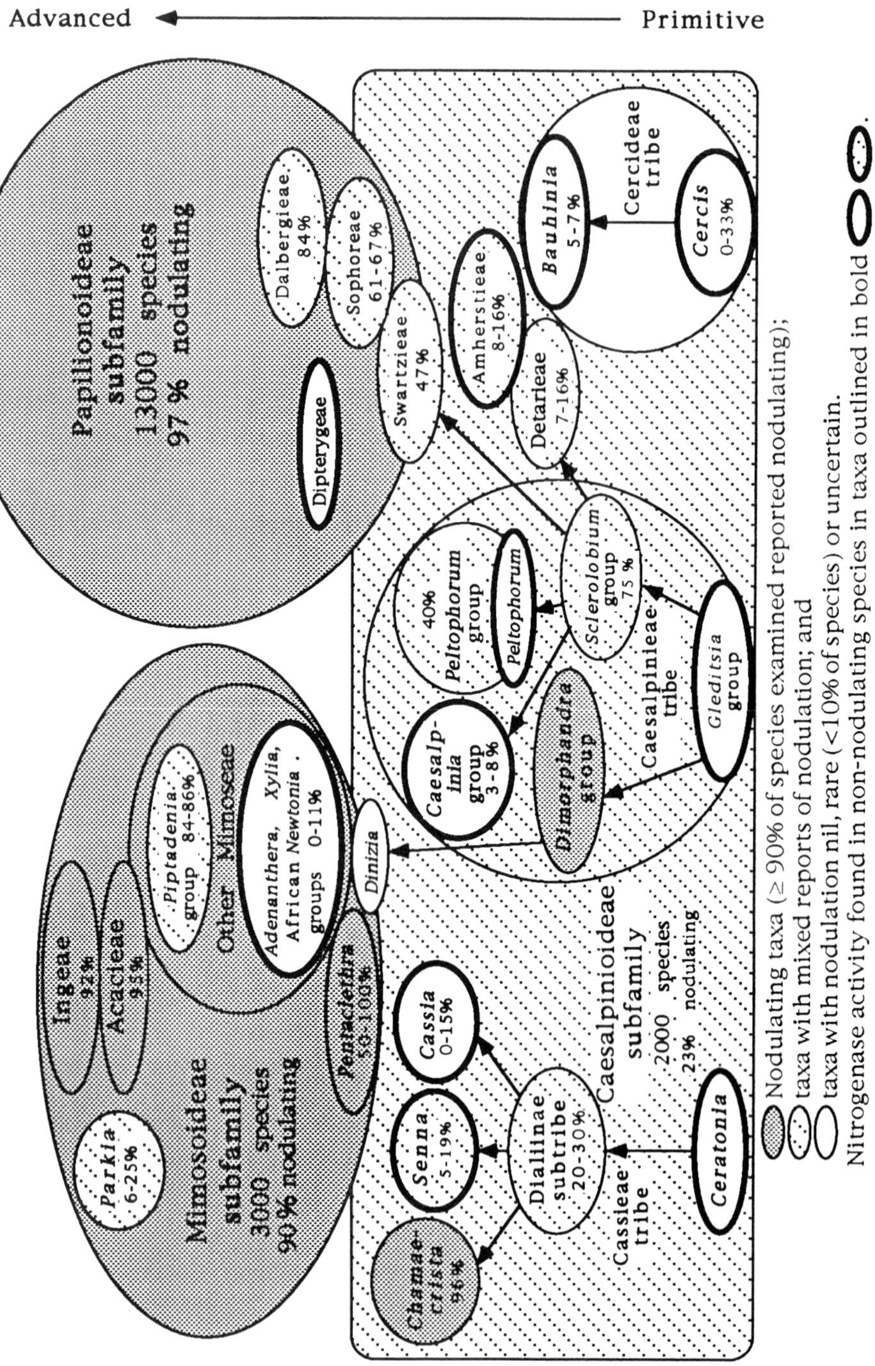

Figure 1. Putative evolutionary pattern of the Leguminosae, derived from Polhill, Raven and Stirton (1981), with nodulation and non-nodulation patterns in the Leguminosae. Percents of nodulating species calculated from Faria et al. (1989), updated. A range of nodulation percents indicates species with conflicting reports of nodulation (updated from Allen and Allen, 1981), calculated here both as negative (lower %) and as positive (higher %). Not to scale.

Lechtova-Trnka, 1931; Lim, 1977; Quiñones as cited in Domingo, 1983; Rothschild, 1970).

Multiple separate evolutions of the nitrogen-fixing symbioses in the Leguminosae appear implausible, however, particularly because, in each of the three major nodulating groups of the Leguminosae, nodulation occurs with both *Rhizobium* and *Bradyrhizobium* (Barnet et al., 1985; Dreyfus and Dommergues, 1981; Jenkins et al., 1988; Martínez-Romero, 1994; Moreira et al., 1993; Moreira and Franco, 1994; Turk and Keyser, 1992; Zhang et al., 1991). *Rhizobium* and *Bradyrhizobium*, distinct genera of bacteria, appear to have ancestors that diverged before the evolution of land plants and long before the legume-rhizobia symbioses (Young and Johnston, 1989; Young, 1993).

An alternative view of the evolution of the symbiosis is suggested by the taxonomy of nodulation. Several generic and supergeneric groups of species (striped in Figure 1) have rare or mixed capacity for nodulation. The wide range of taxa with rare or mixed nodulation capacity, taxonomically positioned between several non-nodulating and nodulating parts of the family, suggests a continuum from non-nodulation to rare, perhaps circumstantial nodulation, to sparse nodulation, often with only partial rhizobial occupation of infected cells, and finally to the abundant nodulation of advanced Mimosoideae, advanced Papilionoideae and some of the *Chamaecrista*. The nodular symbiosis itself appears to have evolved (Faria and Sprent, 1994; Naisbitt et al., 1992; Sprent, 1994; Sutherland et al., 1994) along with a suite of other advanced floral, seed and vegetative characteristics.

Several lines of evidence are now accumulating to suggest that non-nodulating legumes may also form a symbiotic relationship with rhizobia, at the beginning of this symbiotic continuum (Bryan, 1995). Rhizobia have been observed to form infection threads in the root hairs of the non-nodulating legumes *Gleditsia triacanthos* L. (Allen and Allen, 1981; Bryan, 1995), *Cassia fistula* and *C. grandis* and *Senna tora* (as *Cassia tora* Allen and Allen, 1981). Allen and Allen (1981) did not believe that the observed root hairs were part of a symbiosis, the infection threads they observed having grown only to the outer root cortices. However, forms resembling bacteroids have long been reported inside the roots of non-nodulating *Gleditsia* species (Fehér and Bokor, 1926; Friesner, 1926; Rothschild, 1967). Recently, apparent bacteroids were observed in the roots of *G. triacanthos* and five other species by scanning electron microscopy (see discussion) and ethylene evolution (acetylene reduction), an indication of nitrogenase activity, was detected in *G. triacanthos* (Bryan, 1995).

We therefore decided to test whether nitrogenase activity might extend to other non-nodulating species of the Leguminosae and conducted acetylene reduction trials with twelve additional species of non-nodulating legumes. Here, we present the results of that survey and a follow-up assay with one of the species, *Peltophorum pterocarpum* (DC.) Backer.

Materials and methods

Species were selected to represent diverse taxonomic groups of the Leguminosae. In the Caesalpinioideae, the primitive *Gleditsia triacanthos* L. and *G. caspica* Desf., and the more specialized *Cercidium microphyllum* (Torrey) Rose and I. M. Johnston and *P. pterocarpum* represent the Caesalpinieae tribe, apparently fundamental to both the Caesalpinioideae and the Leguminosae (Polhil! and Raven, 1981). *Ceratonia siliqua* L. is thought to represent the base of the Cassieae tribe; *Cassia fistula* L., *Senna alata* (L.) Roxb. and *S. auriculata* (L.) Roxb. of the Cassieae were also selected because of their affinity with the nodulating *Chamaecrista* (Figure 1). The basic *Cercis siliquastrum* L. and the more advanced *Bauhinia purpurea* L. represent the Cercideae tribe. *Tamarindus indica* L. represents the other major division of the Caesalpinioideae, the Amherstieae-Detarieae. *Adenanthera pavonina* L., one of the non-nodulating or rarely nodulating Mimosoideae, was included in the trial. *Dipteryx panamensis* (Pittier) Record and Mell was also evaluated, representing the Papilionoideae.

Availability of diverse rhizobia could enhance the likelihood of compatibility between potential symbionts and the non-nodulating plants. Many species of nodulating legumes had been grown on a sand- and gravel-covered greenhouse bench; non-nodulating plants collected from various locations were also placed on this bench, to form a possible additional reservoir of potentially compatible bacteria.

Collected non-nodulating plants 1-4 years old and seedlings germinated for these studies were transplanted into potting mix in individual open pots with holes in the bottoms and grown on the sand and gravel surface; seedlings were also grown in open plastic (PVC) pipes filled with patting mix, with their bottoms covered with plastic screening. Nodulating legumes continued to be rotated on the greenhouse bench to assure abundant and diverse rhizobia. Plants were periodically provided dilute solutions of low N fertilizer (Peters Blossom Booster, 10-30-20). To maximize the photosynthetic surface and to concentrate the roots into a small volume for these trials, plants had not been recently pruned, and had been grown in small pots.

To permit effective mixing of gas in the root zones during the initial trial, plants of four species of legumes had been grown directly in plastic (PVC) pipes (25 cm height, 10 cm inner diameter) with bulkhead fittings near the bottom and reducing couplings at the top, to permit the stems to be enclosed by their insertion into slotted rubber stoppers, which were then pushed into the reducing couplings and sealed with modeling clay. Bulkhead fittings were also inserted in the rubber stoppers; screening was removed from the bottoms of the tubes, which were then sealed with plastic caps and silicone stopcock grease; and gas was circulated by a peristaltic pump through plastic (PVC) tubing and sampled through a rubber port in the tubing. When the first trial had positive results, two additional trials were conducted, each with four species of non-nodulating leguminous plants grown in plastic pots. Plants in potting mix were removed from their pots with intact root balls. The root balls were placed in 7.5 L (trial 2) or 3.8 L (trial 3) plastic (HDPE) containers with gasket-fitted lids, which had been drilled and slit to fit loosely around the stems. Modeling clay was then molded around the stems to provide gas-tight seals. Tubes were attached to bulkhead fittings near the bottom of the containers and in the lids, to permit circulation of gas through the chambers by a peristaltic pump.

Two plants of each species were separated into initial treatment and control groups. Acetylene (10% of free container volume) was injected into each treatment container, which was briefly vented at the pump; the gas was then circulated through the container. Gas samples (0.1 mL) were drawn by gas-tight glass syringes from rubber ports in the pump tubing. Each plant was sampled with a separate syringe throughout the experiment, to avoid gas contamination between treatments. Standard methods were used for the measurement of ethylene evolution (Turner and Gibson, 1980; Upchurch, 1987). A Hewlett Packard 5890A gas chromatograph was equipped with a 120 cm stainless steel column with 3.2 mm outer diameter, packed with Porapak T 80/100. The column was heated to 100 °C, the injection port to 120 °C and the detector to 200 °C. Nitrogen (40 mL min^{-1}) was used as the carrier gas; hydrogen (25 mL min^{-1}) and air (400 mL min^{-1}) were used for the flame ionization detector. In the second and third trials, plants initially used as

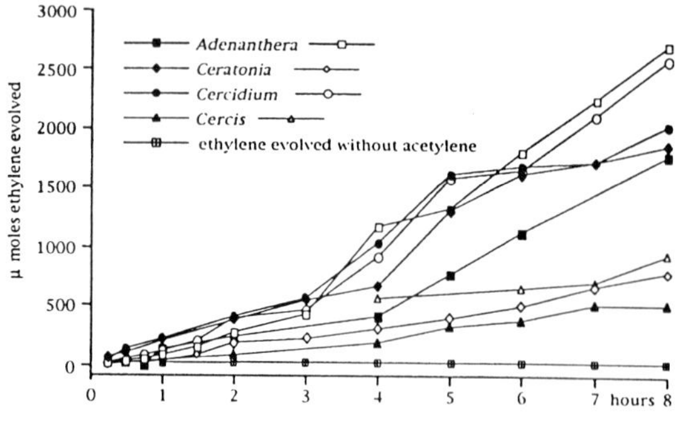

Figure 2. Ethylene evolution in root balls of intact plants after injection of 10% acetylene. Plants of the first test group (black symbols) were tested on a cloudy day, those of the second test group (initial controls; white symbols) the following day, in full sun.

controls were subsequently assayed in the same manner on the following day. Somewhat delayed ethylene evolution had been anticipated based on a previous experiment (Bryan, 1995); however, in these trials, with improved circulation of gas, ethylene evolution was detected immediately; the sampling schedule was then modified in the third trial, reported here, to permit more prompt sampling. To provide an initial estimate of the extent of ethylene evolution on a plant tissue basis, the roots of the plants from the second day of Trial 3 were harvested and oven dried 24 h at 70 °C, 24 h at 80 °C and 24 h at 90 °C prior to weighing.

As a follow-up trial, the potting mix was removed, in November, 1995, from the roots of one of the tested *P. pterocarpum* plants; both a water spray and clippers were needed to remove the potting mix from the tightly grown roots. The plant was repotted in sand, to permit subsequent easy and trauma-free bare-rooting, and returned to the greenhouse to recover; by April, 1996, root growth appeared to be vigorous and many leaves had regrown. Three tests were then conducted: the plant in the sand; the plant bare-rooted, and the sand in which the plant had been growing. To avoid possible pre-incubation, assays were begun an hour after sealing the roots in the pots. Temperature was maintained constant near 25 °C (± 0.25 °C); light, provided by a Sylvania 1000 watt metal arc lamp, also remained constant. Gas was sampled at 10 minute intervals for 2 hours. Samples were taken in replicates of three; the means of three samples at each 10 minute interval is reported here.

Table 1. Ethylene evolved calculated by root weight of plants (sunny day in Figure 2)

Species	Shoot dry wt. (g)	Root fresh wt. (g)	μmol C_2H_4 g^{-1} root fresh wt. hr^{-1}	Root dry wt. (g)	μmol C_2H_4 g^{-1} root dry wt. hr^{-1}
Adenanthera pavonina	12.9	51	6.61	11.1	30.35
Ceratonia siliqua	28.1	44	2.21	14.4	6.74
Cercidium microphyllum	16.4	17	18.97	6.9	46.74
Cercis siliquastrum	6.3	15	7.76	4.5	25.86

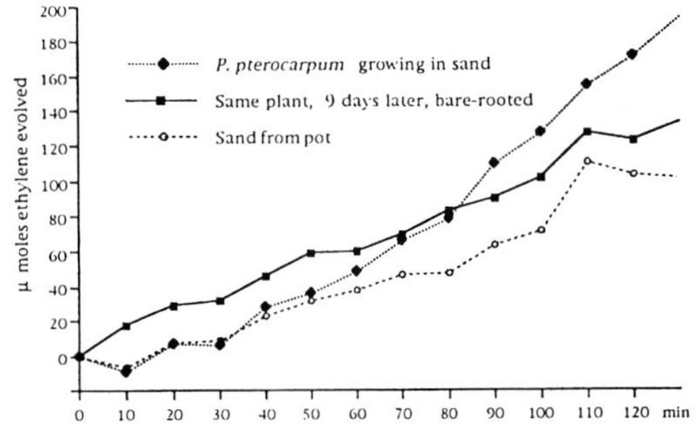

Figure 3. Ethylene evolution by *P. pterocarpum* plant (86 cm shoot height, 1560 g fresh weight): the roots and the sand they were growing in, the bare roots (780 g fresh weight) and the sand (4.4 L) washed from the roots, each after injection of 10% acetylene.

Results

All of the non-nodulating leguminous plants tested evolved ethylene after injection of acetylene. Ethylene evolution began within 15 min for three of the four species sampled immediately after injection of acetylene (*Adenanthera pavonina*, *Ceratonia siliqua* and *Cercidium microphyllum*), within less than an hour in *Cercis siliquastrum* (Figure 2). No ethylene was evolved by any of the control plants (0 acetylene on the frst day). As a reverse control, the original 0-acetylene control plants were injected with 10% acetylene the following day, and all evolved ethylene (Figure 2). The rates of ethylene evolved ranged from 2.21 μmoles g^{-1} root fresh weight hr^{-1} (*C. siliqua*) to 18.97 μmoles C_2H_4 g^{-1} root fresh weight hr^{-1} in *C. microphyllum* (Table 1). In three of the four species in this series of tests, (*A. pavonina*, *C. microphyllum* and *C. siliquastrum*), plants evolved more ethylene on the sunny day than on the cloudy day (Figure 2). No ethylene was detected in plants in the absence of acetylene.

In the follow-up assay with the bare-rooted *P. pterocarpum* plant, ethylene was evolved by the plant growing in sand, by the bare-rooted plant, and by the sand from which the plant had been removed (Figure 3); no ethylene was detected during the hour prior to injection of acetylene in any of the treatments. A lag period (< 20 min.) after injection of acetylene was observed both in the potting-sand and in the plant growing in sand, during which time both the acetylene concentrations and the background ethylene decreased; by the 20 minute measurements, ethylene detected exceeded background ethylene in both cases. In the bare-rooted plant, a slight increase in ethylene was detected within 10 minutes of acetylene injection, and continued at a slow rate (61.3 = μmoles C_2H_4 hr^{-1} $plant^{-1}$, or 0.079 μmoles C_2H_4 hr^{-1} g fresh root $weight^{-1}$) throughout the 2 hour trial (Figure 3).

Discussion

The long-standing assumption that non-nodulating species of the Leguminosae do not fix nitrogen has led to ecological anomalies. Non-nodulating species of the Leguminosae grow well in many nitrogen-limited soils, as do nodulating legumes. Non-nodulating leguminous trees dominate many poor sites and outproduce nodulating species on a variety of sites (Bryan, 1995; Felker and Clark, 1981; Szott et al., 1991).

Tissues of both nodulating and non-nodulating species tend to be rich in nitrogen compared with most non-legumes (Bryan, 1995; McKey, 1994; Waterman, 1994). While nodulating species tend to have greater concentrations of nitrogen, no clear distinction can be drawn between nodulating and non-nodulating species on the basis of their nitrogen content (Bryan, 1995). Several *Bauhinia* species, for example, have higher seed N contents (up to 32.5%) than most nodulating species of legume trees (Barclay and Earle, 1974; Bower et al., 1988; Bryan, 1995; Earle and Jones, 1962; Jones and Earl, 1966). McKey (1994) surveys recent comparisons of the leaf nitrogen concentrations of the nodulating Mimosoideae and Papilionoideae and

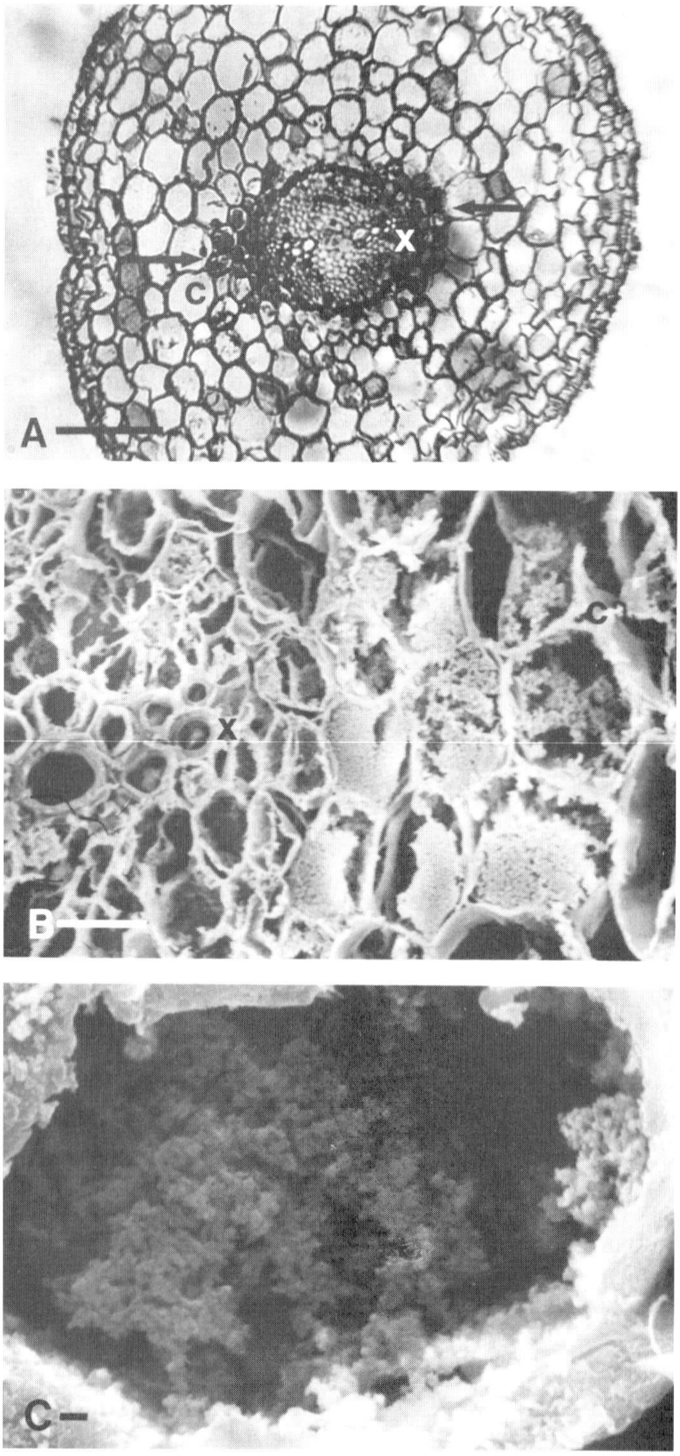

Figure 4. Root of the non-nodulating legume *Peltophorum pterocarpum*, with clusters of forms resembling rhizobial bacteroids (arrows) in cortex (c), outside xylem poles (x). **A**. cross section of root, stained with alum hematoxylon; *Bar* = 100 μ. **B**. scanning electron micrograph of clusters of forms resembling rhizobial bacteroids (arrows) concentrated in inner cortex, outside the xylem pole; *Bar* = 10 μ. **C**. scanning electron micrograph of enlargement of cluster of possible bacteroids in cortical cell outside xylem pole; *Bar* = μ.

non-nodulating Caesalpinioideae (which he assumes not to fix nitrogen); in diverse environments, nodulating mimosoid and papilionoid species had higher leaf nitrogen concentrations than non-nodulating species, which, in turn, had higher leaf nitrogen concentrations than plants of most other families in the sites studied. Nitrogen fixation in non-nodulating species would explain much of the ecological success and high nitrogen tissue contents of these species. Such non-nodular nitrogen-fixation would have provided an environmentally based predisposition for the evolution of the nodular symbioses.

We have conducted other studies that provide a mechanism by which this evolution may have occurred. Using scanning electron microscopy (SEM), we have observed a mechanism previously reported by Fehér and Bokor (1926), Friesner (1926) and Rothschild (1967), which may explain the nitrogenase activity reported here. Forms resembling rhizobial bacteroids were found within the cells of swollen roots of six of the non-nodulating species used in these acetylene reduction trials: *A. pavonina*, *B. purpurea*, *C. siliqua*, *D. panamensis*, *G. triacanthos* and *P. pterocarpum* (Figure 4); apparent bacteroids were observed by SEM in every plant we examined, representing all three subfamilies of the Leguminosae (Bryan, 1995).

In *P. pterocarpum*, a taxonomically advanced caesalpinoid species closely related to several nodulating trees (Faria et al., 1989; Polhill et al., 1981), the forms resembling bacteroids were densely packed and concentrated in the inner root cortex, opposite the xylem poles, as in the initial stages of nodule development of nodulating species (Figure 4). In the other species examined, especially the putatively primitive *G. triacanthos*, the apparent bacteroids only partially filled the infected cells and were widely scattered throughout the root cross section including the stele. Thus, our initial investigations suggest a possible suite of precursors leading to concentration of rhizobial bacteroids outside the xylem poles and then to nodule development.

These acetylene reduction trials were not designed to quantify or compare the nitrogen-fixing potential of different non-nodulating species and no comparison is given. In all of the trials variation between plants of a given species as well as between species was considerable. The previous November, prior to being bare-rooted, the same *P. pterocarpum* root zone, then in potting mix, had evolved ethylene at a considerably higher rate (529 μmoles C_2H_4 hr^{-1} $plant^{-1}$, or approximately 0.68 529 μmoles C_2H_4 hr^{-1} g fresh root $weight^{-1}$) under the same experimental conditions. The lower rates in the trials reported here could have had various causes, such as incomplete regrowth of leaves and roots after the bare-rooting, depletion of available photosynthate over winter in the greenhouse, poorer plant vitality in sand than in potting mix, or perhaps greater activity of free-living nitrogen-fixing bacteria in the potting mix, which was unfortunately not assayed at the time of bare-rooting.

Similarly, comparison of acetylene reduction rates of non-nodulating legumes with those of nodulating legume species cannot be made on the basis of these trials; based on preliminary observations we hypothesize that the rates of acetylene reduction in non-nodulating plants may be between 1 and 2 orders of magnitude lower in than in nodulating plants.

The hypothesis that symbiotic nitrogen-fixation occurs in the roots of non-nodulating legume species was supported, by the acetylene reduction experiments reported here. The evolution of ethylene indicated probable nitrogenase activity in thirteen non-nodulating species of the Leguminosae throughout the family, the taxa outlined in bold in Figure 1. It is clear (Figure 3) that a portion of the ethylene detected was generated in the potting medium, not within the plant roots; the reported rates of ethylene evolution of the plants in potting mix include any ethylene generated by the plant roots, on the plant root surfaces and in the potting medium; however, it is also apparent from the ethylene evolved by the bare-rooted *P. pterocarpum* plant (Figure 3) that part of the ethylene evolved originated in the roots or on the root surfaces. Since no wound response was detected and sand and potting mix residues were slight, nitrogenase activity may be presumed.

Based on the indications of nitrogenase activity reported here, whatever mechanism or mechanisms may ultimately prove responsible, it appears that nitrogen fixation extends to non-nodulating species of the Leguminosae and that non-nodular nitrogen fixation provides a basis for the evolution of the nodular legume-rhizobial symbioses.

Acknowledgements

We acknowledge helpful suggestions by Bernard T Bolmann for detecting low rates of acetylene reduction; by Peter van Berkum for decreasing the likelihood of confusing ethylene evolved by nitrogenase with ethylene derived from extraneous sources; and by two anonymous reviewers. Barry Piekos provided generous guidance with the scanning electron microscope. Mark S Ashton, Florencia Montagnini, B M P Singhakumara and the USDA National Clonal Germplasm Repository of Miami provided seeds and/or plant material for these studies. Brett Hartman, Bret Elderd and Peter Wilshusen helped set up experiments and enter data. This research was supported by a grant from the Andrew W Mellon Foundation.

References

Allen O N and Allen E K 1981 The Leguminosae: A Source Book of Characteristics, Uses, and Nodulation. Univ. Wisc. Press, Madison, WI, USA. 812 p.

Barclay A S and Earle F R 1974 Chemical analysis of seeds III: Oil and protein content of 1253 species. Econ. Bot. 28, 178–236.

Barnet Y M, Catt P C and Hearne D H 1985 Biological nitrogen fixation and root-nodule bacteria (*Rhizobium* sp. and *Bradyrhizobium* sp.) in two rehabilitating sand dune areas planted with *Acacia* spp. Aust. J. Bot. 33, 595–610.

Bonnier C 1957 Symbiose *Rhizobium*-Légumineuses en Région Équatoriale. Publicatians de l'Institut National pour l'Étude Agronomique du Congo Belge. Séries Scientifique, N° 72. 67 p.

Bonnier C 1958 Discussion following Chapter 13 by G B Masefield, Some factors affecting nodulation in the tropics. *In* Nutrition of the Legumes. Ed. E G Hallsworth. p 212. Academic Press, New York, USA.

Bower N, Hertel K, Oh J and Storey R 1988 Nutritional evaluation of marama bean (*Tylosema esculentum*, Fabaceae): Analysis of the seed. Econ. Bot. 42, 533–540.

Bryan J A 1995 Leguminous trees with edible beans, with indications of a rhizobial symbiosis in non-nodulating legumes. Ph.D. thesis, Yale University, New Haven, Connecticut, USA. 310 p.

Ding M, Yi Wei-min and Liao Lan-yu 1986 A survey on the N_2-ase activities of nodules of tree legumes, including *Tamarindus indica*, a species not widely known to nodulate, in artificial forests in Dainbai, Guangdong, China. NFTRR 4, 9–10.

Domingo I L 1983 Nitrogen fixation in Southeast Asian forestry: research and practice. *In* Biological Nitrogen Fixation in Forest Ecosystems: Foundations and Applications. Eds. J C Gordon and C T Wheeler. pp 295–315. Martinus Nijhoff/Dr W Junk Publishers, The Hague, the Netherlands.

Doyle J J 1994 Phylogeny of the legume family: an approach to understanding the origins of nodulation. Annu. Rev. Ecol. Syst. 25, 325–349.

Dreyfus B L and Dommergues Y R 1981 Nodulation of *Acacia* species by fast- and slow-growing tropical strains of *Rhizobium*. Appl. Environ. Microbiol. 41, 97–99.

Earle F R and Jones Q 1962 Analyses of seed samples from 113 plant families. Econ. Bot. 16, 221–250.

Faria S M de, Lewis G P, Sprent J I and Sutherland J M 1989 Occurrence of nodulation in the Leguminosae. New Phytol. 111, 607–619.

Faria S M de, Cavalcante de Lima H, Carvalho A M, Conçalves V F and Sprent J I 1994 Occurrence of nodulation in legume species from Bahia, Minas Gerais and Espirito Santo states of Brazil. *In* Advances in Legume Systematics 5: The Nitrogen Factor. Eds. J I Sprent and D McKey. pp 17–23. Royal Botanic Gardens, Kew, UK.

Faria S M de and Sprent J I 1994 Legume nodule development: an evolutionary hypothesis. *In* Advances in Legume Systematics 5: The Nitrogen Factor. Eds. J I Sprent and D McKey. pp 33–39. Royal Botanic Gardens, Kew, UK.

Fehér D and Bolcor R 1926 Untersuchungen über die bakterielle Wurzelsymbiose einiger Leguminosenhölzer. Planta (Berlin) 2, 406–413.

Felker P and Clark P R 1981 Nodulation and nitrogen fixation (acetylene reduction) in desert ironwood (*Olneya tesota*). Oecologia (Berlin) 48, 292–293.

Fred E B, Baldwin I L and McCoy E 1932 Root Nodule Bacteria and Leguminous Plants. Univ. Wisc. Press, Madison, WI, USA. 343 p.

Friesner G M 1926 Bacteria in the roots of *Gleditsia triacanthos* L. Proc. Indiana Acad. Sci. 34, 215–224.

Hellriegel H and Wilfarth H 1888 Untersuchungen über die Stickstoffnahrung der Gramineen und Leguminosen. Beilageheft zu der Zeitschrift des Vereins für die Rübenzücker-Industrie des deutschen Reiches. Kayssler and Co., Berlin, Germany.

Jenkins M B, Virginia R A and Jarrell W M 1989 Ecology of fast-growing and slow-growing mesquite-nodulating rhizobia in Chihuahuan and Sonoron desert ecosystems. Soil Sci. Soc. Am. J. 53, 543–549.

Jones Q and Earle F R 1966 Chemical analyses of seeds II: Oil and protein content of 759 species. Econ. Bot. 20, 127–156.

Lachmann J 1858 Über Knollen an den Wurzeln der Leguminosen. Landwirthschaftliche Mitteilungen. Zeitschrift der Königlichen höheren landwirthschaftlichen Lehranstalt und der damit vereinigten landwirthschaftlichen Versuchsstation zu Poppelsdorf. Adolph Marcus, Bonn, Germany. pp 34–52.

Lechtova-Trnka M 1931 Étude sur les bactéries des légumineuses et observations sur quelques champignons parasites des nodosités. Botaniste 23, 301–531.

Leonard L T 19:25 Lack of nodule-formation in a subfamily of the Leguminuosae. Soil Sci. 20, 165–167.

Lim G 1977 Nodulation of tropical legumes in Singapore. Trop. Agric. (Trinidad) 54, 135–141.

Martínez-Romero E 1994 Recent developments in *Rhizobium* taxonomy. Plant and Soil 161, 11–20.

McKey D 1994 Legumes and nitrogen: the evolutionary ecology of a nitrogen-demanding lifestyle. *In* Advances in Legume Systematics 5: The Nitrogen Factor. Eds. J I Sprent and D McKey. pp 211–228. Royal Botanic Gardens, Kew, UK.

Moreira, F M S and Franco A A 1994 Rhizobia-host interactions in tropical ecosystems in Brazil. *In* Advances in Legume Systematics 5: The Nitrogen Factor. Eds. J I Sprent and D McKey. pp 63–74. Royal Botanic Gardens, Kew, UK.

Moreira, F M S, Gillis M, Pot B, Kersters K and Franco A A 1993 Characterizatian of rhizobia isolated from different divergence groups of tropical Leguminosae by comparative polyacrylamide gel electrophoresis of their total proteins. Syst. Appl. Microbiol. 16, 135–146.

Naisbitt T, James E K and Sprent J I 1992 The evolutionary significance of the legume genus *Chamaecrista*, as determined by nodule structure. New Phytol. 122, 437–492.

Nobbe, F, Schmid E, Hiltner L and Hotter E 1891 Versuche über die Stickstoff-Assimilation der Leguminosen. Landwirtsch. Vers. Stn. [Tharand] 39, 327–359.

Polhill R M and Raven P H 1981 Advances in Legume Systematics. Royal Botanic Gardens, Kew, UK. 1049 p.

Polhill, R M, Raven P H and Stirton C H 1981 Evolution and systematics of the Leguminosae. *In* Advances in Legume Systematics, Part 1. Eds. R M Polhill and P H Raven. pp 1–26. Royal Botanic Gardens, Kew, UK.

Rothschild D I de 1967 Estudio preliminar sobre la presencia de bacterias en la raiz de *Gleditsia amorphoides* (Gris.) Taub. y *G. triacanthos* L. (Leguminosae). Comunicaciones del Museo Argentino de Ciencias Naturales "Bernardino Rivadivia" e Instituto Nacional de Investigacion de las Ciencias Naturales. Botánica 2, 9–14.

Rothschild D I de 1970 Nodulación en leguminosas subtropicales de la flora Argentina. Revista del Museo Argentino de Ciencias Naturales "Bernardino Rivadivia" e Instituto Nacional de Investigacion de las Ciencias Naturales. Botánica 3, 267–286.

Sprent J I 1994 Evolution and diversity in the legume-rhizobium symbiosis: chaos theory? Plant and Soil 161, 1–10.

Sutherland J M, McInroy S G, James E K and Naisbitt T 1994 Nodule structure with special reference to the tribes Sophoreae, Genisteae and Thermopsideae. *In* Advances in Legume Systematics 5: The Nitrogen Factor. Eds. J I Sprent and D McKey. pp 41–55. Royal Botanic Gardens, Kew, UK.

Szott L T, Palm C A and Sanchez P A 1991 Agroforestry in acid soils of the humid tropics. Adv. Agron. 45, 275–301.

Turk D and Keyser H H 1992 Rhizobia that nodulate tree legumes: specificity of the host for nodulation and effectiveness. Can. J. Microbiol. 38, 451–460.

Turner G L and Gibson A H 1980 Measurement of nitrogen fixation by indirect means. *In* Methods for Evaluating Biological Nitrogen Fixation. Ed. F J Bergersen. pp 111–184. John Wiley and Sons, Ltd., New York, USA.

Upchurch R G 1987 Estimation of bacterial nitrogen fixation using acetylene reduction. *In* Symbiotic Nitrogen Fixation Technology. Ed. G H Elkan. pp 289–305. Marcel Dekker, Inc., New York, USA.

Young J P W 1993 Molecular phylogeny of rhizobia and their relatives. *In* New Horizons in Nitrogen Fixation. Eds. R Palacios, J Mora and W W Newton. pp 587–592. Kluwer Academic Publishers, Dordrecht, the Netherlands.

Young J P W and Johnston A W B 1989 The evolution of specificity in the legume-*Rhizobium* symbiosis. TREE 4, 341–349.

Zhang X, Harper R, Karsisto M and Lindström K 1991 Diversity of *Rhizobium* bacteria isolated from the root nodules of leguminous trees. Int. J. Syst. Bacteriol. 41, 104–113.

Section editor: H Lambers
Guest editors: G.H. Elkan and R.G. Upchurch

Plant and Soil **186:** 161–166, 1996.

Lipopolysaccharide core components of *Rhizobium etli* reacting with a panel of monoclonal antibodies

Elmar L. Kannenberg, L. Scott Forsberg and Russell W. Carlson
*Complex Carbohydrate Research Center and Department of Biochemistry and Molecular Biology, University of Georgia, 220 Riverbend Road, Athens, GA 30602-4712, USA**

Key words: lipopolysaccharide, plant-microbe interaction, *Rhizobium*, symbiosis

Abstract

Monoclonal antibodies that react with *Rhizobium leguminosarum* lipopolysaccharide core antigens (LPS-2) have been used to investigate LPS-2 structure in *Rhizobium etli*. The panel of antibodies (JIM 32 – JIM 35, JIM 37, JIM 38) specific for LPS-2 of *R. leguminosarum* strain 3841 and its core components displays similar reactivities towards isolated LPS-2 from *R. etli* CE109 (a mutant of wild-type strain *R. etli* CE3 that displays LPS-2 as its main LPS form on the cell surface). This result suggests the antibodies bind to similar epitopes on both strains and, hence, that *R. leguminosarum* and *R. etli* have very similar LPS core and lipid A antigen structures. More detailed analysis of the antibody binding sites with isolated LPS-2 and lipid A from *R. etli* suggests that some of the antibodies (JIM 32, 33, 34, and MASM-I) bind some part of the core oligosaccharides, while others (JIM 35 and JIM 38) involve lipid A. These antibodies have already proven useful in the biochemical analysis of the LPS antigen forms. For example, the loss of reactivity of certain LPS forms with antibody JIM 37 has led to the discovery of a hitherto unnoticed form of the LPS antigen in a precipitate formed during the phenol/water extraction procedure. This new form reacts with the JIM 37 antibody. Furthermore, the positive reaction of some of the antibodies with only sonicated wild-type *R. etli* cells suggests that either an effective way of masking the display of core antigens on whole bacterial cells is occurring or that core forms of the LPSs are never displayed on the surface of the bacterial cells. Either possibility, once confirmed, could be important for our picture of the *Rhizobium* cell surface and could also have some bearing on symbiotic nodule infection and development.

Abbreviations: LPS – lipopolysaccharide.

Introduction

Gram-negative soil bacteria belonging to the genus *Rhizobium* form symbiotic root nodules on legumes and fix nitrogen within cells of the nodule tissue. For this process to occur, the rhizobia must invade plant tissue and cells, all the while acclimating to the endophytic environment, and must avoid host plant defence responses. Both of these capabilities probably involve special features of the bacterial cell wall and the bacterial cell surface.

The lipopolysaccharide (LPS) is a major component of the *Rhizobium* cell surface and outer membrane. Based on the electrophoretic mobility of LPS on polyacrylamide gels, two major classes of LPS have been defined (LPS-1, LPS-2) for *Rhizobium leguminosarum* and for *Rhizobium etli* (Carlson et al., 1987). LPS-1 macromolecules are formed from three components, lipid A, the core oligosaccharide, and the O-antigen, whereas LPS-2 lacks the O-antigen. Both classes of molecules are anchored by lipid A in the outer layer of the bacterial outer membrane. The strain-specific O-antigen component is composed of highly variable oligosaccharide units, commonly containing methylated glycosyl residues (Carlson et al., 1995). Evidence to date suggests that the LPS core is relatively conserved between related strains and composed of the following two oligosaccharide components: (i) a tetrasaccharide, α-Gal-(1 $\rightarrow$ 6)[α-GalA-(1 $\rightarrow$ 4)]-α-Man-(1 $\rightarrow$ 5)-Kdo, and (ii) a trisaccharide, a-GalA-(1 $\rightarrow$ 4)[α-GalA-(1 $\rightarrow$ 5)]-Kdo, bound in an unknown fashion to lipid A (Bhat

* FAX No: + 17065424412. E-mail: elmark@ccrc.uga.edu

et al., 1991a; Hollingsworth et al., 1990). The lipid A moiety consists of a carbohydrate backbone of GalA, GlcN, and GlcN-onate acylated with diverse hydroxy fatty acids of variable lengths (C14 to C18), one of which is a unique 27-hydroxyoctacosanoic acid (27-OH C28:0) which may span the entire lipid bilayer (Bhat et al., 1991b, 1994). Because of the complexity of the LPS macromolecule, the exact structures of whole molecules are still unknown and generalizations are further hampered by the fact that to date detailed structural analysis has been conducted on only a few *Rhizobium* strains.

Immunochemical and immunocytological analyses of LPS expression (LPS comprise a strongly immunogenic class of molecules) in *R. leguminosarum* have shown that rhizobia modify LPS-1 antigens according to the micro-environmental conditions of the nodule (Kannenberg et al., 1994). In cases in which mutant rhizobia express LPS forms whose antigen structures cannot be adapted, nodule development is severely affected (Kannenberg et al., 1992; Perotto et al., 1994). For example, *Rhizobium* mutants that lack the O-antigen of the LPS-1 cannot establish normal symbiosis in peas, beans, and other legumes (for review see Noel, 1992). In peas cytological evidence has been presented that a host plant defence response is mounted in response to LPS-defective mutants of *R. leguminosarum* and that bacterial development to the endophytic, nitrogen-fixing stage of the rhizobial bacteroids is affected (Perotto et al., 1994). In beans a *R. etli* mutant that also lacks the O-antigen portion of the LPS molecule causes an arrested infection process, and bacteria are never released into nodule cells (Noel et al., 1986).

Therefore, a complete LPS-1 structure is apparently essential for plant cell and tissue invasion by bacteria. Furthermore, the LPS structure seems also important for bacteroid fitness (Perotto et al., 1994). It has been suggested that the LPS could be either directly involved in nodule development (e.g., the core oligosaccharide may trigger the host defence response and the O-antigen may play a role in preventing this), or indirectly involved by affecting rhizobial cell wall stability and thereby releasing other signal compounds. However, the complexity and variability of the LPS molecules have so far hampered complete elucidation of LPS biological functions.

A powerful approach that can help to elucidate the characteristics and structural relatedness of LPS macromolecules is to use monoclonal antibodies as sub-molecular probes. To date, only the O-antigen components of LPSs have been studied in some detail using monoclonal antibodies and LPS mutants (Kannenberg and Brewin, 1989; Kannenberg et al., 1992; Tao et al., 1992). Comparatively little information is available about the core and lipid A antigen components of different *Rhizobium* strains. Therefore, prior to characterizing the LPS core antigen of various *Rhizobium* strains in more detail, we investigated the cross-reactivity of a panel of monoclonal antibodies specific for the LPS core components of *R. leguminosarum* bv. *viciae* strain 3841 with the LPS of strain *R. etli* CE3 and strain CE109, a LPS mutant derivative of strain CE3, lacking the LPS-1 (Carlson et al., 1987). We studied in some detail the epitopes on these strains recognized by the panel of antibodies and characterized the reactivity with the core antigens of these strains.

Materials and methods

The LPS was affinity-purified for sugar competition experiments and the antibody reactivity tested as described previously (Wood et al., 1989). The LPS was released from *Rhizobium* cells by sonication and the dissolved LPS separated from cell debris by centrifugation. The LPS was bound to 1 mL of matrix-supported polymixin B. The column was washed with 10 mL of water. The LPS was released with 5 mL of an aqueous solution of deoxycholate (1% w/v), subjected to extensive dialysis, and freeze-dried.

The monoclonal antibodies were raised against LPS mutants of strain *R. leguminosarum* 3841 (Lucas et al., 1996). Hybridoma supernatants of the different antibodies (all JIM antibodies) were used in a dilution of 1:100, while MASM-1 ascites fluid was diluted 1:10000. The reactivity of these antibodies with the LPS was tested in duplicate in dot immunoassays with approximately 1 μg of LPS (strain CE109) and tenfold dilutions thereof in the absence and presence of sugar haptens (at a concentration of 100 m*M* in all cases with the exception of Kdo which was used at 5 m*M*).

The reactivity was also tested with purified LPS from *R. etli* CE109 and well-characterized lipid A from strain CE3 (Bhat et al., 1994). For LPS purification, cells from strain CE3 or CE109 were extracted in a standard hot phenol/water extraction method, followed by extensive dialysis of the water phase and size fractionation on Sepharose 4B columns (Carlson et al., 1987). LPS-2 was isolated from strain CE109, while lipid A was released from purified LPS of strain CE3 by mild acid hydrolysis and isolated by centrifugation and extensive washing (Bhat et al., 1994). After hot phenol/water extraction and phase separation, the dialysis of phenol fractions led to the formation of precipitates in the dialysis tubing. The precipitate stemming from strain CE109 was saved and under vigorous stirring further extracted overnight in a triethylamine/EDTA buffer (100 m*M*/25 m*M*) and the extract tested for reactivity with antibodies in dot immunoassays.

Results and discussion

Antigen structure of Rhizobium etli *as revealed by reactivity with a panel of antibodies*

A panel of antibodies (JIM 31, JIM 32, JIM 33, JIM 34, JIM 35, JIM 37, and JIM 38) raised against LPS mutants of strain *R. leguminosarum* 3841 (these mutants partially or completely lacked the O-antigen) has been shown to recognize the LPS-2 of these strains (Lucas et al., 1996). So far, the LPS core components and the lipid A of *R. leguminosarum* 3841 have not been thoroughly characterized chemically. On the other hand, the LPSs from *R. etli* CE109, an LPS mutant of wild-type strain CE3, is one of the best characterized, being exclusively of the LPS-2 class (lacking the O-antigen moiety of LPS-1; Carlson et al., 1987).

As reported (Lucas et al., 1996), the antibodies react positively with whole cells and affinity-purified LPS-2 of strain B660 (strain B660 is an LPS mutant derivative of strain *R. leguminosarum* 3841, lacking LPS-1). In dot immunoassays, whole cells of strain CE109 showed identical reactivity to that reported for strain B660. The LPS-2 of CE109 (and as a positive control that of strain B660) was affinity-purified and its reactivity found to be identical. This suggests a great deal of structural similarity in the LPS-2 antigens of these two strains.

In order to confirm that the antibodies indeed recognize similar epitopes in *R. etli* CE3 and its derivatives, the reactivity with affinity-purified LPSs from strain CE109 was tested in a competition experiment with monomeric sugars to show that the antibodies exhibit the same pattern of inhibition that has been previously found for the LPSs of strain 3841 and its derivatives (Lucas et al., 1996). Table 1 shows the reactivity of the LPSs from strain CE109 with the panel of core antibodies. MASM-I is an antibody raised against the LPS from *Salmonella minnesota* which is known to be specific for Kdo (Lind et al., 1985) and served in this experiment as a positive control. Table 1 shows that the isolated LPSs of strain CE109 (the "blank" column indicates no sugar competitor) are reacting with the panel of antibodies. Strong inhibition of MASM-I reactivity with Kdo was observed, confirming that this antibody recognizes a Kdo-containing epitope. The reactivity of the different antibodies was strongly reduced by GalA (a sugar component of the core oligosaccharides and the lipid As), and the reactivity of JIM 33 and JIM 35 additionally by GlcN. This pattern of inhibition is almost identical to what has been reported previously for the reactivity of mutant deviates of strain *R. leguminosarum* 3841 lacking the O-antigen portion of the LPSs (Lucas et al., 1996). The almost identical reactivity of the LPS-2 antigens isolated from strains *R. leguminosarum* and *R. etli* (i) suggests that the antibodies recognize similar epitopes on both strains, and (ii) re-emphasizes that the two antigens share a high degree of structural similarity in their LPS core components and lipid As.

Reactivity of the antibodies with isolated LPS antigen fractions

In order to characterize further the antigen component recognized by the antibodies, we employed a better defined LPS core antigen from strain CE109 (isolated in a standard hot phenol/water extraction procedure; Carlson et al., 1987) and purified lipid A that had been isolated after mild acid hydrolysis of the LPS from strain CE3 (Bhat et al., 1994). Table 2 shows the reactivity of the panel of antibodies with whole cells and with isolated fractions. While all antibodies reacted with whole cells, two antibodies (JIM 35 and 37) did not react with the isolated LPS from strain CE109. JIM 35 and JIM 38, on the other hand, did react with purified lipid A, while JIM 37 did not react with either fraction. The antibodies were grouped into four classes on the basis of this reactivity. Class I antibodies (JIM 32, JIM 33, JIM 34, and MASM-I) react with the LPS-2 from CE109 but not with lipid A. Class II antibodies (JIM 38) react with the LPS-2 and with isolated lipid A from CE109, and Class III antibodies (JIM 35) react only with isolated lipid A. These patterns of reactivity suggest that the Class I antibodies recognize epitopes that are localized, at least partly, on the LPS core oligosaccharide of strain CE109, while Classes II and III antibody binding sites involve the lipid A portion of the molecule. It is still unclear why the recognition of the strain CE109 LPS is impaired with Class III antibodies. The fact that JIM 33 antibody reactivity was reduced by GlcN (Table 1), a sugar of

Table 1. Characterization by sugar competition of antibody binding sites on the LPS of strain *Rhizobium etli* CE109

Antibody	Blank	GalA	Gal	Man	GlcN	GlcAN	Kdo
MASM I	+	+	+	+	+	+	-
JIM 32	+	(-)	+	+	+	+	+
JIM 33	+	-	+	+	(-)	+	+
JIM 34	+	(-)	+	+	+	+	+
JIM 35	+	-	+	+	(-)	+	+
JIM 37	+	(-)	+	+	+	+	+
JIM 38	+	(-)	+	+	+	+	+

Affinity-purified LPS of strain *R. etli* CE109 was tested in dot immunoassays (both methods adapted from Wood et al., 1989). The reactivity of aliquots at a range of concentrations was tested in the absence and presence of sugar haptens (in all cases at a concentration of 100 m*M* with the exception of Kdo which was used at 5 m*M*). GalA – galacturonic acid, Gal – galactose, Man – mannose, GlcN – glucosamine, GlcAN – amino-deoxygluconic acid, Kdo – ketodeoxyoctonate, + – positive reaction, - – negative reaction, (-) – reaction significantly reduced or trace amounts.

Table 2. Reactivity of the panel of monoclonal antibodies with the LPS antigen from strain *Rhizobium etli* CE109

Monoclonal antibody[b]	Class	Sources of LPS from strain CE109[a] Whole cells[c]	Isolated LPS[d]	Isolated lipid A[e]	Extraction precipitate[f]
MASM-I[g]	I	+	+	-	(+)
JIM 32	I	+	+	-	+
JIM 33	I	+	(+)	-	-
JIM 34	I	+	+	-	(+)
JIM 38	II	+	+	+	+
JIM 35	III	+	-	+	- (tr)
JIM 37	IV	+	-	-	(+)

[a]+ – Positive reaction, (+) – weak positive reaction, - – negative reaction, tr – trace amounts.
[b]Reactivity with the monoclonal antibodies was tested in dot immunoassays using aliquots of different dilutions of the various antigens (1 μg mL^{-1} and tenfold serial dilutions thereof).
[c]Strain *R. etli* CE109 is an LPS mutant derivative of *R. etli* CE3. CE109 lacks the O-antigen-containing LPS-1 forms, and LPS-2 is the main form of LPS found on its cell surface.
[d]LPS-2 was isolated from strain CE109 cells by a standard hot phenol/water extraction procedure, followed by dialysis and gel filtration of the water-soluble material (Carlson et al., 1987).
[e]Lipid A was released from the purified strain CE3 LPS by mild acid hydrolysis and further purified by centrifugation and extensive washing (Bhat et al., 1994).
[f]After phase separation of hot phenol/water extracts, a precipitate formed during dialysis of the phenol phase. This precipitate was separated and further extracted with a triethylamine/EDTA buffer and the extract tested for reactivity with the panel of antibodies.
[g]MASM-I is a monoclonal antibody with specificity for Kdo-containing epitopes (Lind et al., 1985).

the lipid A backbone (Bhat et al., 1994), indicates that this antibody may also bind partly to the lipid A.

Surprisingly, the reactivity of the affinity-purified LPS of strain CE109 differed from that found with LPS purified in the phenol/water extraction (see above and Table 2). The JIM 37 (class IV) antibody, on the other hand, did not seem to react with either of the characterized LPS fractions (Table 2), indicating that either the isolated fractions were modified during the isolation procedure or that the epitope recognized by this antibody was located on a form of the LPS antigen that had separated during the isolation procedure. The hot phenol/water extraction method employed in this purification leads to two fractions (after removal of cell debris and phase separation into a water and phenol phase): water-, and phenol-solubilized LPSs. Normally, during the extensive consecutive dialysis steps performed on the phenol fraction, a water-insoluble precipitation is formed whose contents have not been further investigated. This extraction precipitate was further extracted (with a triethylamine/EDTA buffer) and analyzed to determine if it contains forms of the LPS which were not found in solution. Table 2 shows additional immunogenically distinct LPSs that were identified in this fraction.

The molecular basis for these differential reactivities is not yet clear. Some of the LPS antigen may be modified during the isolation procedure, affecting the antibody reactivity of JIM 37 and perhaps of JIM 35, but more fundamental differences in LPS antigen structures of the various isolates may also be responsible.

Reactivity of the panel of antibodies with cells from Rhizobium etli *CE3*

We wanted to test if the LPS antigens recognized by the panel of antibodies could also be detected on the cell surface or with whole cells of the wild strain *R. etli* CE3. We employed sonicated and unsonicated cell suspensions (at a range of different concentrations) of strain CE3 to test in dot immunoassays for the presence of the antigens. Surprisingly, only JIM 32 from the panel of antibodies reacted strongly with the LPS displayed on the cell surface of CE3 (a weak reaction was also found with antibody JIM 38). After sonication, some of the antibodies reacted weakly with the preparation of disrupted cells (JIM 34, JIM 33, and JIM 36) or showed an increased reactivity, as in the case of JIM 38.

It has been reported that the JIM 32 antibody can react with LPS-1 and LPS-2 molecules (Lucas et al., 1996), suggesting that the reactivity observed with the cells of CE3 may be due to the reactivity with LPS-1 class molecules of these bacteria. On the other hand, the almost complete lack of reactivity of the other antibodies with whole CE3 cells is intriguing. It could mean that the truncated LPS-2 forms are not displayed on the cell surface. In that case, reactivity with the sonicated cells could be explained as reactivity of the antibodies with core precursors, probably bound to the lipid carrier. It could also mean that the truncated forms of the LPSs are effectively masked by an intact cell surface and become exposed only after the cell architecture has been disrupted either mechanically or (as in the case of strain CE109) by a mutation in the chromosome of the bacterium thus producing a modified LPS structure. The latter assumption would indicate that the O-antigen portion of the LPS is involved in the masking effect and would corroborate the conclusions from other papers that suggest that the LPS O-antigen serves a masking function in legume symbiosis (Kannenberg et al., 1992; Perotto et al., 1994). This assumption also indicates that the LPS-2 class is directly or indirectly involved in abnormal nodule development observed with certain LPS mutants lacking the O-antigen.

Acknowledgements

We thank Rosemary Nuri (CCRC) for her help in preparing the manuscript. The work was supported in part by a National Institutes of Health grant GM 395832 to RWC and the US Department of Energy-funded (DE-FG05-93ER20097) Center for Plant and Microbial Complex Carbohydrates (at the CCRC).

References

Bhat U R, Bhagyalakshmi S K and Carlson R W 1991a Re-examination of the structures of the lipopolysaccharide core oligosaccharides from *Rhizobium leguminosarum* biovar *phaseoli*. Carbohydr. Res. 220, 219–227.

Bhat U R, Mayer H, Yokota A, Hollingsworth R I and Carlson R W 1991b Occurrence of lipid A variants with 27-hydroxyoctacosanoic acid in lipopolysaccharides from membranes of the family Rhizobiaceae. J. Bacteriol. 173, 2155–2159.

Bhat U R, Forsberg L S and Carlson R W 1994 Structure of lipid A component of *Rhizobium leguminosarum* bv. *phaseoli* lipopolysaccharide. J. Biol. Chem. 269, 14402–14410.

Carlson R W, Kalembasa S, Turowski D, Pachori P and Noel, K D 1987 Characterization of the lipopolysaccharide from a *Rhi-*

zobium phaseoli mutant that is defective in infection thread development. J. Bacteriol. 169, 4923–4928.

Carlson R W, Forsberg S L, Price N P J, Bhat U R, Kelly T M and Raetz R H 1995 The structure and biosynthesis of *Rhizobium leguminosarum* lipid A. *In* Bacterial Endotoxins: Lipopolysaccharides; From Genes to Therapy. Eds. J Levin, C R Alving, R S Munford and H Redl. pp 25–31. Wiley-Liss, Inc., New York, USA.

Hollingsworth R I, Carlson R W, Garcia F and Gage D A 1990 A new core tetrasaccharide component from the lipopolysaccharide of *Rhizobium trifolii* ANU 843. J. Biol. Chem. 264, 9294–9299 (Erratum 265, 12752).

Kannenberg E L and Brewin N J 1989 Expression of a cell surface antigen from *Rhizobium leguminosarum* 3841 is regulated by oxygen and pH. J. Bacteriol. 171, 4543–4548.

Kannenberg E L, Rathbun E A and Brewin N J 1992 Molecular dissection of structure and function in the lipopolysaccharide of *Rhizobium leguminosarum* strain 3841 using monoclonal antibodies and genetic analysis. Mol. Microbiol. 6, 2477–2487.

Kannenberg E L, Perotto S, Vianciotto V, Rathbun E A and Brewin N J 1994 Lipopolysaccharide epitope expression of *Rhizobium* bacteroids as revealed by in situ immunolabelling of pea root nodule sections. J. Bacteriol. 176, 2021–2032.

Lind S M, Carlin N I A and Lomdberg A A 1985 Production and characterisation of KDO-specific monoclonal antibodies recognising lipopolysaccharides from heptoseless mutants of *Salmonella*. FEMS Microbiol. Lett. 28, 45–49.

Lucas M M, Peart J L, Brewin N J and Kannenberg E L 1996 Isolation of monoclonal antibodies reacting with the core component of lipopolysaccharide from *Rhizobium leguminosarum* strain 3841 and mutant derivatives. J. Bacteriol. 178, 2727–2733.

Noel K D, VandenBosch K A and Kulpaca B 1986 Mutations in *Rhizobium phaseoli* that lead to arrested development of infection threads. J. Bacteriol. 168, 1392–1401.

Noel K D 1992 Rhizobial polysaccharides required in symbiosis with legumes. *In* The Molecular Signals in Plant-Microbe Communication. Ed. D P S Verma. pp 341–357. CRC Press, Boca Raton, USA.

Perotto S, Brewin N J and Kannenberg E L 1994 Cytological evidence for a host defence response that reduces cell and tissue invasion in pea nodules by lipopolysaccharide-defective mutants of *Rhizobium leguminosarum* strain 3841. Mol. Plant-Microbe Interact. 7, 99–112.

Tao H, Brewin N J and Noel K D 1992 *Rhizobium leguminosarum* CFN42 lipopolysaccharide antigenic changes induced by environmental conditions. J. Bacteriol. 174, 2222–2229.

Wood E A, Butcher G W, Brewin N J and Kannenberg E L 1989 Genetic derepression of a developmentally regulated lipopolysaccharide antigen from *Rhizobium leguminosarum* 3841. J. Bacteriol. 171, 4549–4555.

Section editor: H Lambers
Guest editors: G.H. Elkan and R.G. Upchurch

Plant and Soil **186:** 167–171, 1996.

Diversity in symbiotic specificity of cowpea rhizobia indigenous to Zimbabwean soils

S. Mpepereki[1], A.G. Wollum II[2] and F. Makonese[1]

[1]*Department of Soil Science and Agricultural Engineering, University of Zimbabwe, PO Box MP 167, Mt Pleasant, Harare, Zimbabwe* and* [2]*Department of Soil Science, North Carolina State University, Box 7619, Raleigh, NC 27695-7619, USA*

Key words: cowpea rhizobia, diversity, host range, symbiotic specificity

Abstract

Tropical cowpea rhizobia are often presumed to be generally promiscuous but poor N fixers. This study was conducted to evaluate symbiotic interactions of 59 indigenous rhizobia isolates (49 of them from cowpea (*Vigna unguiculata*)), with up to 13 other (mostly tropical) legume species. Host ranges averaged 2.4 and 2.3 legume species each for fast- and slow-growing isolates respectively compared to 4.3 for slow-growing reference cowpea strains. An average of 22% and 19% of fast- and slow-growing cowpea isolates respectively were effective on each of 12 legume species tested. We conclude that the indigenous cowpea rhizobia studied have relatively narrow host ranges. The ready nodulation of different legumes in tropical soils appears due to the diversity of indigenous symbiotic genotypes, each consisting of subgroups compatible with a limited number of legume species.

Introduction

Symbiotic specificity and host range of the "cowpea miscellany" rhizobia indigenous to African soil environments has not been well characterized. These bacteria are often described as promiscuous, nodulating a wide range of legumes but with poor effectiveness (e.g. Singleton et al., 1992). The ability of many rhizobia to nodulate different legumes has led to the abandonment of cross-inoculation groups as a classification criterion (Wilson, 1944), although nodulation and symbiotic effectiveness with specific legumes still remain important characteristics in rhizobia taxonomy (Graham et al., 1991). The capacity of indigenous strains to effectively nodulate a legume crop or, alternatively, to act as a barrier to the establishment of more efficient inoculant strains has important agronomic implications. Evaluation of host range and symbiotic effectiveness of indigenous rhizobia populations also serves to identify effective adapted strains with potential for use as inoculants.

Despite decades of widespread use of commercial rhizobia inoculants on soybean in Zimbabwe (Marufu et al., 1995), very little is known about the symbiotic characteristics of indigenous rhizobia populations which occur widely but at low population densities (Mpepereki and Makonese, 1994). The present study determined the extent of symbiotic specificity of indigenous fast- and slow-growing rhizobia isolated from a range of physiographic environments in Zimbabwe.

* FAX No: + 2634333407. E-mail: smpepe@zimbix.uz.zw.

Materials and methods

Organisms and media

A total of 59 rhizobium isolates (49 from cowpea (*Vigna unguiculata*) and the remainder from 13 legume species) from soils from different locations in Zimbabwe were studied. The isolates have previously been described (Mpepereki, 1994). Reference strains used included CB 756, TAL 1000, TAL 471, MAR 411 (all slow-growing) and the fast-growing TAL 1797 and MAR 1380 (the latter isolated from *Phaseolus*), and all obtained from the Grasslands *Rhizobium* Collection Marondera, Zimbabwe; and the slow-growing strains NC 83.2, NC 92, and TAL 169 from the culture collection of A G Wollum. Cultures were maintained on yeast extract mannitol (YEM) agar slants at 4°C (Vincent, 1970).

Host range and symbiotic effectiveness tests

The following 13 legumes were used: *Arachis hypogaea* (cv. Natal Common); *Cajanus cajan*, *Crotalaria juncea*, *Crotalaria ochraleuca*, *Glycine max* (cv. SCS-1), *Indigofera astragalina*, *Indigofera tinctoria*, *Pisum sativum*, *Phaseolus vulgaris* (cv. Carioca), *Macroptilium atropurpureum* (cv. Siratro), *Sesbania sesban*, *Vigna unguiculata* (cv. Local Mixed) and *Vigna subterranea*. Seed for *C. ochraleuca*, *I. astragalina*, *I. tinctoria*, and *S. sesban* was obtained from M Piha, University of Zimbabwe. All other seed was obtained from the Seed Services Section, Ministry of Agriculture, Zimbabwe.

Prior to planting the seeds were surface-disinfected or scarified with sulphuric acid if hard-coated, using recommended protocols (Somasegaran and Hoben, 1985). Seeds were sown four per pot in 0.5-L plastic containers filled with sterile horticultural vermiculite moistened with N-free nutrient solution (McClure and Israel, 1979). Each seed was inoculated with 0.25 mL of a late-log phase YEM broth culture for each strain. Pots were arranged in a randomized complete block design with three replicates. Plants were thinned to two per pot at one week after emergence. Greenhouse temperatures ranged from 16°C to 35°C. For each plant species, positive controls consisted of plants either inoculated with an appropriate effective reference strain (i.e. CB 756 for *A. hypogaea*, *M. atropurpureum*, and *V. unguiculata*; MAR 1380 for *P. vulgaris* and MAR 411 for *V. subterranea*), or fertilized once weekly with N (0.70 g L^{-1} KNO_3), if a homologous strain was unavailable.

Plants were harvested 42 days after planting. Symbiotic effectiveness was estimated using a 10-point index, subdivided as follows: shoot biomass, 4 pts., plant colour, 3 pts., and nodule colour, 3 pts. of plants inoculated with individual isolates compared with the controls. Plant tops were dried at 70°C for 24 h and biomass ranked using Duncan's Multiple Range Test at the 5% significance level. Mean biomass was assigned a score of 1 to 4 as follows: mean biomass significantly higher than positive control = 4, higher = 3, lower = 2 and significantly lower = 1. Since our objective was to focus on the range of legume hosts that individual isolates can nodulate effectively, we chose to use a rating system for biomass and combining it with other effectiveness data to demonstrate the symbiotic profile of select isolates/strains. Plant colour was scored visually prior to harvesting, where 0 represented yellow and 3 represented green plants. A random sample of 20 nodules from each set of replicate pots was sliced open and scored for apparent N-fixing activity based on visual evidence of red pigment haemoglobin on a scale of 0 to 3: red = 3, pink = 2, partly pink = 1, white/green = 0. The scores for the three parameters (plant biomass, leaf colour and nodule colour) were combined. Isolates with a combined score of 7 and above were recorded as "effective" (E) and those producing scores < 7 were considered poorly effective to ineffective (I) if nodules were present or non-nodulating (0) when nodules were absent.

Results

Ten indigenous isolates obtained from six different host legumes and growing in different soil types, along with two additional isolates (Z-55 and Z-77) from a larger collection were found to be highly specific (Table 1), each being effective on an average of 1.7 species.

Three of the isolates nodulated only their original trap hosts (Table 1). Symbiotic effectiveness profiles of another 49 indigenous cowpea isolates tested on 13 legume species (data not shown) showed that fast- and slow-growing isolates were each effective on an average of 2.4 and 2.3 legume hosts respectively. Except for cowpea and bambarra (*V. subterannea*) which appeared to be compatible with many strains, each isolate tended to effectively nodulate with a different but narrow set of legume hosts. Cowpea reference strains on the other hand were effective on an average of 4.3 hosts each (data not shown).

Table 2 summarizes the distribution of fast- and slow-growing indigenous rhizobia isolates in the categories "effective" (E), "poorly effective/ineffective" (I) and "non-nodulating" (0) on 12 legumes. Four effectiveness groups were identified for the fast-growing cowpea rhizobia (Table 2). In the first group, less than 10% of fast-growing isolates were effective on *S. sesban* (0%), *G. max.* (5%), *C. ochraleuca* (L.) (7%) and *I. astragalina* (8%). The second group were effective with 14–20% of the isolates and included *A. hypogaea* (18%), *C. cajan* (14%), and *C. juncea* (20%). The third group included *M. atropurpureum* (23%) and *P. vulgaris* (27%). The fourth group had significantly higher proportions of effective associations, with *V. subterranea* at 50% and *V. unguiculata* at 73%. The proportion of non-nodulating fast-growing isolates ranged

Table 1. Symbiotic effectiveness profiles of indigenous rhizobia from twelve different Zimbabwean soils

Isolate no.	Trap host[a]	Effectiveness[b] ranking on test legumes							
		Ager	Acsi	Alsp	Crsp	Lesp	Maat	Muse	Sesp
As-gl	Ager	E	E	O	O	E	O	O	O
Ac-sl	Acsi	O	E	O	O	O	O	O	O
Ac-1026	Acsp	E	E	O	O	O	O	O	O
Ac-999	Acsp	I	O	O	O	O	O	O	O
Cr-1	Crsp	I	O	O	O	O	O	O	O
Cr-2	Crsp	O	O	I	E	O	E	O	O
Sir-1	Maat	O	I	I	E	I	E	E	O
Sir-2	Maat	O	E	O	E	I	I	O	E
Ses-1	Sesp	I	O	O	O	E	O	O	E
Ses-2	Sesp	O	O	I	O	E	O	I	E
Z-55	Viun	O	O	O	E	O	nd	nd	O
Z-77	Viun	O	O	O	E	O	E	nd	O

[a]Legumes used: **Ager** = *Acacia gerardii*; **Acsi** = *Acacia sieberana*; **Acsp** = *Acacia* sp; **Alsp** = *Aalbizia* sp; **Crsp** = *Crotalaria* sp; **Lesp** = *Leucena* sp; **Maat** = *Macroptilium atropurpureum*; **Muse** = *Mundulea sericea*; **Sesp** = *Sesbania* sp; **Viun** = *Vigna unguiculata*.
[b]Effectiveness ranking: **E** – nodulating, effective; **I** – nodulating, poorly effective/ineffective; **O** – non-nodulating.

Table 2. Symbiotic effectiveness profiles of fast- and slow-growing indigenous cowpea rhizobia from Zimbabwe selected from an original population of 59 isolates on 12 legumes

Legume	No. of strains tested	Fast-growing			No. of strains tested	Slow-growing		
		E[a]	I	O		E	I	O
		(%)[b]				(%)		
Arachis hypogaea	22	18	9	73	26	12	4	84
Cajanus cajan	21	14	0	86	25	12	24	64
Crotalaria juncea	20	20	5	75	20	5	5	90
Crotalaria ochraleuca	15	7	0	93	9	0	0	100
Glycine max	22	5	0	95	24	13	8	79
Indigofera astragalina	13	8	0	92	15	7	0	93
Pisum sativum	nd[c]	nd	nd	nd	26	0	4	96
Phaseolus vulgaris	22	27	51	23	24	17	50	33
Macroptilium atropurpureum	22	23	41	36	27	26	33	41
Sesbania sesban	20	0	5	95	16	0	0	100
Vigna unguiculata	22	73	27	0	27	85	15	0
Vigna subterranea	22	50	27	23	26	46	35	19
Averages		22	15	63		19	15	67

[a]E – effective; I – nodulating, poorly effective/ineffective; O – non-nodulating.
[b]Based on total number of stains tested.
[c]nd – not determined.

from 0% with *V. unguiculata* to 95% with *S. sesban* and *G. max.*, and averaged 63% across all legumes (Table 2). Unequal numbers of strains reported as tested resulted from logistic problems associated with poor germination or insufficient seed to test all isolates.

Slow-growing cowpea isolates failed to nodulate effectively with three of the 12 legumes i.e., *C. ochraleuca*, *P. sativum* and *S. sesban* (Table 2). Four legumes nodulated effectively with between 10 and 20% of slow-growing isolates, while *V. subterranea* with 46% and *V. unguiculata* with 85% were signif-

Table 3. Symbiotic effectiveness profiles of cowpea reference rhizobia strains

Test legume	No. of strains[a] tested	Effective[b]	Ineffective (%)	Non-nodulating
Arachis hypogaea	8	25	25	50
Cajanus cajan	5	0	80	20
Crotalaria juncea	8	100	0	0
Crotaliaria ochraleuca	6	17	50	33
Indigofera tinctoria	8	13	75	12
Indigofera astragalina	7	14	43	43
Phaseolus vulgaris	8	88	0	12
Macroptilium atropurpureum	8	75	25	0
Sesbania sesban	8	100	0	0
Vigna unguiculata	8	13	87	0
Averages		45	39	17

[a]Cowpea reference strains included: NC 92, NC 83.2, CB 756, TAL 1000, TAL 471, TAL 377, MAR 411, TAL 169.
[b]Based on the total number of strains tested.

icantly more compatible with the test isolates. Non-nodulating combinations with slow-growing cowpea isolates ranged from 0 to 100% and averaged 67% of the total strains tested. An average of 45% of reference strains were effective on the 10 test legumes with only 17% failing to nodulate (Table 3). Only *C. cajan* was totally ineffective with reference strains. Five legumes including *A. hypogaea* and *V. unguiculata* were effectively nodulated by 25% or less of the reference strains.

Discussion

The average number of legume species nodulated by fast- and slow-growing indigenous cowpea rhizobia, 2.4 and 2.3, respectively, suggests that the two groups have a similar but narrow host range. This similarity in size of the host range reflects a close adaptation by rhizobia from different genera to local legumes, perhaps through a process of convergent evolution of symbiotic genotypes. The observed narrow host range pattern was somewhat unexpected given that eight of the test legumes, including *A. hypogaea*, have a long history of cultivation in or are indigenous to Zimbabwean soils and therefore each could be expected to nodulate effectively with more of the indigenous rhizobia. This would be consistent with the hypothesis put forward by Lie et al. (1987) that rhizobia and legumes from a given geographical area may become adapted and form local complementary gene pools that result in effective symbioses. It is possible that the diversity of physiographic environments where the isolates originated within Zimbabwe precluded co-evolution of similar broad host ranges. The indigenous rhizobia tested appear to be a heterogenous mixture of different symbiotic types, each with its own relatively narrow host range. Cowpea (the original host of most of the isolates) and *V. subterranea* (a close relative) had, as expected, the largest number of effective associations with both fast-and slow-growing indigenous isolates emphasizing the two legumes' relative promiscuity. However, although cowpea nodulated with all the reference strains, it was effective with only 13% of them. The diversity in host range and symbiotic effectiveness patterns of indigenous rhizobia observed in this study are consistent with earlier evidence of diversity within the same indigenous cowpea rhizobia collection, which segregated into several groups based on serology, protein profiles and DNA homology (Mpepereki et al., 1994).

Peanut, (*A. hypogaea*) was compatible with a relatively small sub-population of cowpea-nodulating rhizobia, averaging 15% for both fast- and slow-growing indigenous isolates. Thies et al. (1991) also found that only 18% of cowpea-derived isolates formed effective nodules on peanut. Interestingly, peanut nodulates heavily and effectively in most African soils, an indication that it probably associates with a highly compatible sub-group of the indigenous rhizobia population which does not nodulate cowpea. Although *S. sesban* was the least compatible with both fast- and slow-growing indigenous cowpea isolates, it nodulated effectively with both fast- and slow-growing reference strains. Of the eight reference strains tested, the six cowpea strains NC 92, CB 756, TAL 1000, TAL 471, MAR 411 and TAL 169 averaged 60% effective associations with the test legumes.

Phaseolus vulgaris (cv. Carioca was used), which is widely reported as relatively promiscuous, but only with fast-growing rhizobia, was effectively nodulated by all the slow-growing cowpea reference strains except TAL 1000. Indigenous slow-growing isolates also nodulated *P. vulgaris*, with up to 17% of the associations being effective. The wider host range of reference versus "wild" strains may partly explain why the myth about promiscuity of cowpea rhizobia has tended to persist. Cowpea rhizobia reference strains used appear to have limited diversity and wide host range in contrast to the little-studied "wild" populations in most African soils. For example, the cowpea reference strains CB 756, 3G4b20 and 32H1 have been shown to be very closely related on the basis of high DNA homology (Mpepereki et al., 1994).

Our results provide a contribution towards dismissing the myth about general nodulation promiscuity of tropical rhizobia (Singleton et al., 1992) and support the existence of subgroups based on symbiotic specificity (Thies et al., 1991).

It appears that indigenous cowpea rhizobia have relatively narrow host ranges compared to common reference strains and that the ready nodulation of different legumes in tropical soils is less a function of the broad host range of individual strains and more a reflection of the diversity in symbiotic types in the total rhizobia population. Future studies should be conducted to confirm this observation.

Acknowledgement

Funding for this study was provided by the University of Zimbabwe Research Board.

References

Graham P H, Sadowsky M J, Keyser H H, Barnet Y M, Bradley R S, Cooper J E, De Ley D J, Jarvis B D W, Roslycky E B, Strijdom B W and Young J P W 1991 Proposed minimal standards for the description of new genera and species of root-and stem-modulating bacteria. Int. J. Syst. Bacteriol. 41, 582–587.

Lie T A, Goktan D, Engin M, Pijnenborg J and Anlarsal E 1987 Co-evolution of the legume-*Rhizobium* association. Plant and Soil 100, 171–181.

Marufu L, Karanja N and Ryder M 1995 Legume inoculant production and use in East and Southern Africa. Soil Biol. Biochem. 27, 735–738.

McClure P R and Israel D W 1979 Transport of nitrogen in the xylem of soybean plants. Plant Physiol. 64, 411–416.

Mpepereki S 1994 Characterization studies of rhizobia in some Zimbabwean soils. Ph.D. Thesis. University of Zimbabwe, Harare, Zimbabwe.

Mpepereki S and Makonese F 1994 Inoculation response of cowpea (*Vigna unguiculata* [L.]) soyabean (*Glycine max* [L.] Merr.) and dry bean (*Phaseolus vulgaris* [L.]) in soils with low populations of rhizobia. *In* Recent Developments in Nitrogen Fixation Research in Africa. Eds. M Sadiki and A Hilali. pp 32–44. IAV Hassan II, Rabat, Morocco.

Mpepereki S, Wollum A G and Kloos W E 1994 DNA–DNA hybridization, serology and gel electrophoresis studies of rhizobia from Zimbabwean soils. *In* Nitrogen Fixation with Non-Legumes. Eds. N A Hegazi, M Fayes and M Monib. pp 385–386. American University in Cairo Press, Cairo, Egypt.

Singleton P W, Bohlool B B and Nakao P L 1992 Legume rhizobia inoculation in the tropics: myths and realities. *In* Myths and Science of Soils in the Tropics. pp 135–155. ASA, SSSA Special Publication No. 29. Madison, WI, USA.

Somasegaran P and Hoben H 1985 Methods in Legume-*Rhizobium* Technology. NifTAL Project, University of Hawaii, Paia, Honolulu.

Thies J E, Singleton P W and Bohlool B B 1991 Sub-groups of the cowpea miscellany: symbiotic specificity within *Bradyrhizobium* spp. for *Vigna unguiculata*, *Phaseolus lunatus*, *Arachis hypogaea*, and *Macroptilium atropurpureum*. Appl. Environ. Microbiol. 57, 1540–1545.

Wilson J K 1944 Over five hundred reasons for abandoning the cross inoculation groups of legumes. Soil. Sci. 58, 61–69.

Section editor: H Lambers
Guest editors: G.H. Elkan and R.G. Upchurch

Plant and Soil **186:** 173–187, 1996.

Host genetic control of symbiosis in soybean (*Glycine max* L.)

T.E. Devine and L.D. Kuykendall
*Plant Molecular Biology Laboratory and Soybean and Alfalfa Research Laboratory, USDA, ARS, Beltsville, MD 20705, USA**

Key words: *Bradyrhizobium*, legumes, nitrogen fixation, Nod Factors, nodulation, *Rhizobium*

Abstract

Genes controlling nitrogen-fixing symbioses of legumes with specialized bacteria known as rhizobia are presumably the products of many millions of years of evolution. Different adaptative solutions evolved in response to the challenge of survival in highly divergent complexes of symbionts. Whereas efficiency of nitrogen fixation appears to be controlled by quantitative inheritance, genes controlling nodulation are qualitatively inherited. Genes controlling nodulation include those for non-nodulation, those that restrict certain microsymbionts, and those conditioning hypernodulation, or supernodulation. Some genes are naturally occurring polymorphisms, while others were induced or were the result of spontaneous mutations. The geographic patterns of particular alleles indicate the role of coevolution in determining symbiont specificities and compatibilities. For example, the *Rj4* allele occurs with higher frequency (over 50%) among the soybean (*G. max*) from Southeast Asia. DNA homology studies of strains of *Bradyrhizobium* that nodulate soybean indicated two groups so distinct as to warrant classification as two species. Strains producing rhizobitoxine-induced chlorosis occur only in Group II, now classified as *B. elkanii*. Unlike *B. japonicum*, *B. elkanii* strains are characterized by (1) the ability to nodulate the *rj1* genotype, (2) the formation of nodule-like structures on peanut, (3) a relatively high degree of ex planta nitrogenase activity, (4) distinct extracellular polysaccharide composition, (5) distinct fatty acid composition, (6) distinct antibiotic resistance profiles, and (7) low DNA homology with *B. japonicum*. Analysis with soybean lines near isogenic for the *Rj4* versus *rj4* alleles indicated that the *Rj4* allele excludes a high proportion of *B. elkanii* strains and certain strains of *B. japonicum* such as strain USDA62 and three serogroup 123 strains. These groups, relatively inefficient in nitrogen fixation with soybean, tend to predominate in soybean nodules from many US soils. The *Rj4* allele, the most common allelic form in the wild species, has a positive value for the host plants in protecting them from nodulation by rhizobia poorly adapted for symbiosis.

Introduction

Although the two bacterial genera, *Rhizobium* and *Bradyrhizobium*, probably diverged about 500 million years ago, presumably prior to the evolution of nitrogen-fixing symbioses with legumes, (Wilson et al., 1987), symbiosis between legumes and rhizobial microsymbionts is thought to have originated during the Cretaceous period. Evidence for this hypothesis is that symbiosis is found in two major taxonomic subgroups of the Leguminosae: the Mimosoideae and the Caesalpinioideae. Symbiosis likely evolved before these legume groups diverged, rather than symbiosis having evolved independently in both groups. Since these groups had diverged by the end of the Cretaceous period, the Cretaceous period would be the likely time for the evolution of the symbiosis (Polhill et al., 1981). Some view the complex and diverse array of plant/bacterial interactions as a sort of biological chaos that defies interpretation. For example, Sprent (1994) stated "Environmental constraints superimposed on initial evolutionary developments result in an apparently chaotic situation where there is no obvious pattern of co-evolution between hosts and rhizobia." If, however, evolution and co-evolution have played a role in the origin of present biological forms, it would seem that, through critical analysis, the possibility of deciphering these forms and understanding

* FAX No: +13015045320.
E-mail: dkuykendall@asrr. arsusda.gov

the processes that shaped them is a feasible, although challenging, objective.

The biological advantage to the legume host derived from symbiosis is clear – the fixed nitrogen supplied by the bacteria provides an essential nutrient that is usually a limiting factor in the environment. For over 100 years, an advantage to the bacteria derived from the symbiosis was surmised, but never proven. However, Kuykendall et al. (1982) and Kuykendall (1989) established that a biological advantage accrued to the bacteria from symbiosis. In field studies, the logarithmic multiplication within the protected ecological niche of the nodule resulted in increased recovery from nodules of a subsequent crop and increased persistence in the soil.

Different adaptative solutions evolved in response to the challenge of survival in taxonomically divergent complexes of symbionts. Of particular interest is the range of rhizobia with which a host population is able to (a) nodulate, and (b) fix nitrogen efficiently, and conversely, the ability of a strain of microsymbiont to nodulate and efficiently fix nitrogen with an array of host genotypes, species or genera. Some host plants have the ability to nodulate with a wide array of rhizobia. An example is soybean which nodulates and fixes nitrogen with both fast-growing *Rhizobium* and slow-growing *Bradyrhizobium*, two very distinct genera that differ greatly in their morphology, physiology and DNA homology (Jordan, 1982). In contrast, other species such as alfalfa nodulate only with a specific species of *Rhizobium*. Understanding the biological factors and environmental forces that governed the canalization of genetic control mechanisms toward these different biological solutions to adaptation is an intriguing and formidable challenge to research biologists.

Questions for research

In order to stimulate and guide the research needed to develop a comprehensive understanding of the biology of symbiotic nitrogen fixation, we will pose the questions that, as geneticists, we see as significant.

- In an ecological niche populated simultaneously by a diversity of legume species or genera, is there an advantage to the bacteria in possessing the capacity to nodulate the diverse population of host species present?
- To what extent is it feasible for a single bacterial strain to retain the capacity for highly efficient nitrogen fixation in the substrate milieu provided by different host species?
- Does survival demand that the bacteria specialize to maximum efficiency of fixation with a particular host species?
- Does retention of the capacity for fixation with a broad array of host species diminish the capacity for maximum fixation with a single species and what is the trade-off in efficiency?
- Does a diminished efficiency of fixation with a host result in a lower selective opportunity for the host and its consequent loss from the ecosystem?
- What relative population frequency of such bacteria in the soil is tolerable for the survival of the host species?
- For the host, is there an advantage to carrying genes which prevent nodulation with particular rhizobial strains that may not have the capacity for efficient fixation with a specific host genotype?
- If the localized ecosystem has only a single host species rather than several, may evolution canalize selection of the bacteria toward specificity of nodulation and efficiency of fixation with that host population?
- Then, subsequently, should reproductive barriers isolating the host population or a series of host populations from several ecological niches be eliminated by some environmental modification (domestication by man, climatic change, topographical change, pollen vector or seed vector change), what would be the impact on symbiosis of introgressing these populations?
- Is there an advantage to the host in developing or maintaining a broad adaptability to a diverse population of soil microsymbionts in a soil containing a highly heterogeneous population of *Rhizobium* and *Bradyrhizobium*?
- Or, on the contrary, in such an ecosystem is there an advantage to the host in developing genetic mechanisms for excluding nodulation with specific groups of rhizobia and favoring nodulation with others?
- What is the role of polyploidy, particularly allopolyploidy, in the host in altering both nodulation and efficiency relationships in the case of highly restrictive specificity and broad-based symbiosis?

Based on genetic evidence of duplicated quantitative genes (Palmer and Kilen, 1987), evolutionary studies and haploid gene analysis (Hadley and Hymowitz, 1973) and molecular mapping evidence

Table 1. Soybean genes controlling nodulation response

Gene	Phenotype	Source
Spontaneous nodulation-restrictive mutant		
rj1	Non-nodulating	segregate in a cross of Lincoln×Richland
Indigenous asiatic polymorphisms		
Rj2	Ineffective nodulation with many strains of *B. japonicum*	Hardee, CNS, and others
Rj4	Ineffective nodulation primarily with certain *B. elkanii* strains	Hill and others
Rfg1	Ineffective nodulation with *R. fredii* strains USDA 205 and 257	Kent and McCall
Induced nodulation-restrictive mutations		
rj5	Non-nodulating	Harosoy 63
rj6	Non-nodulating	nod 139, NN5
Induced supernodulation mutations		
rj7	Hypernodulating	NOD3-7
*nts1**	Supernodulating	nts382

* (*rj7* and *nts1* are allelic).

(Shoemaker et al., 1995), soybean is currently understood to be an ancient polyploid that has been heavily diploidized by chromosomal rearrangments. Meiosis is characterized by preferential bivalent pairing, thus simplifying genetic analysis. Since soybean is an ancient tetraploid, it is not surprising that multiple loci would govern common or related processes. Several host plant genetic factors have been identified in soybean that govern nodulation with *Rhizobium* and *Bradyrhizobium* (Table 1). Can these genes provide insights into some of the questions we have posed? Let us examine this genetic menagerie.

Spontaneous nodulation-restrictive mutant

The *rj1* allele was discovered in a plant breeding program as a simple recessive gene in a single line segregating in the cross of the cultivars 'Lincoln' and 'Richland' (Williams and Lynch, 1954). The *rj1* allele, when present in homozygous mode, produces the "non-nodulating" phenotype, a failure of nodulation with most microsymbionts. In field tests, with over 200,000 soybean plants, less than one nodule was recovered per thousand plants (Devine, 1984). Some strains of *Bradyrhizobium* have been shown to have the ability to induce a low frequency of nodulation on *rj1*, plants (Devine and Weber, 1977; Murphy and Elkan, 1965). The *rj1* locus has been genetically mapped to classical linkage group 11, forty recombination units from the *f* locus conditioning fasciated stem (Devine et al., 1983). Hedges et al. (1990) established that the *Idh 1* locus, controlling isocitrate dehydrogenase, was located approximately midway between the *rj1* and *f* loci, 26.9 centimorgans from *rj1*.

Induced nodulation-restrictive mutations

Several loci governing nodulation have been identified by chemically induced mutation. Mathews et al. (1989) reported the induction of 3 non-nodulation mutants, nod49, nod139 and nod772, by ethyl methanesulphonate mutagenesis of the cultivar Bragg. Allelism tests demonstrated that non-nodulation genes in nod49 and nod772 were allelic with the *rj1* gene. Later Pracht et al. (1993) established that the non-nodulating gene in nod139 is allelic with the non-nodulation allele designated *rj6*. The *rj6* gene was identified in the non-nodulating mutant NN5, isolated from *N*-nitroso-*N*-methylurea-mutagenized seed of the cultivar Williams. The *rj5* and *rj6* alleles are recessive alleles which, when both are simultaneously present in homozygous mode in the same plant, produce a non-nodulating phenotype (Pracht et al., 1993). These genes are located at loci distinct from the *rj1* locus.

Induced supernodulation mutations

Several researchers have used induced mutation to produce supernodulating or hypernodulating lines of soybeans. In pioneering research, Gresshoff and coworkers applied chemical mutagenesis to the maturity group VII cultivar 'Bragg' and produced nine supernodulating mutants (Carroll et al., 1985). These mutants were characterized by a profusion of root nodules, a pattern that was clearly distinct from the normal, restrained nodulation pattern. Tests for genetic allelism indicated that these mutant lines carried a recessive mutation at the same locus, which was designated *nts1* (Delves et al., 1988). Gremaud and Harper (1989) also produced similar hypernodulating mutants using chemical mutagenesis on the maturity group III cultivar 'Williams'. Akao and Kouchi (1992) reported the generation of a supernodulating mutant, designated En6500, by chemical mutagenesis of the Japanese cultivar 'Enrei 87'. The supernodulating trait induced in this cultivar was controlled by a single recessive gene (Kokubun and Akao, 1994). Buzzell et al. (1990) reported obtaining five supernodulating mutants by chemical mutagenesis of the cultivar 'Elgin 87'. Allelism tests have shown that each of the five mutants involved the same locus with a recessive allele conditioning the supernodulation response. Anomalous segregations in one of the mutant lines has been attributed to a second mutation linked to the recessive supernodulation gene (Buzzell, personal communication, 1995). Tests for allelism with Nod 1-3 and Nod 4 have shown that the locus that mutated to produce the hypernodulating trait in 'Williams' is the same locus as En6500 (Vuong et al., 1996). The En6500 locus is also allelic with the locus identified in 'Bragg' by Delves et al. (1988) (Gresshoff, personal communication, 1995). This suggests that this particular locus either is especially responsive to chemical mutagenesis and/or to the efficiency of the screening methods employed to isolate the mutants, or that this locus has a particularly significant role in controlling this trait.

The *nts1* gene has been genetically mapped to linkage group H in the Iowa State/USDA molecular marker genetic map. Harper and Nickel (1995) have designated this locus *rj7*. The *rj1* non-nodulation gene or the combination of the *rj5* and *rj6* genes epistatically repress expression of the *rj7* gene (Harper and Nickel, 1995; Matthews et al., 1990). Grafting experiments with the supernodulating or hypernodulating mutants and their respective non-mutagenized sources demonstrated that nodule number was regulated by the plant shoot (Akao and Kouchi, 1992; Cho and Harper, 1991; Delves et al., 1986, 1992)

The initial agronomic performance of the supernodulating or hypernodulating mutants was most unsatisfactory (Pracht et al., 1994). Supernodulation restricted plant growth and yield (Gresshoff, 1995). Gresshoff (1995) has attributed the physiological growth limitation to extensive nodule development, rather than an ancillary deleterious mutation. Other researchers have suggested that the yield loss is most likely due to additional induced mutations (Song et al., 1995).

Although the same genetic locus is involved, individual mutational events may produce alleles with variable intensity in expression of the supernodulation phenotype. Additionally, Harper suggests that environmental conditions may play a role in the expression of this trait (Harper, personal communication, 1996). Gresshoff (1995) reported that soybean genotypes derived by backcrossing the *nts382* allele, which expressed clear supernodulation, into cultivars adapted to northern Australia had increased yield compared to the parents, Bragg and nts. He suggested that the increased plant size of the tropical cultivar or the longer growth season permitted the support of the extra nodule complement and good root growth, thereby enabling expression of the increased nitrogen fixation capacity and thus affecting plant growth and seed set. Gresshoff (personal communication, 1995) reports that the cultivar 'Nitrobean 60', developed from his supernodulating mutants, produced a 15% yield increase over the 'Bragg' source. Song et al. (1995) reported that, in results from tests conducted over several years at two sites in southern Australia grain yield with supernodulating soybean ranged from 25% less to equal to the yield of the normally nodulating cultivars 'Centaur' or 'Bragg'. However, at one site, the dry matter of oats, planted immediately after soybean harvest, was greater following the supernodulators than following the normally nodulated cultivars. They suggested that a major advantage for the supernodulating trait may be the contribution of fixed nitrogen to a subsequent cereal crop in the rotation. Harper (personal communication, 1995) reports that breeding efforts with the mutants developed in his laboratory have resulted in the recovery of only 80% of the yield of the source cultivar 'Williams'.

Gresshoff (1993) proposed an auxin burst control (ABC) hypothesis to account for supernodulation. The hypothesis assumes that the Nod Factor produced by the bacteria is perceived as an elicitor of flavone changes in the epidermis, root hairs and cortex. Such changes are presumed to cause an inhibition of radial auxin transport from the phloem parenchyma. The auxin burst control (ABC) hypothesis assumes that the plant responds to this alteration of auxin transport by increasing the shoot-derived auxin supplied to the root. This condition blocks, or slows, further cell division or reinitiation of cell divisions in younger tissues. The ABC hypothesis proposes that supernodulation in soybean is caused by a lowering of the plant's ability to generate the auxin burst.

Indole biosynthesis by the bacterial microsymbiont may be involved in controlling nodulation since mutants of the *Bradyrhizobium japonicum* strain lacking this ability are unable to nodulate soybean (Kummer and Kuykendall, 1989; Wells and Kuykendall, 1983). Hunter and Kuykendall (1990), Kuykendall and Hunter (1991), and Kuykendall et al. (1992) described the development of a *B. japonicum* strain with 40% greater nodulation. This USDA-patented strain, now available from Urbana Labs (St Joseph, MO) as a commercial inoculant, was derived as a prototrophic revertant of a mutant specifically deficient in indole glycerol phosphate synthase (Kuykendall and Hunter, 1991). Recently, a *trpCD* deletion mutant was constructed and shown to be ineffective, conclusively establishing that the ability to synthesize indole is a prerequisite for symbiosis (Kuykendall and Hunter, 1995). Selection of 5-methyltryptophan-resistant mutants may offer a new means of selecting symbiotically improved strains (Hunter and Kuykendall, 1995).

Naturally occuring polymorphisms for nodulation restriction

Genetically undefined traits

In 1980, Nangju reported that US–bred soybean cultivars did not nodulate with the indigenous bradyrhizobia in Nigeria, however the Southeast Asian cultivars 'Malayan', 'Orba' and 'TGm686' nodulated and fixed nitrogen with indigenous *Bradyrhizobium*. Unfortunately, the Southeast Asian cultivars were not agronomically suitable for production in Nigeria. Plant breeding efforts were undertaken to develop new soybean lines combining desirable agronomic characteristics with the ability to form a successful symbiosis with indigenous *Bradyrhizobium*. The International Institute for Tropical Agriculture (IITA) annual report for 1987-88 indicated that researchers screened 250,000 breeding lines for nodulation promiscuity and desirable seed storage characteristics. The 1993 IITA annual report indicated that the breeding program had produced soybean lines whose yield perfomance with the uninoculated treatment was similar to the treatment with applied nitrogen or inoculation with commercial inoculum.

In an effort to exclude *B. japonicum* strains of serogroup 123, which occur in high frequency in some midwestern states, attempts were made to construct soybean genotypes that restricted nodulation with this serogroup (Cregan and Keyser, 1986). Such a host genotype has not been produced and these efforts appear to have been abandoned.

Genetically defined traits

The Rj4 *gene*

Several naturally occuring genetic polymorphisms have been identified in soybean that control nodulation restriction with groups of *Rhizobium* or *Bradyrhizobium*. Vest and Caldwell (1972) identified, in the soybean cultivars 'Hill' and 'Dunfield', the dominant allele *Rj4* that conditioned an ineffective nodulation response with strain USDA61 of the Beltsville *Rhizobium* Culture Collection. The ineffective nodulation response produced by the *Rj4* allele was described as the presence of numerous small cortical proliferations, or protuberances, on the roots. Evidently, nodulation was initiated, but aborted at an early stage. As a result, the ineffectively nodulated plants were reduced in growth and had the same appearance as nitrogen-deficient, uninoculated control plants. The cultivar 'Hill' nodulates normally with other strains of *Bradyrhizobium*. Since strain USDA61 nodulates and fixes nitrogen with soybean cultivars homozygous for the recessive allele *rj4*, the ineffective nodulation response is a specific genetic interaction of the host genotype and microsymbiont genotype. The occurrence of a gene conditioning such an apparently disadvantageous phenotype prompted the question of why such a gene would be encountered. The answer coming most quickly to mind was that it was a mutation or inborn metabolic error. However, this explanation

Table 2. Prevalence of the Rj4 phenotype in *G. max.* introductions from Asia

Country of origin	Number	Percent frequency
USSR	14	0
China	248	35.7
Korea	171	19.4
Japan	243	24.2
Taiwan	5	60.0
Vietnam	5	97.4
Myanmar and Malaya	14	71.7
Thailand	26	66.9
Indonesia	33	64.8
India and Pakistan	88	15.8
Total or proportion of total	847	29.7

was not satisfactory. The *Rj4* allele and several other naturally occuring alleles that apparently frustrate the normal process of nitrogen fixation were dominant alleles. The dominant nature of these genes was not congruent with the usual pattern that mutations are recessive. Therefore, studies were undertaken to determine if other factors were involved. In a survey of over 800 plant introductions from 12 Asian countries, the *Rj4* phenotype occurred in 29% of all introductions and was most frequent in the populations from Southeast Asia, i.e. Myanmar Indonesia, Malasia, Thailand, and Vietnam (all over 60%) and lower in frequency in Northern Asia, i.e. 0% in USSR (Table 2) (Devine and Breithaupt, 1981). In a survey of over 500 Asiatic accessions of *Glycine soja*, the putative wild progenitor species of soybean, for nodulation by strain USDA61, Devine (1987) found that the ineffective nodulation response that is characteristic of plants carrying the *Rj4* allele was the most common response (63.4%). The high frequency of the *Rj4* phenotype in the wild progenitor species and the domesticated soybeans from Southeast Asia suggest that there exists a selection pressure exerted in favor of the *Rj4* phenotype to maintain it as the predominant form in these populations. Several lines of evidence, that initially appear unconnected, converge to resolve our understanding of the role of the *Rj4* allele.

An important key to understanding the significance and role of the *Rj4* gene is found in the research conducted with the *rj1* gene by Dr Elkan and his students at North Carolina State University in Raleigh, NC. These researchers established that the *Bradyrhizobium* strains that had the ability to induce limited nodulation on *rj1* plants shared several characteristics, including high capsule formation, inability to metabolize nitrite and failure to reduce triphenyl tetrazolium chloride (TTC) (Table 3) (Hubbell and Elkan, 1967; Murphy and Elkan, 1965). Pueppke and Paine (1987) documented an observation that ruled out the hypothesis that nodulation of *rj1* was by entry through cracks in the root, since such strains were shown to nodulate *rj1* via infection threads.

Some strains of *Bradyrhizobium* that nodulate soybean also produce the phytotoxin rhizobitoxine (Johnson et al., 1959; Owens et al., 1972). Rhizobitoxine interferes with chlorophyll synthesis in developing leaves producing a distinctive interveinal foliar chlorosis. This foliar chlorosis has been reported in producers' fields in the states of Alabama, Arkansas, California, Florida, Georgia, Mississippi, North and South Carolina, and Tennessee (Erdman et al., 1957). Chlorosis-inducing strains of *Bradyrhizobium* nodulate cowpea and fix nitrogen in symbiosis with cowpea with such efficiency as to produce very vigorous growth, but do not produce chlorosis symptoms on cowpea (Table 4) (Devine, 1989). These chlorosis-inducing strains of *Bradyrhizobium* are able to induce a limited degree of nodulation on the nodulation-restrictive soybean genotype *rj1* (Devine and Weber, 1977; Murphy and Elkan, 1965). They also induce nodule-like structures on peanut (Devine et al., 1983). The evidence, thus, indicated that a subgroup of bradyrhizobia nodulating soybean share three distinctive characteristics related to host affinity: a) the ability to nodulate the nodulation-restrictive soybean *rj1*, b) the propensity to induce chlorosis symptoms on soybean, and c) the ability to induce nodule-like swellings on peanut. Devine et al. (1983) postulated that the clustering of these three characteristics affecting symbiosis in a common population suggested that these bradyrhizobia may represent a taxonomically distinct group.

Researchers in Professor Elkan's laboratory used DNA::DNA hybridization analyses to characterize genetic diversity among *Bradyrhizobium* strains nodulating soybean and established two major groups, Group I and Group II (Table 5) (Hollis et al., 1981). The differences in DNA homology between these two groups were of sufficient magnitude as to warrant classification as separate taxonomic species. Devine et al. (1988) established that the *Bradyrhizobium* strains inducing chlorosis in planta were not found in DNA homology Group I, but only in DNA homology Group II. Therefore, a correlation was established between the DNA homology characteristics and host affinity

Table 3. Characteristics of strains with the ability to nodulate vs. inability to nodulate *rj1* soybeans

Non-nodulating	*rj1* nodulating
Moderate capsule formation*	High capsule formation
Inability to metabolize nitrite**	Inability to metabolize nitrite
Reduction of triphenyl tetrazolium chloride*	Non-reduction of triphenyl tetrazolium chloride

*Hubbell and Elkan (1967).
**Murphy and Elkan (1965).

Table 4. Rhizobitoxine-induced chlorosis symptoms

Strain	Soybean	Cowpea
76	+	0
84	+	0
94	++	0
100	+++	0
Uninoculated	0	0

Table 5. Classification, by DNA homology, of *Bradyrhizobium* strains nodulating soybean*

Group I *B. japonicum* subspecies 1	Group Ia *B. japonicum* subspecies 2	Group II *B. elkanii*
ATCC 10324	USDA 110	USDA 76
USDA 24	8°	61 A76
D193	USDA 140	USDA 86
D209	5631	USDA 29
THA6	USDA 62	USDA 46
USDA 58	61A50	USDA 31
USDA 123		USDA 117
USDA 38		USDA 130
5633		USDA 94
USDA 115		

*Hollis, Kloos and Elkan (1981).

characteristics. From the data of Young et al. (1991), we may deduce that there are clearly two species of soybean-nodulating *Bradyrhizobium* differing in 16S rRNA sequence. In 1982, Jordan transferred *R. japonicum* to the new genus *Bradyrhizobium* with only one species being named. Thus, based on characteristics of symbiotic affinity, numerous genetic and biochemical criteria, and most importantly, DNA divergence, we deemed it appropriate to recognize a new and distinct species of *Bradyrhizobium*, designated as *B. elkanii* in recognition of the seminal work, previously cited, of Dr Gerald H Elkan and his students (Table 6) (Kuykendall et al., 1992).

A survey by Fuhrmann (1990) of 360 isolates from soybean nodules on 18 farms in Delaware indicated that 37% of the nodules sampled contained bradyrhizobia with the colony morphology characteristic of the chlorosis-inducing strains. Nitrogen content of plant top growth was lower for strains with the colony morphology characteristic of chlorosis-inducing strains.

Serogroups can be assigned to *B. elkanii* or *B. japonicum*; serogroups 31, 46, 76 and 94 are *B. elkanii* and serogroups 6, 110, 122, 123 and 135 are *B. japonicum*. Two major surveys (Keyser et al., 1984; Weber et al., 1989) of the serological types of bradyrhizobia nodulating soybean in the US production area were summarized by Streeter (1994) (Table 7). The combined studies were based on 1568 nodules from sites in 25 states. Classifying the strains based on serological response indicates that 40% of the nodules contained *B. elkanii* strains. The *B. elkanii* serogroups occur with especially high frequency in the southeastern United States (Weber et al., 1989). Presumably *B. elkanii* strains tolerate higher soil temperatures than strains of *B. japonicum* (Miller and Weber, 1972). Major phenotypic differences between *B. elkanii* and *B. japonicum* are summarized in Table 6.

A ten-year program of backcrossing and selfing was employed to produce a complimentary set of near isogenic lines differing specifically in the presence of

Table 6. Differences between *B. japonicum* and *B. elkanii*

Characteristic	Species		Reference
	B. japonicum	*B. elkanii*	
Nod factors	2 Nod factors	8 Nod factos	Stacey et al. (1995)
Rhizobitoxine chlorosis	Negative	Positive	Devine et al. (1988)
Antibiotic resistance	Sensitive	Multiple	Kuykendall et al. (1988)
Hydrogenase expression	Positive	Negative	Minamasawa (1989)
EPS	Contains glucose mannose, galacturonic acid, and galactose	Contains rhamnose and 4-o-methyl-glucuronic acid	Huber et al. (1984)
Nodule polysaccharides	Contains rhamnose and 2-o methyl-glucuronic acid	Contains rhamnose and 4-o-methyl glucuronic acid	An et al. (1995) and Streeter et al. (1992)
Fatty acid content	Less than 3% 19:0 cyclopropane	Between 20 and 30% 19:0 cyclopropane	Kuykendall et al. (1988)
Host range	Lacks the ability to nodulate both *rj1* soybean and peanut	Generally forms some nodules on *rj1* soybean and peanut	Devine and Weber (1977); Murphy and Elkan (1965)
Rj4 compatibility	Restriction appears to be limited to strain 62 and some serogroup 123 strains	A significant proportion of strains are interdicted	Devine et al. (1990); Sadowsky et al. (1992)
Hemoprotein content	A CO-reactive hemoprotein was present in significant amounts in bacteroids	The major terminal oxidase cytochrome aa3 was present in significant amounts in bacteroids	Keister and Marsh (1990)

Table 6. Continued

Characteristic	Species		Reference
	B. japonicum	*B. elkanii*	
Nitrite reductase	Positive	Lacks this enzyme activity; thus nitrite accumulates	Murphy and Elkan (1965); Van Berkum and Keyser (1985)
IAA production	Low	High	Minamisawa and Fukai (1991)

Table 7. Prevalence of the two *Bradyrhizobum* species in nodules from US soils

Species /Serogroup	Number of isolates		
	Keyser '84	Weber '89	% of Grand total
B. japonicum			
123	132	183	20.1
6	74	23	6.2
122	41	48	5.7
135	59	nd	3.8
110	15	25	2.5
B. elkanii			
31	209	107	20.1
76	99	34	8.5
46	84	27	7.1
94	49	32	5.2
Unknown	95	118	13.5
cross reaction	nd	114	7.3
Grand total	857	711	100.0
B. elkanii	441	200	40.9

Adapted from Streeter (1994).
nd = not determined.

Table 8. Definition of *Rj4* compatibility by use of near isogenic lines

	Rj4 compatible	*Rj4* incompatible
BARC-2 (*Rj4*)	+	-
BARC-3 (*rj4*)	+	+

+ = effective nodulation.
− = ineffective nodulation.

the *Rj4* vs. the *rj4* allele (Devine and O'Neill, 1986). These lines are calculated to be 99.95% identical in their nuclear DNA and completely identical in their cytoplasmic DNA. With these precise biological tools it was possible to ascertain whether other strains, in addition to USDA61, specifically elicit the *Rj4* ineffective nodulation response. Since the near isogenic lines essentially differ only in the allele present at the *Rj4* locus, any difference in nodulation response with individual bradyrhizobial strains can be attributed specifically to the *Rj4* allele (Table 8). Devine et al. (1990) found eight strains in addition to strain USDA61 that elicited the nodulation restriction of the *Rj4* allele (Table 9). Most interestingly, although only 16% of the 119 *Bradyrhizobium* strains tested induced chlorosis, seven of the nine strains (78%) restricted by the *Rj4* allele were chlorosis-inducing strains. Furthermore, when classified by antibiotic resistance profile, eight of the nine restricted strains (89%) were classified as *B. elkanii.* The evidence can be interpreted as indicating that the *Rj4* allele, rather than being merely a deleterious mutation frustrating the normal process of symbiosis, instead has a positive effect on the host plant by shielding it from nodulation by many strains of *B. elkanii.* Such a view is congruent with the high frequency of the *Rj4* allele found in soybean lines in southeastern Asia where, as in the southeastern US, warm temperatures during much of the year provide an environment conducive to strains of *B. elkanii.* This understanding of the role of the *Rj4* allele provides guidance to soybean breeders in the southern US and other areas with similar climatic conditions. By incorporating the *Rj4* allele in their breeding lines and new cultivars breeders, can reduce nodulation with many of the strains of *B. elkanii* that are undesirable for symbiosis with soybean.

The *Rj4* allele did not interdict nodulation by all chlorosis-inducing bradyrhizobia (only 36%), suggesting that the *Rj4* exclusion mechanism is not elicited specifically by the rhizobial gene(s) conditioning

Table 9. Bradyrhizobium strains rejected by the *Rj4* allele

Strain	Origin	Chlorosis-induction
USDA61	North Carolina, USA	+
USDA62	North Carolina, USA	-
USDA83	Maryland, USA	+
USDA94	North Carolina, USA	+
USDA238	Hokkaido, Japan	+
USDA252	Ventai, China	+
USDA259	Shanghai, China	+
USDA260	Shanghai, China	+
USDA340	Japan	-

Table 10. Relationship of chlorosis-induction and reaction to the *Rj4* allele among bradyrhizobial strains

Rj4 reaction	Chlorosis-inducing number of strains	Non chlorosis-inducing number of strains
Reactive	7	2
Non reactive	12	98
Total	19	100
% of strains *Rj4* reactive	36%	2%

rhizobitoxine production in symbiosis, per se, but by some other genetic component carried in high frequency in strains of *B. elkanii* (Table 10). As reported by Devine et al. (1990) this genetic component is apparently not exclusive to *B. elkanii*, since strain USDA62 of *B. japonicum* was evidently recognized and excluded by the *Rj4* allele. Later, Sadowsky et al. (1992) described a subset, composed of three strains, of *B. japonicum* serogroup 123 that was also restricted by the *Rj4* allele. These strains were a small fraction of all available serogroup 123 strains, which had earlier been screened for poor nodulation of PI 377578. PI 377578 was earlier identified, in 1981, by Devine and Breithaupt as expressing the *Rj4* phenotype. Thus from the report of Sadowsky et al. (1992) it can be deduced that the *nolA* locus of *B. japonicum* strain USDA110, when transferred, permits serocluster 123 strain USDA438 to nodulate *Rj4* plants. The bacterial *nolA* locus presumably represses the expression of another bacterial locus responsible for the production of a signal molecule that functions in the *Rj4* recognition process. Dockendorf et al. (1994) reported that *nolA* is a repressor of nod gene expression in *B. japonicum*. In *Bradyrhizobium* sp. (*Arachis*), *nolA* is required for effective nodulation of siratro and cowpea (Gillette and Elkan, personal communication, 1995). Devine et al. (1993) described mutants of *B. elkanii* strain USDA61 that overcame the *Rj4* barrier. Table 6 lists differences between *B. elkanii* and *B. japonicum.*

Earlier in this paper we raised the question whether there was an advantage to the host plant in developing a broad adaptability to a diverse population of microsymbionts or, on the contrary, developing genetic mechanisms for excluding nodulation with specific groups of microsymbionts. The populations of soybeans in Southeast Asia with a high frequency of the *Rj4* allele may be an example of achieving an adaptive solution through exclusion.

The Rj2 *gene*

Caldwell (1966) discovered the dominant allele *Rj2*, carried in the soybean cultivars 'Hardee' and 'CNS', that produces an ineffective nodulation response, forming cortical proliferations or rudimentary nodules with bradyrhizobial strains USDA7, USDA14, and USDA122. Plants homozygous for the recessive allele *rj2* nodulate normally with these strains. Plants carrying the *Rj2* allele nodulate normally with other strains of *Bradyrhizobium* such as USDA110 and USDA142. Caldwell et al. (1966) reported that 'Hardee' also produced a ineffective nodulation response with the 22

Table 11. Prevalence of the *Rj2* restriction phenotype in Asiatic lines of soybean, *G. max*

Region	Number of isolates	% restricted
China, northeast	50	62
China, central	21	43
China	7	29
Japan	8	38
Thailand	3	0
Total	89	51

Table 12. Prevalence of *Rj2* reactivity among *Bradyrhizobium* strains classified by serogroup

Serogroup	Number of isolates	% restricted
6	25	84
31	3	0
76/123	10	0
123	9	44
Other	42	48
Total	89	51

strains tested from serogroup 3-24-44 (later designated serogroup c1, Vest et al. (1973), and subsequently as serogroup 6, Van Berkum and Keyser (1985)) and the five strains then known to comprise serogroup 122. A survey of Asiatic accessions of the soybean host plant in the USDA plant introduction collection established a center of gene frequency for the *Rj2* response in the adjacent Chinese province of Jiangsu and Zhejiang and the Shanghai Municipality in the lower Yangzi region (Devine, 1987). Devine et al. (1991) established the close genetic linkage of the *Rj2* locus with the *Rps2* locus, one of several loci conditioning resistance to phytophthora root rot disease of soybean. These two loci, designated linkage group 19, were estimated to be separated by only 2.7 centimorgans.

Using soybean lines near isogenic for the *Rj2* alleles (Devine and O'Neill, 1987) in an experimental protocol similar to that described previously for the *Rj4* near isogenic lines, Devine et al. (1991) tested 89 Asiatic isolates of *B. japonicum* principally from northeast China. Approximately half, 51%, of the isolates were restricted from nodulation by the *Rj2* allele (Table 11). No geographic differentiation in the frequency of *Rj2*-restricted strains was apparent in the sample tested. The analysis by serogroup indicated that a predominance of the serogroup 6 strains tested were restricted by the *Rj2* allele, however, four of the nine serogroup 123 strains tested were also restricted by the *Rj2* allele (Table 12). No *Rj2*-restricted isolates were found among the 13 cultures tested from serogroup 31 and serocluster 76/123. Forty-two of the strains tested did not fit the criteria for established serogroups and are presently of unknown serotype. Almost half, 48%, of these serologically undefined strains were restricted by the *Rj2* gene. Since several of the isolates in serogroup 6 were not restricted by the *Rj2* allele and only 21 of the 45 isolates restricted by the *Rj2* allele were serogroup 6, specific serological reaction per se is not essential to, or necessary for, action of the *Rj2* allele. It appears more likely that the *Rj2* rejection is activated by some other factor(s) that occurs in high frequency in some serogroups.

The Rfg1 *gene*

Until 1982, soybean was thought to be nodulated only by slowgrowing bacteria. Then, several fast-growing strains were isolated that successfully nodulated the cultivar 'Peking', an introduction from China (Keyser et al., 1982). However, these early isolates did not form a successful nitrogen-fixing symbiosis with several US-bred cultivars. Devine (1985) established that the majority (56%) of a sample of 285 plant introductions of *G. max* from nine Asian countries nodulated effectively with strain USDA205, the type strain for the fast-growing rhizobia nodulating soybean, *Rhizobium fredii* (Scholla and Elkan, 1984). Although the frequency of effective nodulation with USDA205 was low in northeast China, the source of much of the germplasm base for North American-bred cultivars, the frequency of such lines was over 80% in several Southeast Asian 3 countries (Table 13). Dowdle and Bohlool (1985) later isolated other strains of *Rhizobium fredii* from Asia, notably HH 303, that nodulate North American cultivars and fix nitrogen with high efficiency. The *Rfg1* allele is a dominant gene identified in the cultivar 'Kent' that produces an ineffective nodulation response with strain USDA205 (Devine, 1984; Devine and Kuykendall, 1994). Heron and Pueppke (1984) reported that *Rhizobium fredii* strain USDA2S7 produced an ineffective nodulation response with the cultivar 'McCall'. Trese (1995) demonstrated, by a thorough genetic analysis, that the gene conditioning the ineffective nodulation response in McCall is the same *Rfg1* gene demonstrated by Devine in the cultivar 'Kent'. Research in Pueppke's laboratory has identified genetic loci in *R. fredii* strain USDA257 that are involved in cultivar-specific incompatibility with 'McCall' soybean (Heron et al., 1989). The rhizobial genes eliciting the *Rfg1* response are: (1) a chromosomal locus, *nolC*, (Krishnan and Pueppke, 1992) and (2) another locus of six open reading frames, *nolBTUVWX*, located on a Sym plasmid (Meinhardt et al., 1993). Inactivation of any of these open reading frames permits strain USDA257 to nodulate 'McCall' (Meinhardt et al., 1993) and *Erythrinia* species (Krishnan and Pueppke, 1994).

Conclusions

None of the nodulation control genes of the host plant have been molecularly sequenced nor have their gene products been described. The availability of near isogenic lines for the *Rj1*, *Rj2*, and *Rj4* loci provides unique biological tools for creative experimentation to permit such determinations. When this is accomplished, there will be an opportunity to address the intriguing questions of the relationship of these genes to one another. For example, were the *Rj2*, *Rj4* and *Rj1* genes derived from a common ancestral locus and what is the relationship, if any, to the *Rj1*, *Rj5* and *Rj6* loci? Has the relict polyploidy in soybean provided a base for the elaboration of genes controlling specificity with different populations of microsymbionts?

Research previously cited has established ecogeographic variation in the distribution of the nodulation controlling alleles *Rj2*, *Rj4*, and *Rfg1*. The predominance of the *Rj4* allele in soybean populations in Southeast Asia can be understood in terms of its function with respect to *Bradyrhizobium elkanii* and climatic conditions. The explanation for the center of gene frequency for the *Rj2* allele in the lower Yangzi River Valley is unknown. However, it is interesting that the *Rj2* allele is quite closely linked to the *Rps2* allele conditioning resistance to phytophthora root rot, a disease favored by wet or poorly drained soils. Whether this linkage association has a functional or evolutionary advantage, or is purely a fortuitous association, is not known. The low frequency of the *Rfg1* allele in Southeast Asia parallels the high frequency of the *Rj4* allele in this area. The reason for this negative correlation is unknown, but is intriguing and invites further investigation. We anticipate that future studies of both the microsymbionts and host populations of Asiatic germplasm will provide fruitful insights for understanding ecogeographic relationships and evolutionary processes governing the nitrogen-fixing symbiosis in soybean. In this paper, we have endeavored to provide, through a series of questions and through the information available from soybean, an outline of the research needs for an understanding of the biology of the nitrogen-fixing symbiosis. It is hoped that such an understanding will provide the basis for control of the symbiosis for agricultural improvement.

References

Akao S and Kouchi H 1992 A supernodulating mutant isolated from soybean cultivar Enrei. Soil Sci. Plant Nutr. 38, 183–187.

An J, Carlson R W, Glushka J and Streeter J G 1995 The structure of a novel polysaccharide produced by *Bradyrhizobium* species within soybean nodules. Carbohydr. Res. 269, 303–317.

Buzzell R I, Buttery B R and Ablett G R 1990 Supernodulation mutants in Elgin 87 soybean. *In* Nitrogen Fixation: achievements and objectives. Eds. P M Gresshoff, L E Roth, G Stacey and W E Newton. p. 726. Chapman and Hall, New York, USA.

Caetano-Anolles G and Gresshoff P M 1990 Early induction of feedback regulatory responses governing nodulation in soybean. Plant Sci. 71, 69–81.

Caetano-Anolles G and Gresshoff P M 1991 Efficiency of nodule initiation and autoregulatory responses in a supernodulating soybean mutant. Appl. Environ. Microbiol. 57, 2205–2210.

Caldwell B E 1966 Inheritance of a strain specific ineffective nodulation in soybeans. Crop Sci. 6, 427–428.

Carlson R W, Sanjuan J, Bhat U R, Glushka J, Spaink H P, Wijfjes A H M, van Brussel A A N, Stokkermans T J W, Peters N K and Stacey G 1993 The structures and biological activities of the lipo-oligosaccharide nodulation signals produced by type I and II strains of *Bradyrhizobium japonicum*. J. Biol. Chem. 268, 18372–18381.

Carroll B J, McNeil D L and Gresshoff P M 1985 A supernodulation and nitrate-tolerant symbiotic (*nts*) soybean mutant. Plant Physiol. 78, 34–40.

Cho M J and Harper J E 1991 Root isoflavanoid response to grafting between wild-type and nodulation-mutant soybean plants. Plant Physiol. 96, 1277–1282.

Cregan P B and Keyser H H 1986 Host restriction of nodulation by *Rhizobium japonicum* strain USDA123. Crop Sci. 26, 911–916.

Delves A C, Carroll B J and Gresshoff P M 1988 Genetic analysis and complementation studies on a number of mutant supernodulatin soybean lines. J. Genet. 67, 1–8.

Delves A C, Higgins A and Greshoff P M 1992 Shoot apex removal does not alter autoregulation of nodulation in soybean. Plant Cell Environ. 15, 249–254.

Delves A C, Mathews A, Day D A, Carer A S, Carroll B T and Greshoff P M 1986 Regulation of soybean - *Rhizobium* symbiosis by shoot and root factors. Plant Physiol. 82, 588–590.

Devine T E 1984 Genetics and breeding of nitrogen fixation. *In* Biological Nitrogen Fixation. Ed. M Alexander. pp 127–154. Plenum Press, New York, USA.

Devine T E 1985 Nadulation of soybean (*Glycine max* L. Merr.) plant introduction lines with the fast-growing rhizobial strain USDA205. Crop Sci. 25, 354–356.

Devine T E 1987 A comparison of rhizobial strain compatibilities of *Glycine max* and its progenitor species *Glycine soja*. Crop Sci. 27, 635–639.

Devine T E 1989 The role of co-evolution in the soybean/microsymbiont interaction *In* World Soybean Research Conference IV Proceedings. Ed. A J Pascale. pp 1119–1124. Orientacion Grafica Editora S R L, Buenos Aires, Argentina.

Devine T E and Breithaupt B H 1981 Frequencies of nodulation response alleles, *Rj2* and *Rj4*, in soybean plant introductions and breeding lines. USDA Tech. Bull., No. 1628.

Devine T E, Kilen T C and O'Neill J J 1991 Genetic linkage of the phytophthora resistance gene *Rps2* and the nodulation response gene *Rj2* in soybean. Crop Sci. 713–715.

Devine T E and Kuykendall L D 1994 *Rfg*1, a soybean gene controlling nodulation with fast growing *Rhizobium fredii* strain 205. Plant and Soil 158, 47–51.

Devine T E, Kuykendall L D and Breithaupt B H 1983 Nodule-like structures induced on peanut by chloriosis producing strains of *Rhizobium* classified as *R. japonicum*. Crop Sci. 23, 394–397.

Devine T E, Kuykendall L D and O'Neill J J 1988 DNA homology group and the identity of bradyrhizobial strains producing rhizobitoxine-induced foliar chlorosis on soybean. Crop Sci. 28, 939–941

Devine T E, Kuykendall L D and O'Neill J J 1990 The *Rj4* allele in soybean represses nodulation by chlorosis-inducing bradyrhizobia classified as DNA homology group II by antibiotic resistance profiles. Theor. Appl. Genet. 80, 33–37.

Devine T E, Kuykendall L D and O'Neill J J 1991 Nodulation interaction of the soybean allele *Rj2* with asiatic isolates of *Bradyrhizobium japonicum*. Crop Sci. 31, 1129–1131.

Devine T E and O'Neill J J 1986 Registration of BARC-2 (*Rj4*) and BARC-3 (*rj4*) soybean germplasm. Crop Sci. 26, 1263–1264.

Devine T E and O'Neill J J 1987 Registration of BARC-4 (*Rj2*) and BARC-5 (*rj2*) soybean germplasm. Crop Sci. 27, 1322–1323.

Devine T E, O'Neill J J and Kuykendall L D 1993 Near isogenic lines of soybeans as tools to identify nodulation specific mutants of *Bradyrhizobium elkanii*. Plant and Soil 149, 205–209.

Devine T E, Palmer R G and Buzzell R I 1983 Analysis of genetic linkage in the soybean. J. Heredity 74, 457–460.

Devine T E and Weber D F 1977 Genetic specificity of nodulation. Euphytica 26, 527–535.

Dobert R C, Breil B T and Triplett E W 1994 DNA sequence of the common nodulation genes of *Bradyrhizobium elkanii* and their phylogenetic relationship to those of other nodulating bacteria. Mol. Plant-Microbe Inter. 7, 564–572.

Dockendorff T C, Sanjuan J, Grob P and Stacey G 1994 *NolA* represses nod gene expression in *Bradyrhizobium japonicum*. Mol Plant-Microbe Inter. 7, 596–602.

Dowdle S F and Bohlool B B 1985 Predominance of fast-growing *Rhizobium japonicum* in a soybean field in the People's Republic of China. Appl. Environ. Microbiol. 50, 1171–1176.

Erdman L W, Johnson H W and Clark F 1957 Varietal responses of soybeans to a bacterial-induced chlorosis. Agron. J. 49, 267–271.

Fuhrmann J J 1990 Symbiotic effectiveness of indigenous soybean bradyrhizobia as related to serological, morphological, rhizobitoxin and hydrogenase phenotype. Appl. Environ. Microbiol. 56, 224–229.

Fukuhara H, Minakawa Y, Akao S and Minamisawa K 1994 The involvement of indole-3-acetic acid produced by *Bradyhizobium elkanii* in nodule formation. Plant Cell Physiol. 35, 1261–1265.

Gremaud M F and Harper J E 1989 Selection and initial characterization of partially nitrate tolerant nodulation mutants of soybean. Plant Physiol. 89, 169–173.

Gresshoff P M 1993 Molecular genetic analysis of nodulation genes in soybean. *In* Plant Breeding Reviews. Ed. J Janick. Vol. 11. pp 275–318. John Wiley and Sons Publishers, New York, NY, USA.

Gresshoff P M 1993 Plant function in nodulation and nitrogen fixation in legumes. *In* New Horizons in Nitrogen Fixation. Eds. R Palacios, J Mora and W E Newton. pp 31–54. Kluwer Academic Publishers, Dordrecht, The Netherlands.

Hadley H H and Hymowitz T 1973 Speciation and cytogenetics. *In* Soybeans: Improvement, Production, and Uses. Ed. B E Caldwell. Agronomy 16, 97–116.

Harper J E and Nickell C D 1995 Genetic analysis of nonnodulating soybean mutants in a hypernodulated background. Soybean Genet. Newsl. 22, 185–190.

Hedges B R, Sellner J M, Devine T E and Palmer R G 1990 Assigning isocitrate dehydrogenase to linkage group 11 in soybean. Crop Sci. 30, 940–942.

Hollis A B, Kloos W E and Elkan G H 1981 DNA:DNA hybridization studies of *Rhizobium japonicum* and related Rhizobiaceae. J. Genet. Microbiol. 123, 215–222.

Hubbell D H and Elkan G H 1967 Correlation of physiological characteristics with nodulating abilty in *Rhizobium japonicum*. Can. J. Microbiol. 13, 235–241.

Huber T A, Agarwal A K and Keister D L 1984 Extracellular polysaccharide composition, ex planta nitrogenase activity, and DNA homology in *Rhizobium japonicum*. J. Bacteriol. 158, 1168–1171.

Hunter W J and Kuykendall L D 1990 Enhanced nodulation and nitrogen fixation by a revertant of a nodulation-defective *Bradyrhizobium japonicum* tryptophan auxotroph. Appl. Environ. Microbiol. 56, 2399–2403.

Hunter W J and Kuykendall L D 1995 Symbiotic properties of 5- methyltryptophan-resistant mutants of *Bradyrhizobium japonicum*. Plant and Soil 173, 293–298.

Jordan D C 1982 Transfer of *Rhizobium japonicum* Buchanan 1980 to *Bradyrhizobium* gen. nov.,a genus of slow-growing, root nodule bacteria from leguminous plants. Int. J. System. Bacteriol. 32, 136–139.

Keister D L and Marsh S S 1990 Hemoproteins of *Bradyrhizobium japonicum* cultured cells and bacteroids. Appl. Environ. Microbiol. 56,2736–2741.

Keyser H H and Cregan P B 1987 Nodulation and competition for nodulation of selected soybean genotypes among *Bradyrhizobium japonicum* serogroup 123 isolates. Appl. Environ. Microbiol. 53, 2631–2635.

Keyser H H, Weber D F and Uratsu S L 1984 *Rhizobium: japonicum* serogroup and hydrogenase phenotype distribution in 12 states. Appl. Environ. Microbiol. 47, 613–615.

Kokubun M and Akao S 1994 Inheritance of supernodulation in soybean mutant En6500. Soil Sci. Plant Nutr. 40, 715–718.

Krishnan H B and Pueppke S G 1992 Inactivation of *nolC* conditions developmental abnormalities in nodulation of Peking soybean by *Rhizobium fredii* USDA257. Mol Plant-Microbe Inter. 5 14–21.

Krishnan H B and Pueppke S G 1994 Cultivar-specific genes of the nitrogen-fixing soybean symbiont,*Rhizobium fredii* USA257, also regulate nodulation of *Erythrina* spp. Am. J. Bot. 8, 38–45.

Kummer R M and Kuykendall L D 1989 Symbiotic properties of amino acid auxotrophs of *Bradyrhizobium japonicum.* Soil Biol. Biochem. 21, 779–782.

Kuykendall L D 1989 Influence of*Glycine max* nodulation on the persistence in soil of a genetically marked strain of *Bradyrhizobium japonicum.* Plant and Soil 116, 275–277.

Kuykendall L D, Devine T E and Cregan P B 1982 Positive role of nodulation on the establishment of *Rhizobium japonicum* in subsequent crops of soybean. Current Microbiol. 7, 79–81.

Kuykendall L D, Hahn M, Hennecke H and Hunter W J 1992 Genetically improved rhizobia and their use in agriculture. *In* Biological Nitrogen Fixation and Sustainability of Tropical Agriculture. Eds. K Mulongoy, M Gueye and D S C Spencer. pp 211–210. John Wiley and Sons, London, UK.

Kuykendall L D and Hunter W J 1991 Enhancement of nitrogen fixation with *Bradyrhizobium japonicum* mutants. US Patent 5,021,076.

Kuykendall L D and Hunter W J 1995 Symbiotic ineffectiveness of *trpCD* deletion mutants of *Bradyrhizobium japonicum.* Soil Biol. Biochem. 27, 1035–1039.

Kuykendall L D, Roy M A, O'Neill J J and Devine T E 1988 Fatty acids, multiple antibiotic resistance, and DNA homology among *Bradyrhizobium japanicum* strains. Int. J. Syst. Bacteriol. 37, 358–361.

Kuykendall L D, Saxena B, Devine T E and Udell S E 1992 Genetic diversity in *Rhizobium japonicum* and a proposal for *Bradyrhizobium elkanii* sp. nov. Can. J. Microbiol. 38, 501–505.

Lohrke S M, Orf J H, Martinez-Romero E and Sadowsky M J 1995 Host-controlled restriction of nodulation of *Bradyrhizobium japonicum* strains in Serogroup 110. Appl. Environ. Microbiol. 61, 2378–2383.

Mathews A, Carroll B J and Gresshoff P M 1989 A new nonnodulation gene in soybean. J. Heredity 80, 357–360.

Mathews A, Carroll B J and Gresshoff P M 1990 The genetic interaction between non-nodulation and supernodulation in soybean: an example of developmental epistasis. Theor. Appl. Genet. 79, 125–130.

Meinhardt L W, Krishnan H B, Balatti P A and Pueppke S G 1993 Molecular cloning and characterization of a sym plasmid locus that regulates cultivar-specific nodulation of soybean by *Rhizobium fredii* USA257. Mol. Microbiol. 9, 17–29.

Minamisawa K 1989 Extracellular polysaccharide composition, rhizobitoxine production, and hydrogenase phenotype in *Bradyrhizobium japonicum.* Plant Cell Physiol. 30, 877–884.

Minamisawa K and Fukai K 1991 Production of indole-3-acetic acid by *Bradyrhizobium japonicum.* A correlation with genotype grouping and rhizobitoxine production. Plant Cell Physiol. 32, 1–9.

Murphy S G and Elkan G H 1965 Nitrogen metabolism of some strains of *Rhizobium japonicum* having different nodulating capacities. Can. J. Microbiol. 11, 1039–1041.

Nangju D 1980 Soybean response to indigenous rhizobia as influenced by cultivar origin. Agron. J. 72, 403–406.

Palmer R G and Kilen T C 1987 Qualitative genetics and cytogenetics. *In* Soybeans: Improvement, Production and Uses. Ed. J R Wilcox. Agronomy 16, 135–209.

Polhill R M, Raven P H and Stirton C H 1981 Evolution and Systematics of the Leguminosae. *In* Advances in Legume Systematics. Part I. Eds. R M Polhill and R H Raven. pp 1–26. Royal Botanic Gardens, Kew, UK.

Pracht J E, Nickell C D and Harper J E 1993 Genes controlling nodulation in soybean: *Rj5* and *Rj6*. Crop Sci. 33, 711–713.

Pracht J E, Nickell C D, Harper J E and Bullock D G 1994 Agronomic evaluation of non-nodulating and hypernodulating mutants of soybean. Crop Sci. 34, 738–740.

Pueppke S G and Payne J H 1987 Responses of *Rj1* and *rj1* soybean isolines to inoculation with *Bradyrhizobium japonicum.* Plant Physiol. 84, 1291–1295.

Sadowsky M J and Cregan P B 1992 The soybean *Rj4* allele restricts nodulation by *Bradyrhizobium japonicum* serogroup 123 strains. Appl. Environ. Microbiol. 58, 720–723.

Scholla M H and Elkan G H 1984 *Rhizobium fredii* sp. nov., a fast growing bacterium that effectively nodulates soybeans. Int. J. Syst. Bacteriol. 34, 484–486.

Shoemaker R, Polzin K, Lorenzen L and Specht J 1995 Molecular genetic mapping of soybean: Map utilization. Crop Sci. 32, 1091–1098

Song L, Carroll B J, Gresshoff P M and Herridge D F 1995 Field assessment of supernodulating genotypes of soybean for yield, N2 fixation and benefit to subsequent crops. Soil Biol. Biochem. 27, 563–569.

Sprent J I 1994 Evolution and diversity in the legume-*Rhizobium* symbiosis: chaos theory? Plant and Soil 161, 1–10.

Stacey G, Sanjuan S, Luka S, Dockendorff T and Carlson R W 1995 Signal exchange in the *Bradyrhizobium*-soybean symbiosis. Soil Biol. Biochem. 27, 473–483.

Streeter J G 1994 Failure of inoculant rhizobia to overcome the dominance of indigenous strains for nodule formation. Can. J. Microbiol. 40, 513–522.

Streeter J G and Salminen S O 1993 Distribution of the two types of polysaccharide formed by *Bradyrhizobium japonicum* bacteroids in nodules on field-grown soybean plants (*Glycine max* (L,.) Merr.) Soil Biol. Biochem. 25, 1027–1032.

Streeter J G, Salminen S O, Whitmoyer R E and Carlson R W 1992 Formation of novel polysaccharides by *Bradyrhizobium japonicum* bacteroids in soybean nodules. Appl. Environ. Microbiol. 58, 607–613.

Trese A T 1995 A single dominant gene in McCall soybean prevents effective nodulation with *Rhizobium fredii* USDA257. Euphytica 81, 279–282.

Van Berkum P and Keyser H H 1985 Anaerobic growth and denitrification among different serogroups of soybean rhizobia. Appl. Environ. Microbiol. 49, 772–777.

Vest G and Caldwell B E 1972 *Rj4* a gene conditioning ineffective nodulation in soybeans. Crop Sci. 12, 692–694.

Vest G, Weber D F and Sloger C 1973 Nodulation and nitrogen fixation. *In* Soybeans: Improvement, Production, and Uses. Ed. B E Caldwell. Agronomy 16, 363–390.

Vuong T D, Nickell C D and Harper J E 1996 Genetic and allelism analyses of hypernodulating soybean (*Glycine max* (L.) Merr.) mutants from two genetic backgrounds. Crop Sci. (*In press*).

Weber D F, Keyser H H and Uratsu S L 1989 Serological distribution of *Bradyrhizobium japonicum* from US soybean production areas. Agron. J. 81, 786–789.

Wells S E and Kuykendall L D 1983 Tryptophan auxotrophs of *Rhizobium japonicum*. J. Bacteriol. 156, 1356–1358.

Williams L F and Lynch D L 1954 Inheritance of a nonnodulating character in the soybean. Agron. J. 46, 28–29.

Wilson A C, Ochman H and Prager E M 1987 Molecular time scale for evolution. Trends Genet. 3, 241–247.

Young J P W, Downer H L and Eardly B D 1991 Phylogeny of the phototrophic *Rhizobium* strain BTAil by polymerase chain reaction-based sequencing of a 16S rRNA gene segment. J. Bacteriol. 173, 2271–2277.

Section editor: H Lambers
Guest editors: G.H. Elkan and R.G. Upchurch

G.H. Elkan and R.G. Upchurch (eds.), Current Issues in Symbiotic Nitrogen Fixation, 189–192.

A comparison of the binding of *Agrobacterium tumefaciens* and *Rhizobium meliloti* to alfalfa roots, carrot suspension culture cells, and *Arabidopsis thaliana* root segments

Ann G. Matthysse
*Department of Biology, University of North Carolina, Chapel Hill, NC 27599-3280 USA**

Key words: Agrobacterium tumefaciens, alfalfa, *Arabidopsis thaliana*, attachment, binding, carrot, *Rhizobium meliloti*

Abstract

Although *A. tumefaciens* and *R. meliloti* are closely related, the interactions of these bacteria with cells of non-leguminous dicots differ markedly. When *R. meliloti* was added to carrot suspension culture cells less than 2% of the added bacteria bound to the plant cells. The addition of the *nod* gene inducer luteolin had no effect on the binding of the bacteria to carrot cells. When *A. tumefaciens* was mixed with the same cells more than 30% of the added bacteria bound to the carrot cells. *A. tumefaciens*, but not *R. meliloti*, bound to root segments of *A. thaliana*. Both species of bacteria bound to alfalfa roots.

Introduction

Rhizobium meliloti and *Agrobacterium tumefaciens* are closely related to each other. However, the two species of bacteria differ markedly in their host range. *R. meliloti* has a narrow host range including alfalfa and excluding all nonlegumes. *A. tumefaciens* has a broad host range including most dicots. *A. tumefaciens* has been reported to be capable of binding to tissue culture cells or cells of wound sites in all dicots tested (Matthysse, 1986). *R. meliloti* has been reported to bind to root hairs of alfalfa (Caetano-Anolles et al., 1989 and Wall and Favelukes, 1991). The binding of these bacteria to each others preferred host cells has not been examined.

Materials and methods

A. tumefaciens virulent wild type strain C58 and *R. meliloti* strain 1021 were grown in Luria broth (L), minimal medium (RK, Roberts and Kerr, 1974), TY, or TY calcium (TYC, Swart et al., 1993) at room temperature. They were used for experiments when they had reached the early stationary phase of growth. Carrot suspension culture cells were grown in Murashige and Skoog (MS) medium as previously described (Matthysse et al., 1981). For incubation with bacteria 0.5 ml of carrot cells were transferred to 5.0 ml of fresh MS or to 5.0 ml of 4% sucrose containing 1 mM $CaCl_2$ and 0.1 ml of an appropriate dilution of the bacterial culture in phosphate-buffered saline (PBS) was added. The per cent of the bacterial inoculum which bound to the carrot cells was measured as previously described (Matthysse et al., 1981). Briefly, carrot cells with attached bacteria were separated from free bacteria by filtration through Miracloth (Calbiochem). The carrot cells with attached bacteria were resuspended in PBS and ground in a Waring blender. The number of free and attached bacteria was determined by viable cell counts.

Alfalfa seeds were surface sterilized and germinated on 1% agar covered with Whatman number 1 filter paper. When the roots were between 1 and 2 cm in length the seedlings were transferred aseptically to sterile glass slides covered with 1% agar containing 4% sucrose and 1 mM $CaCl_2$; 0.05 ml of bacteria were added to the root surface followed by a cover slip. The slide was sealed with melted paraffin and incubated

* FAX No: 919-962-6941. E-mail: matthysse@unc.edu

Table 1. Attachment of various bacteria to carrot suspension cells

Bacterial strain	Per cent bacteria inoculum attached[1]
R. meliloti 1021	2 ± 3[2]
A. tumefaciens C58	35 ± 5
R. meliloti 1021 plus 50 μM luteolin	3 ± 2

[1]Attachment was measured after 60 min incubation of 1-5 $\times$ 10^3 bacteria per ml with approximately 10^5 carrot cells per ml.
[2]Mean and standard deviation of a minimum of 3 experiments

standing on end in a sterile Coplin jar containing a 2-5 mm layer of water in the bottom. The slides were examined in the light microscope after 48 hours. Photomicrographs were made using a Zeiss photoscope 2 and Nomarski optics.

Arabidopsis thaliana ecotype WS seeds were surface sterilized, stored at 4° for 3 days, and grown in MS salts with 2% sucrose for 1 to 5 weeks. Roots were cut asceptically into segments 1 to 3 cm long and incubated with about 10^7 bacteria per ml in a 1/10 dilution of MS salts containing 0.4% sucrose. They were examined for bacterial attachment after 4 and 20 hours.

Results

When *A. tumefaciens* strain C58 grown in L was incubated with carrot cells for 60 min about 35% of the inoculated bacteria bound to the plant cells. During the same time less than 3% of inoculated *R. meliloti* 1021 grown in L bound to the carrot cells (Table 1 and Figure 1). When the incubation was continued for a total of 120 min there was still no detectable binding of *R. meliloti*. The use of 4% sucrose 1 mM $CaCl_2$ rather than MS medium in the incubation had no effect on the binding of either bacterium to carrot cells. Similar results were obtained with bacteria grown in RK, TY, and TYC media.

Flavonoids such as luteolin released from legume roots have been reported to induce the expression of *nod* genes in *R. meliloti* (Fisher and Long, 1992). Therefore the effect of the addition of luteolin to the medium on the binding of *R. meliloti* to carrot cells was examined. The addition of 50 mM luteolin to either MS medium or 4% sucrose failed to increase the binding of *R. meliloti* grown in L to carrot cells (Table 1).

Both *R. meliloti* and *A. tumefaciens* were seen to be associated with the alfalfa roots after 48 hours incubation. In both cases the bacteria appeared to be trapped in the mucigel layer surrounding the root and also to be bound to the root epidermis and to root hairs. Single cells and bacterial aggregates were seen on the root hairs. Most of the bacteria attached to the epidermis were single cells. The mucigel layer surrounding the root contained numerous bacteria largely distributed along the outer edge of the mucigel. The only difference between the two species was seen in areas where root cap cells had sloughed off the root tip and were visible free in the agar. Greater numbers of *R. meliloti* appeared to be associated with these cells than *A. tumefaciens* (Figure 2).

A. tumefaciens grown in all 4 media was observed to bind to root hairs, epidermal cells, and cut ends of *A. thaliana* root segments. Polarly-attached bacteria and bacterial aggregates were visible on most of the root surface after 4 hours. After 20 hours the root surface was largely covered in a layer of bacteria several cells deep. Large bacterial aggregates were seen at the cut ends. The root hairs were covered with a mass of polarly-attached bacteria and bacterial aggregates. When *R. meliloti* grown in each of the 4 media was incubated with *A. thaliana* root segments no bacteria were seen attached to the root hairs or cut ends after 4 or 20 hours incubation. Rare bacteria were seen on the root epidermis, approximately one bacterium on every 10 to 20 epidermal cells. *R. meliloti* grown in TYC formed small bacterial aggregates which floated free in the medium (data not shown).

Discussion

Although *R. meliloti* and *A. tumefaciens* are closely related, it appears that under the conditions used in these experiments the two species of bacteria differ in their ability to bind to plant cells. *R. meliloti* was unable to bind to carrot suspension culture cells or *A. thaliana* root segments; these bacteria bound to alfalfa roots. These results correspond with the host range of *R. meliloti* which includes alfalfa but does not include nonlegumes such as carrot or *A. thaliana*. *A. tumefaciens* bound to carrot suspension culture cells and also to the epidermal cells and root hairs of intact alfalfa roots and *A. thaliana* root segments. These results correspond with the host range of *A. tumefaciens* which includes most dicot species. Both species of bacteria also appeared to adhere to the mucigel layer on the surface of the alfalfa roots.

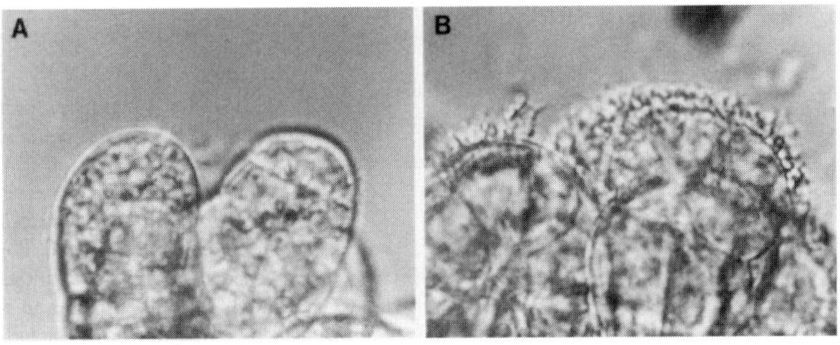

Figure 1. Photomicrograph of the attachment of *R. meliloti* 1021 (A) and *A. tumefaciens* C58 (B) to carrot suspension culture cells. Approximately 10^8 bacteria per ml were inoculated into MS medium containing 10^5 carrot cells per ml and incubated for 20 hours. The living cells were photographed using Nomarski optics. Note that *R. meliloti* 1021 did not bind to carrot cells, while *A. tumefaciens* C58 bound both as individual cells and in large aggregates.

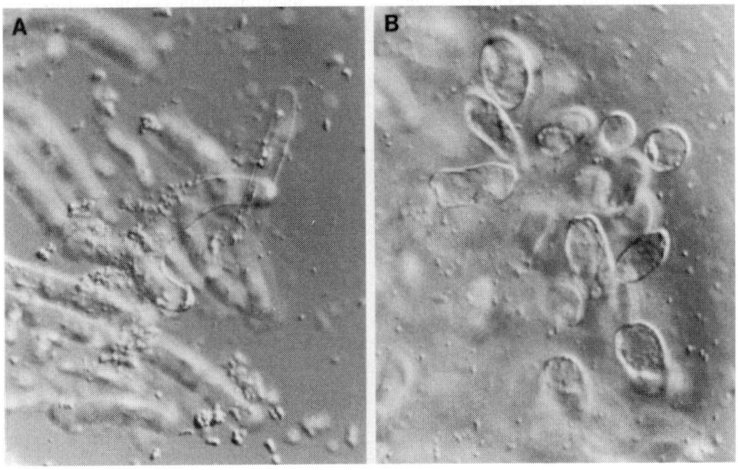

Figure 2. Photomicrograph of the attachment of *R. meliloti* 1021 (A) and *A. tumefaciens* C58 (B) to root cap cells of alfalfa. *R. meliloti* were associated with the root cap cells while *A. tumefaciens* cells were distributed at random. The bacteria were grown with the root in 1% agar for 48 hours. The living cells were photographed using a Zeiss photoscope 2 and Nomarski optics.

Smit et al (1992) reported that the medium in which *R. leguminosarum* was grown affected its ability to bind to pea roots. The ability of *A. tumefaciens* to bind to dicot cells was unaffected by the bacterial growth medium. *R. meliloti* was unable to bind to carrot cells or Arabidopsis root segments regardless of the bacterial growth medium. *R. meliloti* bound to alfalfa roots.

The actual molecules involved in the binding of *A. tumefaciens* to dicot cells or Rhizobia to legume root hairs are unknown. Smit et al (1992) have suggested that the bacterial binding molecule for both Agrobacteria and Rhizobia is a small calcium-binding protein called rhicadhesin. If this is so, then the results presented above suggest that the expression of rhicadhesin (or its accessibility) differ between *A. tumefaciens* and *R. meliloti*. *A. tumefaciens* is capable of expressing the binding molecule when the bacteria are incubated with dicot cells or roots. *R. meliloti* may only express the binding molecule when it is incubated with a potential host plant such as alfalfa. Alternatively, the binding molecule of *R. meliloti* may differ from that of *A. tumefaciens* and the plant receptor to which *R. meliloti* binds may be found only on legumes.

Conclusions

Both *R. meliloti* and *A. tumefaciens* bound to root hairs and to root epidermis of alfalfa. *A. tumefaciens* bound to carrot suspension culture cells and *A. thaliana* root segments. *R. meliloti* grown in L, RK, TY, or TYC media did not show significant binding to carrot cells or to *A. thaliana* root segments.

Acknowledgments

This research was supported by grant number 94-37303 from the US Department of Agriculture.

References

Caetano-Anolles A, Lagares A and Favelukes G 1989 Adsorption of *Rhizobium meliloti* to alfalfa roots: dependence on divalent cations and pH. Plant and Soil 177, 67–74.

Fisher RF and Long SR 1992 Rhizobium-plant signal exchange. Nature 357, 655–660.

Matthysse AG 1986 Initial interactions of *Agrobacterium tumefaciens* with plant host cells. CRC Crit. Rev. Microbiol. 13, 281–307.

Matthysse AG, Holmes KV and Gurlitz RHG 1981 Elaboration of cellulose fibrils by *Agrobacterium tumefaciens* during attachment to carrot cells. J. Bacteriol. 145, 583–595.

Roberts WP and Kerr A 1974 Crown gall induction: serological reactions, isozyme patterns, and sensitivity to mitomycin C and to bacteriocin of pathogenic and non-pathogenic strains of *A. radiobacter*. Physiol. Plant Pathol. 4, 81–91.

Smit G, Swart S, Lugtenberg BJJ and Kijne JW 1992 Molecular mechanisms of attachment of Rhizobium bacteria to plant roots. Mol. Microbiol. 6, 2897–2903.

Swart S, Smit G, Lugtenberg BJJ and Kijne JW 1993 Restoration of attachment, virulence, and nodulation of *Agrobacterium tumefaciens* ChvB mutants by rhicadhesin. Mol. Microbiol. 10, 597–605.

Wall LG and Favelukes G 1991 Early recognition in the Rhizobium meliloti-alfalfa symbiosis: root exudate factor stimulates root adsorption of homologous Rhizobia. J. Bacteriol. 173, 3492–3499.

G.H. Elkan and R.G. Upchurch (eds.), Current Issues in Symbiotic Nitrogen Fixation, 193–196.

Symbiotic effectiveness of marked rhizobial and bradyrhizobial mutants in pea, pigeonpea and mungbean

[1]V P Singh, [2]S B Singh, [2]O Prakash and [1]B Dhar
[1]*Department of Genetics and Plant Breeding* and [2]Department of Biochemistry, Banaras Hindu University, Varanasi – 221 005, India*

Key words: antibiotics, azide, *Bradyrhizobium*, effectiveness, nodulation, pulses, *Rhizobium*

Abstract

The symbiotic effectiveness of rhizobial and bradyrhizobial mutants of the parent strains, namely, P5 (*Rhizobium leguminosarum* bv. *viceae*), AO12 (*Bradyrhizobium* sp.) and MO5 (*Bradyrhizobium* sp.), resistant to streptomycin, kanamycin, erythromycin and sodium azide, was studied in aseptic conditions. The mutants isolated spontaneously were sensitive to their respective phages, RL1, VAR6 and MRV6 and were found to nodulate their host plants pea, pigeonpea and mungbean. Some of these mutants under reference (*str*1, *ery*1, *kan*1, *azr*1 and *azr*2 of P5; *syr*2, *ery*9, *kan*16, *azr*9 and *azr*11 of MO5 and *str*1, *ery*5, *kan*2, *azr*1, *azr*9 and *azr*16 of AO12) exhibited significantly the increased values for the symbiotic traits (nodule number, nodule dry weight, plant dry weight and nitrogen content/plant) as compared to the parents. The azide resistant mutants of strains P5, MO5 and AO12 showed superiority over the parent strains as well as antibiotic resistant mutants for most of the symbiotic traits.

Introduction

The antibiotic resistant mutants of different rhizobial strains were reported to have a wide range of variability for symbiotic properties (Kuykendall, 1981). Ram et al. (1978) observed that symbiotic efficiency can be improved by introducing effective azide resistant mutants. The present study was, therefore, undertaken to characterize different antibiotic- and azide-resistant mutants of *Rhizobium* and *Bradyrhizobium* for their symbiotic effectiveness and phage sensitivity as markers. This was studied by developing mutants and then inoculating them to the host plants and various symbiotic traits were analysed.

Materials and methods

The rhizobial and bradyrhizobial mutants of parent strains, namely, P5 (*Rhizobium leguminosarum* bv. *viceae*), AO12 (*Bradyrhizobium* sp.) and MO5 (*Bradyrhizobium* sp.), resistant to streptomycin, erythromycin, kanamycin and sodium azide and their respective phages, namely, RL1, VAR6 and MRV6 were employed in the present investigation.

Isolation of mutants

The mutants were isolated spontaneously on YM agar medium supplemented with streptomycin (5 and 25 μg mL^{-1}), erythromycin (10 and 50 μg mL^{-1}), kanamycin (5 and 50 μg mL^{-1}) and sodium azide (50, 100, 150 and 200 μg mL^{-1}) individually. Sixteen clones which showed resistance at high levels of inhibitors were selected randomly and tested for sensitivity with their respective phages in broth. The clones sensitive to phages were selected. Growth of individual clones was then compared in the presence and absence of the antibiotics and antimetabolite. Mutants stocked on agar slants containing respective antibiotics and azide were maintained for further studies.

Plant growth conditions

Symbiotic effectiveness of the mutants along with the parent strains P5, AO12 and MO5 were tested on pea (cv. T163), pigeonpea (cv. Bahar) and mungbean (cv. PS16), respectively under aseptic conditions on agar

* FAX No: + 91542312059

Table 1. Spontaneous mutation frequency of antibiotic and azide resistant rhizobial and bradyrhizobial strains in YM medium

Strain	CFU mL^{-1}	Streptomycin (μg mL^{-1}) 5	Streptomycin 25	Erythromycin (μg mL^{-1}) 10	Erythromycin 50	Kanamycin (μg mL^{-1}) 5	Kanamycin 50	CFU mL^{-1}	Sodium azide (μg mL^{-1}) 50	Sodium azide 100	Sodium azide 150	Sodium azide 200
P5	4.8×10^7	16	0	1	0	96	0	1.5×10^8	111	20	3	0
		(3.3)	(0)	(0.2)	(0)	(20)	(0)		(7.4)	(1.3)	(1.7)	(0)
AO12	1.8×10^7	180	2	5×10^3	15	0	29	1.46×10^8	800	150	50	2
		(100)	(1.1)	(2803)	(8.3)	(0)	(16)		(54)	(10.2)	(3.4)	(0.13)
MO5	1.4×10^8	3.5×10^3	5	1.9×10^3	8	0	1.3×10^3	2.8×10^8	1	0.5	0	0
		(250)	(0.35)	(142)	(0.57)	(0)	(95)		(0.04)	(0.02)	(0)	(0)

Figures in parenthesis denote mutation frequency ($\times 10^{-7}$).

Table 2. ANOVA for nodulation traits in pea (cv. T163), pigeonpea (cv. Bahar) and mungbean (cv. PS16) inoculated with rhizobial strain P5 and bradyrhizobial strains A012 and M05, respectively

Host	Source	d.f.	Mean sum squares: Number of nodules/ plant	Nodule dry weight/ plant	Total dry weight/ plant	Total N content/ plant
Pea (cv. T163)	Replications	3	0.09	1.23	2.57	0.11
	Strains	6	128.45*	236.00*	39232.57*	27.74*
	Error	18	6.59	1.46	32.91	1.10
Pigeonpea (cv. Bahar)	Replications	3	1.00	2.58	3.25	0.04
	Strains	7	640.86*	278.57*	3574.79*	10.57*
	Error	21	1.86	1.63	7.06	0.19
Mungbean (cv. PS16)	Replications	3	1.74	1.45	0.37	0.10
	Strains	6	167.48*	138.80*	15844.00*	19.51*
	Error	18	1.63	2.20	30.16	0.35

*$p > 0.01$.

slants. The plants were kept in a culture room at 26 $\pm$ 2°C with light intensity of approximately 3000 lux. Each treatment had four replications containing one plant each. The experiments were terminated after 45 days and the data on nodule number, nodule dry weight, total plant dry weight and total nitrogen content/plant were recorded and analysed statistically following randomized block design and the means were compared by critical difference (CD) (Panse and Sukhatme, 1967). The total nitrogen content of the plants was determined by the Kjeldahl method (Vincent, 1970).

Results and discussion

The spontaneous mutation frequencies of antibiotic and azide resistance of rhizobial and bradyrhizobial strains are given in Table 1. These spontaneously isolated antibiotic and azide resistant mutants of strains P5, AO12 and MO5 exhibited lytic behavior with their respective phages (RL1, VAR6 and MRV6). All the phage sensitive clones retained their antibiotic and antimetabolite resistance level after five successive subculturing. The antibiotic and azide resistant mutants from strains P5, AO12 and MO5 were selected for determination of nitrogen fixing efficiency with their host plants, namely, pea, pigeonpea and mungbean, respectively, under controlled conditions. The analysis

Table 3. In vitro symbiotic effectiveness of antibiotic and azide resistant mutants of *R. leguminosarum* strain P5 in pea cv. T163

Strain	Mean sum squares			
	Number of nodules/plant	Nodule dry weight/plant (mg)	Total dry weight/plant (mg)	Total N content/plant (mg)
Control	0	0	275	6.2
Parent	7	13	486	11.6
*str*1	3	8	320	6.4
*ery*1	5	12	410	8.3
*kan*1	1	4	310	6.5
*azr*1	11	18	513	11.8
*azr*2	17	22	492	11.4
Mean	6	11	401	8.9
SEM ±	2	1	4	0.74
CD (1%)	5.24	2.45	11.68	2.13

of variance showed that the strains, in general, had the significant effect for all the traits under reference (Table 2).

Symbiotic effectiveness of the mutants of strain P5 resistant to streptomycin, erythromycin, kanamycin and sodium azide at 5, 10, 5 and 100 $\mu g\ mL^{-1}$ concentrations, respectively, was studied. These mutants were tested on pea cv. T163. The mean values of four replications of all the characters are summarized in Table 3. Certain antibiotic resistant mutants (*str*1, *ery*1 and *kan*1) had a negative impact on symbiotic traits as they were inferior compared to the parent strain. However, azide resistant mutants (*azr*1 and *azr*2) were statistically superior to their parent with respect to certain symbiotic traits. Nevertheless, none of the resistant mutants could show any significant difference for nitrogen content/plant over the parent strain.

The selected mutants of *Bradyrhizobium* strain MO5 were resistant at 25, 50, 50 and 50 $\mu g\ mL^{-1}$ concentrations of streptomycin, erythromycin, kanamycin and sodium azide, respectively. All the mutants had lyric behaviour with phage MRV6 in broth like their parent. These mutants formed nodules on mungbean cv. PS16. The mean values of four replications for each character are summarized in Table 4. It is clear that all the antibiotic and antimetabolite resistant mutants were statistically superior to the parent for number of nodules/plant; all the mutants, except *strs*2 and *ery*9, for nodules dry weight/plant and all the mutants, excluding *ery*9, for plant dry weight. In this case also, no mutant

Table 4. In vitro symbiotic effectiveness of antibiotic and azide resistant mutants of *Bradyrhizobium* strain MO5 in mungbean cv. PS16

Strain	Mean sum squares			
	Number of nodules/plant	Nodule dry weight/plant (mg)	Total dry weight/plant (mg)	Total N content/plant (mg)
Control	0.0	0.0	157	1.75
Parent	10.0	11.0	298	7.00
*str*2	16.3	14.0	320	7.50
*ery*9	17.0	13.0	304	7.10
*kan*11	15.5	16.5	328	7.75
*azr*9	16.5	15.0	325	7.80
*azr*11	18.4	17.5	340	7.95
Mean	13.3	12.4	296	6.69
SEM ±	0.9	1.1	3.9	0.42
CD (1%)	2.59	3.02	11.17	1.21

Table 5. In vitro symbiotic effectiveness of antibiotic and azide resistant mutants of *Bradyrhizobium* strain AO12 in pigeonpea cv. Bahar

Strain	Mean sum squares			
	Number of nodules/plant	Nodule dry weight/plant (mg)	Total dry weight/plant (mg)	Total N content/plant (mg)
Control	0	0	290	5.82
Parent	22	21	330	9.93
*str*1	28	21	320	9.60
*ery*5	38	25	370	10.24
*kan*2	35	20	345	9.90
*azr*1	30	19	360	10.20
*azr*9	25	20	350	10.45
*azr*16	40	28	384	11.20
Mean	27	19	344	9.67
SEM ±	1	1	2	0.31
CD (1%)	2.72	2.55	5.32	0.88

could exhibit any significant difference for total nitrogen content/plant.

Similarly, selected mutants of *Bradyrhizobium* strain AO12 were resistant to streptomycin, erythromycin, kanamycin and sodium azide at 25, 50, 50 and 200 $\mu g\ mL^{-1}$ concentrations, respectively. The mutants showed lysis in broth as of their parent with the phage VAR6. These mutants were tested on plants of pigeonpea cv. Bahar. It was found that all the mutants had nodulating ability and they produced pink coloured nodules. The values of four replications for all the symbiotic traits are summarized in Table 5. All the six mutants produced significantly more nodules than the parent strain; it was highest for *azr*16 and *ery*5. Moreover, these two mutants had higher nodule dry weight/plant than the parent. In addition, all the mutants, except *str*1, showed their superiority over the parent for total plant dry weight. But none of them, except *azr*16, could show a higher level of total nitrogen content/plant than the parent strain. Interestingly, the mutant *azr*16 showed higher values for all the symbiotic traits significantly over the parent strain in the present investigation.

Since the wild rhizobial strains, viz., P5, AO12 and MO5 were phage sensitive, the spontaneously isolated mutants of these strains were also sensitive to the respective phages and had no difference in lyric behaviour from their wild types in broth. Some of the mutants, however, exhibited increased values for certain symbiotic traits as compared to the parent strains. This is an indication of increased effectiveness of the mutants to exhibit better nodulation traits. The present findings are in accordance with the earlier reports of Subba Rao (1976), Minchin et al. (1978), Mytton (1978) and Gupta et al. (1989). The results also indicated that the occurrence of resistance to antibiotics was not associated with loss of effectiveness as was apparent from various symbiotic traits and are in consonance with the observations made by Schwinghamer (1964, 1967) for intrinsic antibiotic marker in *R. trifolii*, *R. meliloti* and *R. leguminosarum*; Josey et al. (1980) in *R. phaseoli* and Gupta et al. (1989) in *Rhizobium* sp. Azide resistant mutants were symbiotically effective and showed significant results for all the symbiotic characters. These observations are in agreement with the report of Ram et al. (1978) that azide resistant mutant of *R. leguminosarum* fixed more nitrogen symbiotically. Identification of the inoculated strains can be made more accurately if the strains are sensitive to the specifc rhizobiophages as an additional marker (Srivastava et al., 1980).

Acknowledgements

The work was financially supported by the University Grants Commission, New Delhi, through a research project completed by Dr V P Singh.

References

Gupta R P, Kalra M S and Sahoo M S 1989 Intrinsic multiple antibiotic markers for clover cultivar and *Rhizobium* strain interaction. Trop. Agric. (Trinidad) 66, 313–315.

Josey D P, Beyon J L, Johnston A W B and Beringer J E 1980 Strain identification in *Rhizobium* using intrinsic and antibiotic resistance. J. Appl. Bacteriol. 46, 343–350.

Kuykendall L D 1981 Mutants of *Rhizobium* that are altered in legume interaction and nitrogen fixation. Int. Rev. Cytol. Suppl. 13, 299–309.

Minchin F R, Summerfield R J and Eaglesham A R J 1978 Plant genotype in *Rhizobium* strain interaction in cowpea (*Vigna unguiculata*). Trop. Agric. (Trinidad) 55, 107–115.

Mytton L R 1978 Some problems and potentials in breeding for improved white clover (*Trifolium repens–Rhizobium trifolii*). Ann. Appl. Biol. 88, 445–448.

Panse V G and Sukhatme P V 1967 Statistical Methods for Agricultural Workers. ICAR, New Delhi, India.

Ram J, Grover R P, Rewari R B and Kumar S 1978 Improvement in nitrogen fixing effectiveness of *R. leguminosarum* by incorporating genetic resistance to azide. Indian J. Exp. Biol. 16, 1321–1322.

Schwinghamer E A 1964 Association between antibiotic resistance of ineffectiveness in mutant strains of *Rhizobium* spp. Can. J. Microbiol. 10, 221–233.

Schwinghamer E A 1967 Effectiveness of *Rhizobium* as modified by mutation for resistance to antibiotics. Antonie van Leeuwenhoek 33, 121–136.

Srivastava J S, Singh B D, Dhar B, Singh V P, Singh R B and Singh R M 1980 Utilization of streptomycin resistance and phage sensitivity as markers in ecological studies with *Rhizobium leguminosarum*. Indian J. Exp. Biol. 18, 1171–1173.

Subba Rao N S 1976 Field response of legumes in India to inoculation and fertilizer application. *In* Symbiotic Nitrogen Fixation in Plants. Ed. P S Nutman: 255–268. International Biological Programme, Vol. 7. Cambridge Univ. Press, London, UK.

Vincent J M 1970 A Manual for the Practical Study of the Root-Nodule Bacteria. IBH Hand Book No. 15, 135–137. Blackwell Scientific Publications, Oxford, UK.

Developments in Plant and Soil Sciences

29. United Nations Economic Commission for Europe and FAO (eds.): *The Utilization of Secondary and Trace Elements in Agriculture.* 1987 ISBN 90-247-3546-7
30. H.G. v.d. Meer, R.J. Unwin, T.A. van Dijk and G.C. Ennik (eds.): *Animal Manure on Grassland and Fodder Crops.* Fertilizer or Waste? 1987 ISBN 90-247-3568-8
31. N.J. Barrow: *Reactions with Variable-Charge Soils.* 1987 ISBN 90-247-3589-0
32. D.P. Beck and L.A. Materon (eds.): *Nitrogen Fixation by Legumes in Mediterranean Agriculture.* 1988 ISBN 90-247-3624-2
33. R.D. Graham, R.J. Hannam and N.C. Uren (eds.): *Manganese in Soils and Plants.* 1988 ISBN 90-247-3758-3
34. J.G. Torrey and J.L. Winship (eds.): *Applications of Continuous and Steady-State Methods to Root Biology.* 1989 ISBN 0-7923-0024-6
35. F.A. Skinner, R.M. Boddey and I. Fendrik (eds.): *Nitrogen Fixation with Non-Legumes* (Proceedings of the 4th Symposium, Rio de Janeiro, 1987). 1989 ISBN 0-7923-0059-9
36. B.C. Loughman, O. Gašparíková and J. Kolek (eds.): *Structural and Functional Aspects of Transport in Roots.* 1989 ISBN 0-7923-0060-2; Pb 0-7923-0061-0
37. P. Plancquaert and R. Haggar (eds.): *Legumes in Farming Systems.* 1990 ISBN 0-7923-0134-X
38. A.E. Osman, M.M. Ibrahim and M.A. Jones (eds.): *The Role of Legumes in the Farming Systems of the Mediterranean Areas.* 1990 ISBN 0-7923-0419-5
39. M. Clarholm and L. Bergström (eds.): *Ecology of Arable Land – Perspectives and Challenges.* 1989 ISBN 0-7923-0424-1
40. J. Vos, C.D. van Loon and G.J. Bollen (eds.): *Effects of Crop Rotation on Potato Production in the Temperate Zones.* 1989 ISBN 0-7923-0495-0
41. M.L. van Beusichem (ed.): *Plant Nutrition – Physiology and Applications.* 1990 ISBN 0-7923-0740-2
42. N. El Bassam, M. Dambroth and B.C. Loughman (eds.): *Genetic Aspects of Plant Mineral Nutrition.* 1990 ISBN 0-7923-0785-2
43. Y. Chen and Y. Hadar (eds.): *Iron Nutrition and Interactions in Plants.* 1991 ISBN 0-7923-1095-0
44. J.J.R. Groot, P. de Willigen and E.L.J. Verberne (eds.): *Nitrogen Turnover in the Soil-Crop System.* 1991 ISBN 0-7923-1107-8
45. R.J. Wright, V.C. Baligar and R.P. Murrmann (eds.): *Plant-Soil Interactions at Low pH.* 1991 ISBN 0-7923-1105-1
46. J. Kolek and V. Kozinka (eds.): *Physiology of the Plant Root System.* 1992 ISBN 0-7923-1205-8
47. A.U. Mokwunye (ed.): *Alleviating Soil Fertility Constraints to Increased Crop Production in West Africa.* 1991 ISBN 0-7923-1221-X; Pb 0-7923-1222-8
48. M. Polsinelli, R. Materassi and M. Vincenzini (eds.): *Nitrogen Fixation* (Proceedings of the 5th Symposium, Florence, 1990). 1991 ISBN 0-7923-1410-7
49. J.K. Ladha, T. George and B.B. Bohlool (eds.): *Biological Nitrogen Fixation for Sustainable Agriculture.* 1992 ISBN 0-7923-1774-2
50. P.J. Randall, E. Delhaze, R.A. Richards and R. Munns (eds.): *Genetic Aspects of Plant Mineral Nutrition.* 1993 ISBN 0-7923-2118-9
51. K.S. Kumarasinghe and D.L. Eskew (eds.): *Isotopic Studies of Azolla and Nitrogen Fertilization of Rice.* 1993 ISBN 0-7923-2274-6
52. F.A. Bliss and G. Hardarson (eds.): *Enhancement of Biological Nitrogen Fixation of Common Baen in Latin America.* 1993 ISBN 0-7923-2451-X
53. M.A.C. Fragoso and M.L. van Beusichem (eds.): *Optimization of Plant Nutrition.* 1993 ISBN 0-7923-2519-2
54. N.J. Barrow (ed.) *Plant Nutrition – From Genetic Engineering to Field Practice.* 1993 ISBN 0-7923-2540-0
55. A.D. Robson (ed.): *Zinc in Soils and Plants.* 1993 ISBN 0-7923-2631-8

Developments in Plant and Soil Sciences

56. A.D. Robson, L.K. Abbott and N. Malajczuk (eds.): *Management of Mycorrhizas in Agriculture, Horticulture and Forestry*. 1994 ISBN 0-7923-2700-4
57. P.H. Graham, M.J. Sadowsky and C.P. Vance (eds.): *Symbiotic Nitrogen Fixation*. 1994 ISBN 0-7923-2781-0
58. F. Baluška, M. Čiamporová, O. Gašparíkova and P.W. Barlow (eds.): *Structure and Function of Roots*. 1995 ISBN 0-7923-2832-9
59. J. Abadía (ed.): *Iron Nutrition in Soils and Plants*. 1995 ISBN 0-7923-2900-7
60. P.S. Curtis, E.G. O'Neill, J.A. Teeri, D.R. Zak and K.S. Pregitzer (eds.): *Belowground Responses to Rising Atmospheric CO_2*. Implications for Plants, Soil Biota, and Ecosystem Processes. 1994 ISBN 0-7923-2901-5
61. P.C. Struik, W.J. Vredenberg, J.A. Renkema and J.E. Parlevliet (eds.): *Plant Production on the Threshold of a New Century*. 1994 ISBN 0-7923-2903-1
62. L.O. Nilsson, R.F. Hüttl and U.T. Johansson (eds.): *Nutrient Uptake and Cycling in Forest Ecosystems* (CEC/IUFRO Symposium, Halmstad, Sweden, 1993). 1995 ISBN 0-7923-3030-7
63. H.P. Collins, G.P. Robertson and M.J. Klug (eds.): *The Significance and Regulation of Soil Biodiversity* (International Symposium, East Lansing, 1993). 1995 ISBN 0-7923-3138-9
64. R.A. Date, N.J. Grundon, G.E. Rayment and M.E. Probert (eds.): *Plant-Soil Interactions at Low pH: Principles and Management* (3rd International Symposium, Brisbane, 1993). 1995 ISBN 0-7923-3198-2
65. J.K. Ladha and M.B. Peoples (eds.): *Management of Biological Nitrogen Fixation for the Development of More Productive and Sustainable Agricultural Systems*. (15th Congress of Soil Science, Acapulco, Mexico, 1994) 1995 ISBN 0-7923-3413-2; Pb 0-7923-3414-0
66. C. Rodriguez-Barrueco (ed.): *Fertilizers and Environment*. 1996 ISBN 0-7923-3729-8
67. B. Buerkert, B.E. Allison and M. von Oppen (eds.): *Wind Erosion in Niger*. Implications and Control Measures in a Millet-based Farming System. 1996 ISBN 0-7923-3885-5
68. O. van Cleemput, G. Hofman and A. Vermoesen (eds.): *Progress in Nitrogen Cycling*. 1996 ISBN 0-7923-3962-2
69. N. Ahmad (ed.): *Nitrogen Economy in Tropical Soils*. 1996 ISBN 0-7923-4094-9
70. M. Rahman, A. Kumar Podder, C. van Hove, Z.N. Tahmida Begum, T. Heulin and A. Hartmann (eds.): *Biological Nitrogen Fixation Associated with Rice Production*. 1996 ISBN 0-7923-4197-X
71. D. Rosen, E. Tel-Or, Y. Hadar and Y. Chen (eds.): *Modern Agriculture and the Environment*. 1997 ISBN 0-7923-4295-X
72. G.H. Elkan and R.G. Upchurch (eds.): *Current Issues in Symbiotic Nitrogen Fixation*. 1997 ISBN 0-7923-4367-0
73. H.M. Anderson, P.W. Barlow, D.T. Clarkson, M.B. Jackson, and P.R. Shewry (eds.): *Plant Roots – From Cells to Systems*. 1997 ISBN 0-7923-4369-7
74. Zhu Zhao-liang, Wen Qi-xiao and J.R. Freney (eds.): *Nitrogen in Soils of China*. 1997 ISBN 0-7923-4372-7
75. J.K. Ladha, F.J. de Bruin and K.A. Malik (eds.): *Assessing Opportunities of Nitrogen Fixation in Rice*. 1997 ISBN 0-7923-4514-2

Kluwer Academic Publishers – Dordrecht / Boston / London